HIMALAYAN PHYTOCHEMICALS

HIMALAYAN PHYTOCHEMICALS
SUSTAINABLE OPTIONS FOR SOURCING AND DEVELOPING BIOACTIVE COMPOUNDS

SUMIRA JAN
ICAR-Central Institute of Temperate Horticulture, Srinagar, India

NAZIA ABBAS
CSIR-Indian Institute of Integrative Medicine, Jammu, India

ELSEVIER

Elsevier
Radarweg 29, PO Box 211, 1000 AE Amsterdam, Netherlands
The Boulevard, Langford Lane, Kidlington, Oxford OX5 1GB, United Kingdom
50 Hampshire Street, 5th Floor, Cambridge, MA 02139, United States

Library of Congress Cataloging-in-Publication Data
A catalog record for this book is available from the Library of Congress

British Library Cataloguing-in-Publication Data
A catalogue record for this book is available from the British Library

ISBN: 978-0-08-102227-6

Publisher: John Fedor
Acquisition Editor: Anneka Hess
Editorial Project Manager: Michelle W. Fisher
Production Project Manager: Omer Mukthar
Cover Designer: Christian Bilbow

Typeset by SPi Global, India

Dedicated to Sheikh Noor ud-Din Wali (RA)
(Sheikh Ul Alam)

O Himalaya!
Are you an arena at large
that the hand of nature
has created to replicate itself?

Contents

4. Chemistry of Himalayan Phytochemicals

5. Biotechnological Strategies for Improvement of Aromatic and Medicinal Herbs

6. Quality Assurance and Quality Control of Medicinal and Aromatic Herbs

7. Marketing Indigenous Herbs: Connecting Link Between International Trade and Himalayan Livelihood

Preface

Himalaya is loftiest and longest mountain range with diverse microclimatic belts embracing huge diversity. Due to increased proclivity of modern civilization toward herb-based drugs, there is an immense thrust on utilization of botanicals as drugs. Though the majority of medicinal herbs are procured from Himalaya, there is major area of Himalaya that is still uncharted or yet not fully explored such as Trans Himalaya covering northern areas Ladakh, Zanaskar, and Karakoram. Western belt of Indian Himalaya including Jammu and Kashmir is totally neglected in spite of covering 60% of forest cover of Himalayan range. The utilization of Himalayan herbs is still obscure due to lack of interaction between farmers and scientists, scattered market chain, and above all dearth of authenticated data of Himalayan flora. The immense threat to Himalayan diversity is indiscriminate harvesting of herbs at immature stage leading to depletion of their total population and introduction of exotic species, which further endangers their nativity. Modern pharmaceutical industries are now digging on natural resources for procurement of bioactive compounds, which could pave the way toward the sustainable development. The current book emphasizes the data pertaining to biogeographical distribution, botanical aspects, and biotechnological interventions that have been done so far in Himalayan herbs. We have also provided recommendations to cultivators for mass cultivation of Himalayan herbs. Further, we have reviewed market scenario of Himalaya herbs and elaborated strategies for expanding national marketing chains to connect with international market. This book is an effort to present before readers an integrated volume covering taxonomy, ecology, and biotechnology of medicinal plants together. We have tried our best to provide glimpse into huge Himalayan diversity, but there is always scope for further improvement. We will be delighted if students or readers/naturalists could add to our volume by inculcating their findings.

Sumira Jan

Nazia Abbas

Acknowledgments

We are thankful to our parents and mentors for providing us with esthetics for nature and natural beauty. I am thankful to technical staff *Manzoor Ahmad Rather* for the collection of monographs and their beautiful photographs. We are thankful to our institutes ICAR-Central Institute of Temperate Horticulture and Indian Institute of Integrative Medicine wherein we have been working on medicinal herbs. We are thankful to the reviewers and all the members of the production team of Elsevier.

CHAPTER

1

Biogeographic Distribution of Medicinal and Aromatic Herbs

OUTLINE

1.1 INTRODUCTION

Indian Himalaya is the youngest mountain range with breadth of 250–300 km stretched over 2500 km from Jammu and Kashmir in the west to Arunachal Pradesh in the east covering area of about 5,33,604 km^2 (Nandy et al., 2006). Indian Himalaya is well known for its cultural and biological diversity. The Indian Himalayan Region (IHR) spans states of Jammu and Kashmir, Himachal Pradesh, Uttarakhand, Sikkim, Arunachal Pradesh, Assam, and West Bengal. Physiographically, IHR starts from foothills of Siwalik (south) extending up to Tibetan Plateau on Trans-Himalaya (north) (Anand and Kumar, 2008). Geographically, IHR

https://doi.org/10.1016/B978-0-08-102227-6.00001-2

has three major regions; the Himadri (greater Himalaya), the Himanchal (lesser Himalaya), and the Siwaliks (outer Himalaya) extending almost uninterrupted throughout its length are separated by major geologic fault lines as represented in Fig. 1.1. Trans-Himalaya covers the northern part of India spanning Jammu and Kashmir and Himachal Pradesh (Nandy, 2014). The Trans-Himalaya is divided into the following:

(a) Western Himalaya: Himachal Pradesh and Jammu and Kashmir
(b) Central Himalaya: Uttarakhand
(c) Northeast Himalaya: Assam and West Bengal (parts of Darjeeling district)

The major Himalayan ranges are as follows:

1. Pir Panjal Range (part of middle Himalayas)
2. Dhauladhar Range (middle Himalayas)
3. Zanskar Range
4. Ladakh Range
5. East Karakoram Range
6. Mahabharata Range (middle Himalayas in Nepal)

Due to the advancement of biotechnological interventions, biodiversity remained as an obsolete theme of developing countries (Benbrook, 2012). Indian Himalaya in spite of rich floristic diversity still lacks updated biodiversity inventories that provide distribution pattern and

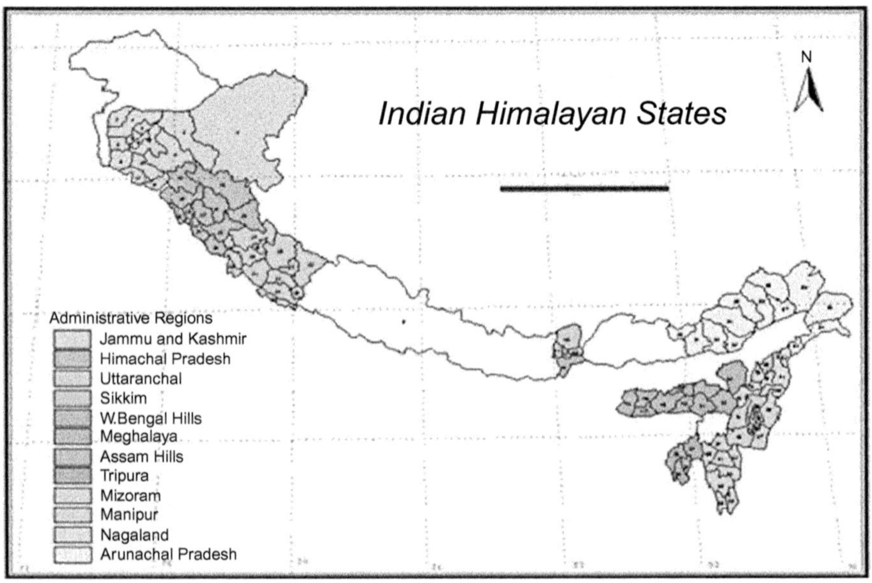

FIG. 1.1 Schematic representation of Indian Himalayan states. Source: http://gbpihedenvis.nic.in/indian_him_reg.htm.

status of surviving species (Khuroo et al., 2010; Jayakumar et al., 2011). Documentation of plant species is indispensable toward conservation of species (Bickford et al., 2007). The Indian Himalayan Region (IHR) covers over 1748 (23.2% of India) plant species (angiosperms 1685, gymnosperms 12, and pteridophytes 51 species) of identified medicinal value (Samant et al., 1998; Dhar et al., 2000). Conservation of medicinal and aromatic plants (MAPs) is increasing due to proclivity of modern civilization toward organic/herbal resources (Rai et al., 2000; Uniyal et al., 2002). The global trade of herbal drugs has registered a steep rise in the recent years (UN, 2004) and now exceeds over US$20,000 million (Dubey et al., 2004). This chapter summarizes information pertaining diverse characteristics of MAPs of the Indian Himalayan Region and identifies gaps that hamper conservation and effective management of MAPs. We will evaluate main priority areas that need to be focused to conserve MAPs based on value index signifying threats imposed by users and sensitivity index that reflects susceptibility of MAPs to ecological pressures (Dhar et al., 2000). Environmental assessment and conservation strategies exclusively rely on updated biodiversity inventories covering information pertaining to both native and naturalized alien species to distinguish between native and alien species distribution (Molnar et al., 2008; Pyšek and Richardson, 2010; Early et al., 2016). Documenting distinctive measures for alien and native species will lead to distinct remarks regarding consideration of native and alien species together or alone. Moreover, the determining factors for native species diversity are unlikely to that of alien species diversity, since alien species is remarkably associated to cultivation practices and socioeconomic factors (Norton, 2014; Measey et al., 2016). So, in order to document alien species in biodiversity inventories, its distribution pattern and species richness are to be determined so as to decipher invasive status given to species (Aranda et al., 2015). Species invasion has detrimental impact on abundance and biogeographic distribution, structure of ecosystem, and economic provisions (Marbuah et al., 2014) together with alien species that exhibit prospects of becoming invasive in the future (Leppäkoski et al., 2002; Hui and Richardson, 2017). Indian Himalaya being a major global biodiversity spot still lacks information on distribution of invasive species (Pandit et al., 2014; Dar et al., 2015; Adhikari et al., 2015). In contrast, several other countries documented illustrative distribution of alien species for China (Weber et al., 2008; Weber, 2017), Japan (Enomoto, 1999), Korea (Koh et al., 2000), Taiwan (Wu et al., 2004), Singapore (Corlett, 1988), and Chonju, Korea (Zerbe et al., 2004). Khuroo et al. (2007a,b) provided illustrative information on alien flora from Kashmir Himalaya but has not covered geographic distribution of native species, and Reddy (2008) assembled an introductory directory of the invasive alien flora for an Indian Himalaya. But all these studies provided alien species without considering native species of Indian Himalaya. In order to reveal the floristic diversity of the Himalayas, we need to analyze both native flora and provide discrepancy among native and alien flora using taxonomic and biogeographic patterns. In order to fill this gap, we have summarized three main issues: (1) documentation of native and woody flora in Indian Himalaya, (2) evaluation of species distribution patterns and taxonomic symphony, and (3) analysis of species invasion by alien woody species.

1.2 HISTORICAL PERSPECTIVE AND CURRENT POSITION OF MAPs

In Indian history, the practice of using medicinal herbs for drug usage dates back 5000 years ago found in Nagpur inscribed on Sumerian clay slab. It enclosed 12 recipes of drug preparations containing 250 plants (Mukherjee and Wahile, 2006; Ameh et al., 2010). The religious scriptures from India, Rig Veda, treat diverse disease with plants that exclusively originate from India, for example, nutmeg, pepper, and clove (Gurib-Fakim, 2006). This book encloses data on the utilization of aromatic and medicinal plants of Indian Himalaya as phytomedicines. Another Indian inscription, Ebers Papyrus, written about 1550 BC ago contained 800 drug formulations prepared from 700 plant species including pomegranate, castor oil plant, aloe, senna, garlic, onion, fig, willow, coriander, juniper, and common centaury (Alves et al., 2013). India is the oldest and seventh largest country using almost 7500 species in ethnomedicine (Mukherjee et al., 2007) and is sixth among twelve mega diversity spots across the world (Chaudhuri and Chaudhuri, 2003). Medicinal and aromatic herbs used in ethnomedicines are subjected to advanced pharmaceutical evaluation and are now used in diverse system of medicines including Ayurveda (2000 spp.), Siddha (1121 spp.), homeopathy (234 spp.), Unani (751 spp.), and Tibetan (337 spp.) (Joshi, 2012). The Himalayan region due its diverse agroclimatic zones including alpine, tropical, subtropical, temperate, and desert harbors metabolically rich herbs (Sati, 2013; Maikhuri et al., 2016). During the recent years, consumption of medicinal and aromatic herbs for therapeutic purposes has been increased. Medicinal and aromatic herbs can be used as phytochemical, nutraceutical, and cosmeceuticals, and their utilization is expected to rise as compared with traditional drug formulations (Bhowmik et al., 2009; Gupta, 2011). Increase in demands of aromatic and medicinal herbs as phytochemicals has been reported in both indigenous and international system of medicines (Barboza et al., 2009; Gupta, 2010). Scientific research documenting ethnomedicinal uses of medicinal and aromatic herbs from Indian Himalaya can be assessed in numerous scientific journals and scientific databases, namely, Google Scholar, SciFinder, and Scopus (Elvin-Lewis, 2011). We have summarized ethnomedicinal reports from surveys in Indian Himalaya carried out during the recent decade. Kaul (1997) in his book entitled *Medicinal Plants of Kashmir and Ladakh: Temperate and Cold Arid Himalaya* was the first book on Indian Himalaya that summarizes scientific data on 111 selected medicinal and aromatic herbs growing in Northwestern Himalaya of India (Kaul, 2010). Later on, Chaurasia et al. (2007) documented the utilization of 329 ethnomedicinal herbs from Trans-Himalaya (Ladakh and Lahul-Spiti) in their book entitled *Ethnobotany and Plants of Trans-Himalaya*. These books exclusively covered flora of Northwestern and Trans-Himalaya without any reports on other regions of Indian Himalaya. Indian Himalaya has grabbed immense attention from global herb trade due to its floristic wealth. According to Singh et al. (1998), there are 4439 species (1220 genera and 189 families) of angiosperms in the Northwestern Himalaya. Numerous primordial native nomadic communities within Indian Himalaya, namely, Gujjars, Bakarwals, Gaddis, Ladakhis, Zanskaris, Padris, Lahulis, Kinnars, and Indo-Tibetan, have no or little exposure to technological advancements (MOEF, 1988). Nomadic communities such as Gujjars, Bakarwals, and Gaddis keep on foraging over large stretches of land (MOEF, 1988). These nomadic communities possess immense traditional knowledge and have been used in folklore medicine. But cultivation practices and technological intervention for integrating traditional knowledge of medicinal and aromatic herbs are still lacking (Karki et al., 2003). The utilization of traditional medicines is discreet limited to particular ethnic group, and cross-cultural analysis has demonstrated varied traditional knowledge of herbs (Leonti et al., 2010; Lardos and Heinrich, 2013).

1.3 DISTRIBUTION PATTERN OF MAPs IN HIMALAYA

The Himalayas are rich in biodiversity exhibiting 21 vegetation types, 10 forest types, and 11 forest formations, as reviewed by Dhar (2002). Northwestern Himalaya encloses dense tropical forest around lowland Bhabhar, Siwalik Hills, deciduous forests covering the center and alpine- and arctic-type vegetation at higher elevations (Nandy et al., 2006; Ghosh, 2008). The lower ranges of Indian Himalaya embrace exclusively vegetation of *Pinus roxburghii* and *Alnus nitida* as slopes having poor nutrient availability (Samant et al., 2007a,b; Devi et al., 2013; Sharma and Raina, 2013). However, slopes bearing adequate soil moisture and relatively higher soil nutrients exhibit luxuriant growth of *Alnus nepalensis*, *Quercus leucotrichophora*, *Rhododendron arboreum*, etc. (Sharma et al., 2009; Singh and Rawat, 2012). The forests of *Quercus semecarpifolia*, *Q. floribunda*, and *Pinus wallichiana* are distinguishing feature of the higher altitude vegetation covering Lesser Himalaya (Chauhan, 1999; Vidyarthi et al., 2013). Great Himalayan ranges are covered predominantly by *Abies* sp., *Betula utilis*, *Rhododendron campanulatum*, and *Juniperus communis*. Indian Himalaya possesses about 46% of total endemic species; approximately 3471 angiosperms plants are enlisted in endemic species (Lanker et al.,

2010). Elevation clines beyond 3600m exhibit luxuriant assortment of aromatic and medicinal herbs growing in alpine meadows and grasslands. Most predominantly, growing herbs include *Aconitum*, *Arnebia*, *Picrorhiza*, *Rheum*, *Meconopsis*, *Rhododendron anthopogon*, and *R. lepidotum* (Singh et al., 2002; Bisht et al., 2013). The data pertaining to the distribution of medicinal and aromatic herbs in Indian Himalaya have been recorded about 1000 BC ago preserved in scriptures. Indian Himalaya covers more than 1789 plants, 330 tree species, 1020 herbaceous species, and 335 shrubs exhibiting high medicinal value (Samant et al., 1998; Kala et al., 2006). Red data reported 15.5% endemism and 31% nativity among biodiversity of medicinal herbs (Pant and Pant, 2011). Most biodiversity in terms of species richness in proportion to geographic area is Sikkim and Darjeeling (0.0996) followed by HP (0.0098) while as for Jammu and Kashmir estimate is still uncovered (Badola and Aitken, 2003). The species richness of medicinal herbs per square kilometer of forestland is recorded highest for Sikkim and Darjeeling at 0.267 (Kala, 2000). Endemism is a prominent feature of Himalayan biodiversity listed endemic species, 6 from Meghalaya state, 8 from Manipur and Nagaland, 7 from Arunachal Pradesh, 11 from Jammu and Kashmir, 51 from Sikkim, and 16 from Himachal Pradesh (Samant and Dhar, 1997; Vetaas and Grytnes, 2002; Kala et al., 2006; Mani, 2012). Indian Himalayan range covers about 45 medicinal taxa including *Acorus calamus (bach)*, *Artemisia nilagirica*, *Bergenia ciliata (pashanbhed; near endemic)*, *Cannabis sativa (bhang)*, *Celastrus paniculatus (mal-kangani)*, *Centella asiatica (brahmi)*, *Chenopodium album (bethu)*, *Cinnamomum tamala (tejpatra; near endemic)*, *Dactylorhiza hatagirea (salam panja; near endemic)*, *Dioscorea bulbifera (barahi)*, *Gloriosa superba (kalihari)*, *Hedychium spicatum (banhaldi or kapurkachri; near endemic)*, *Juglans regia* (walnut; *near endemic*), *Litsea glutinosa*, *Papaver somniferum (post)*, *Picrorhiza kurroa (karu)*, *Rumex nepalensis*, *Solanum nigrum (makoi)*, *Taraxacum officinale (dudhi)*, *Urtica dioica* (stinging nettle), and *Zanthoxylum armatum (timur)* (Kala, 2005; Khare, 2008; Bhat et al., 2013; Rawat et al., 2014, 2016). There are different tree species exhibiting medicinal properties; however, all these tree species, shrubs, and herbs need urgent attention to be conserved (Tables 1.1–1.3).

The three most diversity-rich Himalayan ranges were divided into three high-elevation clines: Great Himalayan National Park (GHNP) (192 species); Lahaul Spiti (L-S), a suggested site for cold-desert biosphere (213 species); and Nanda Devi Biosphere Reserve (NDBR) (95 species) in Uttaranchal Himalaya (Miller, 2010). These three mountain ranges (2000–3000m) provide most suitable habitat to almost 56% of medicinal and aromatic plant species among which 39 species were common (Chakraborty et al., 2017; Ahmad et al., 2017). These species are restricted to particular elevation clines with specific climatic and restricted geographic coordinates (Dar et al., 2002). Such climatic features become a limiting factor for geographic distribution and availability of these herbs. Indian Himalaya

TABLE 1.1 Medicinal Tree Species From Indian Himalaya and Their Status

Tree species	Family	Altitude range	Status
Abies pindrow	Pinaceae	2900–3100 m 3350–3550 m	Vulnerable
Aesculus indica	Hippocastanaceae	1550–1750 m	Alien
Carpinus viminea	Corylaceae	2000–2200 m	Vulnerable
Callicarpa arborea	Verbenaceae	1550–1750 m	–
Hippophae salicifolia	Elaeagnaceae	2450–2650 m	
Lyonia ovalifolia	Ericaceae	1550–1750 m	Rare
Myrica esculenta	Myricaceae	1550–1750 m	Vulnerable
Neolitsea pallens	Lauraceae	1550–1750 m	Vulnerable
Pinus roxburghii	Pinaceae	–	Vulnerable
Rhododendron arboreum	Ericaceae	–	Rare
Sapindus mukorossi	Sapindaceae	–	–
Sorbus aucuparia	Rosaceae	3350–3550 m	Threatened
Sorbus cuspidata	Rosaceae	2450–2650 m	Threatened
Taxus baccata	Taxaceae	2900–3100 m	Endangered
Viburnum cotinifolium	Caprifoliaceae	2450–2650 m	Rare

TABLE 1.2 Medicinal Shrub Species From Indian Himalaya and Their Status

Tree species	Family	Altitude range	Status
Arachne cordifolia	Euphorbiaceae	1550–1750 m	Vulnerable
Buddleja asiatica	Buddlejaceae	1550–1750 m	Alien
Clematis buchananiana	Ranunculaceae	2450–2650 m	Vulnerable
Coriaria nepalensis	Coriariaceae	2900–3100 m	Rare
Cotoneaster microphyllus	Rosaceae	2900–3100 m	Vulnerable
Debregeasia salicifolia	Urticaceae	1550–1750 m	Alien
Lonicera angustifolia	Caprifoliaceae	3350–3550 m	Vulnerable
Leptodermis lanceolata	Rubiaceae	1550–1750 m	–
Leycesteria formosa	Caprifoliaceae	2900–3100 m	
Inula cappa	Asteraceae	2450–2650 m	Rare
Indigofera heterantha	Fabaceae	2450–2650 m	Vulnerable

(Continued)

TABLE 1.2 Medicinal Shrub Species From Indian Himalaya and Their Status—cont'd

Tree species	Family	Altitude range	Status
Hypericum choisianum	Hyperiacaceae	1550–1750 m	Rare
			Vulnerable
Holmskioldia sanguinea	Verbenaceae	1550–1750 m	Rare
Eupatorium odenophorum	Asteraceae	1550–1750 m	–
Elsholtzia fruticosa	Lamiaceae	2450–2650 m	Threatened
Deutzia compacta	Hydrangeaceae	2450–2650 m	Threatened
Desmodium elegans	Fabaceae	2000–2200 m	–
Persicaria polystachya	Polygonaceae	2900–3100 m	Rare
Prinsepia utilis	Rosaceae	1550–1750 m	
Rhamnus virgatus	Rhamnaceae	–	
Rhus javanica	Anacardiaceae	–	
Rosa sericea	Rosaceae	2900–3100 m	Rare
Roylea cinerea	Lamiaceae	1550–1750 m	Vulnerable
Rubia manjith	Rubiaceae	2450–2650 m	Alien
Rubus niveus	Rosaceae	2450–2650 m	Vulnerable
Sarcococca saligna	Buxaceae	1550–1750 m	–
Sorbaria tomentosa	Rosaceae	2450–2650 m	
Viburnum erubescens	Caprifoliaceae	2450–2650 m	Rare
Viburnum grandiflorum	Caprifoliaceae	2900–3100 m	Vulnerable
Viburnum nervosum	Caprifoliaceae	2900–3100 m	–
Woodfordia fruticosa	Lythraceae	1550–1750 m	–
Zanthoxylum armatum	Rutaceae	–	Vulnerable

TABLE 1.3 Medicinal Herb Species From Indian Himalaya and Their Status

Herb species	Family	Altitude range	Status
Aconitum heterophyllum	Ranunculaceae	3350–3550 m	Critically endangered
Ainsliaea aptera	Asteraceae	2900–3100 m	Alien
Ainsliaea latifolia	Asteraceae	2900–3100 m	Vulnerable
Anaphalis contorta	Asteraceae	1550–1750 m	Rare

TABLE 1.3 Medicinal Herb Species From Indian Himalaya and Their Status—cont'd

Herb species	Family	Altitude range	Status
Anaphalis margaritacea	Asteraceae	2450–2650m	Vulnerable
Anaphalis triplinervis	Asteraceae	2000–2200m	Alien
Anemone obtusiloba	Ranunculaceae	3350–3550m	Vulnerable
Anemone rivularis	Ranunculaceae	2450–2650m	–
Arisaema jacquemontii	Araceae	2450–2650m	
Artemisia japonica	Asteraceae	2000–2200m	Rare
Artemisia roxburghiana	Asteraceae	2450–2650m	Vulnerable
Asparagus filicinus	Asparagaceae	–	Rare
Aster peduncularis	Asteraceae	–	Vulnerable
Barleria cristata	Acanthaceae	2000–2200m	Rare
Begonia picta	Begoniaceae	2900–3100m	–
Bergenia ciliata	Saxifragaceae	2450–2650m	Threatened
Bidens bipinnata	Asteraceae	2000–2200	Threatened
Bidens biternata	Asteraceae	2000–2200m	–
Bidens pilosa	Asteraceae	2900–3100m	Rare
Bistorta amplexicaulis	Polygonaceae	2900–3100m	
Bistorta vaccinifolia	Polygonaceae	3350–3550m	Rare
Blumea lanceolaria	Asteraceae	1550–1750m	
Bupleurum falcatum	Apiaceae	2000–2200m	Rare
Calanthe tricarinata	Orchidaceae	3350–3550m	Vulnerable
Cannabis sativa	Cannabaceae	2000–2200m	Alien
Clematis montana	Ranunculaceae	1550–1750m	Vulnerable
Clinopodium umbrosum	Lamiaceae	2000–2200m	–
Corallodiscus lanuginosus	Gesneriaceae	2450–2650m	
Cyathula capitata	Amaranthaceae	2000–2200m	Rare
Cyathula tomentosa	Amaranthaceae		Vulnerable
Cynoglossum glochidiatum	Boraginaceae	2000–2200m	–
Cynoglossum lanceolatum	Boraginaceae	1550–1750m	–
Delphinium vestitum	Ranunculaceae	3350–3550m	Vulnerable
Dicliptera bupleuroides	Acanthaceae	2000–2200m	Vulnerable

(Continued)

TABLE 1.3 Medicinal Herb Species From Indian Himalaya and Their Status—cont'd

Herb species	Family	Altitude range	Status
Dipsacus inermis	Dipsacaceae	2450–2650 m	Alien
Elephantopus scaber	Asteraceae	1550–1750 m	Vulnerable
Elsholtzia strobilifera	Lamiaceae	2900–3100 m	Rare
Euphorbia chamaesyce	Euphorbiaceae	1550–1750 m	Vulnerable
Euphorbia hypericifolia	Euphorbiaceae	2900–3100 m	Alien
Euphorbia pilosa	Euphorbiaceae	2900–3100 m	Vulnerable
Fagopyrum dibotrys	Polygonaceae	2450–2650 m	–
Fragaria nubicola	Rosaceae	1550–1750 m	
Galinsoga parviflora	Asteraceae	2450–2650 m	Rare
Galium aparine	Rubiaceae	2900–3100 m	Vulnerable
Galium asperifolium	Rubiaceae	3350–3550 m	Rare
Gerunium wallichianum	Gerianiaceae	2450–2650 m	Vulnerable
Gerbera gossypina	Asteraceae	1550–1750 m	Rare
Girardania diversifolia	Urticaceae	1550–1750 m	–
Gonatanthus pumilus	Araceae	1550–1750 m	Threatened
Gonostegia hirta	Urticaceae	1550–1750 m	Threatened
Impatiens scabrida	Balsaminaceae	2900–3100 m	–
Jurinea dolomiaea	Asteraceae	3350–3550 m	Endangered
Lamium album	Lamiaceae	2000–2200 m	
Leucas lanata	Lamiaceae	2450–2650 m	
Lindenbergia indica	Scrophulariaceae	2450–2650 m	
Maianthemum purpureum	Liliaceae	3350–3550 m	Rare
Morina longifolia	Morinaceae	2900–3100 m	Rare
Nepeta ciliaris	Lamiaceae	2000–2200 m	Alien
Nomocharis oxypetala	Liliaceae	3350–3550 m	Vulnerable
Origanum vulgare	Lamiaceae	3350–3550 m	–
Paeonia emodi	Paeoniaceae	1550–1750 m	
Parnassia nubicola	Saxifragaceae	2900–3100 m	Rare
Pedicularis hoffmeisteri	Scrophulariaceae	3350–3550 m	Vulnerable
Phalaris minor	Poaceae	1550–1750 m	–

TABLE 1.3 Medicinal Herb Species From Indian Himalaya and Their Status—cont'd

Herb species	Family	Altitude range	Status
Picrorhiza kurroa	Scrophulariaceae	3350–3550 m	Critically endangered
Pimpinella acuminata	Apiaceae	2000–2200 m	Vulnerable
Pimpinella diversifolia	Apiaceae	3350–3550 m	Vulnerable
Plantago depressa	Plantaginaceae	3350–3550 m	Alien
Plantago himalaica	Plantaginaceae	3350–3550 m	Vulnerable
Podophyllum hexandrum	Podophyllaceae	3350–3550 m	Rare
Polygonatum verticillatum	Liliaceae	2450–2650 m	Vulnerable
Primula denticulata	Primulaceae	3350–3550 m	Alien
Ranunculus hirtellus	Ranunculaceae	3350–3550 m	Vulnerable
Reinwardtia indica	Linaceae	2000–2200 m	–
Roscoea alpina	Zingiberaceae	2900–3100 m	Rare
Rubus nepalensis	Rosaceae	2900–3100 m	Rare
Rumex hastatus	Polygonaceae	1550–1750 m	Vulnerable
Rumex nepalensis	Polygonaceae	2900–3100 m	Rare
Salvia hians	Lamiaceae	3350–3550 m	Rare
Salvia nubicola	Lamiaceae	3350–3550 m	Rare
Saussurea albescens	Asteraceae	2000–2200 m	–
Saussurea auriculata	Asteraceae	3350–3550 m	Rare
Saxifraga diversifolia	Saxifragaceae	2900–3100 m	Threatened
Selinum candollei	Apiaceae	3350–3550 m	–
Senecio graciliflorus	Asteraceae	3350–3550 m	Rare
Silene edgeworthii	Caryophyllaceae	2450–2650 m	
Solanum suratteuse	Solanaceae	1550–1750 m	
Solidago virgaurea	Asteraceae	2900–3100 m	
Swertia chirayita	Gentianaceae	2900–3100 m	Rare, vulnerable
Swertia ciliata	Gentianaceae	3350–3550 m	Vulnerable
Synotis alatus	Asteraceae	2450–2650 m	Rare
Taraxacum officinale	Asteraceae	3350–3550 m	Vulnerable

(Continued)

TABLE 1.3 Medicinal Herb Species From Indian Himalaya and Their Status—cont'd

Herb species	Family	Altitude range	Status
Triumfetta rhomboidea	Tiliaceae	1550–1750 m	–
Urena lobata	Malvaceae	–	
Urtica ardens	Urticaceae	–	Rare
Urtica dioica	Urticaceae	–	Vulnerable
Valeriana hardwickii	Valerianaceae	–	–
Verbascum thapsus	Scrophulariaceae	–	–
Vernonia anthelmintica	Asteraceae	2450–2650 m	Vulnerable
Vernonia cinerea	Asteraceae	2000–2200 m	Vulnerable
Veronica anagallis-aquatica	Scrophulariaceae	2450–2650 m	Alien
Viola canescens	Violaceae	2900–3100 m	Vulnerable

exhibits varied microclimates that contribute to endemism that is more pronounced along altitudinal or latitudinal lines (Rao et al., 2004). This might be the reason why endemic medicinal and aromatic herbs are more susceptible to climatic changes than widely distributed taxa. For this reason, endemic species thrive in rocky mountains and/or extremely dry habitats as noticed in alpine taxa (Bhat et al., 2013; Bhatt et al., 2014). Herbs thriving in altitudinal range 1000–3000 m in Himachal Pradesh include C. *asiatica, Dioscorea deltoidea, Jurinea dolomiaea, Podophyllum hexandrum, Taxus baccata,* and *Valeriana jatamansi* (Gairola et al., 2014). To be further detailed, the 1500–2000 m (305 species) and 2500–3000 m (297 species) ranges are the richest reservoirs (Bhat et al., 2012). Himalayan ranges above 5000 m exhibit extreme climatic conditions, snowy and relatively shorter phenological stages (Eaton, 2015; Kumar, 2016). Himalayan taxa exhibiting predominant phytochemical value grow naturally in higher elevation clines and display immense effects development of reproductive organs, for instance, seed and flower development (Sheikh, 2014). In order to overcome developmental constraints, medicinal herbs develop varied modes of multiplication, for example, the development of storage organs likes rhizomes in medicinal herbs including *Angelica glauca, Arctium lappa* (Burdock), *Megacarpaea polyandra, Rheum* (Rhubarb), and *Saussurea*; the stolon formation in *P. kurroa, Gentiana* sp., and *V. jatamansi*; the fragmentation of the genets via deliberate centrifugal extension in herbaceous rosettes, and the formation of root tubers in *Aconitum* (Dhyani et al., 2010; Parihaar et al., 2015). However, all these vegetative modes of production were limited due to poor soil structure, and habitat fragmentation led to the devastation of natural vegetation (Thompson et al., 2014; Leroux et al., 2017). Due to over harvesting of rhizomes and tubers of *A. glauca*, *Aconitum heterophyllum*, *P. kurroa*, and *P. hexandrum*, wild plant species were either extinct or near verge of extinction particularly in Mari-Rohtang mountains of Manali-Kullu (HP) and varied devastation in upper Parvati Valley, Kullu (HP) (Sharma et al., 2014). In order to conserve these medicinal species, strategies both in situ and ex situ conservation are prerequisite steps (Ghosh and General, 2013). Mass multiplication of medicinal herbs in their natural habitats is required to meet high market demands of pharmaceutical drugs (Borowitzka, 2013). Growing herbs in natural agroclimatic conditions like high altitudes ensure high accumulation of active compounds; for example, pronounced accumulation of alkaloids is immensely affected by altitude (Jan et al., 2016). High-altitude Himalayan range embraces Himalayan rhubarbs (*Rheum emodi* and *R. nobile*), and secondary metabolites decreased with lower altitudinal regimes (Malik et al., 2010; Rokaya et al., 2012). In addition to the investigation of climatic variation in active metabolites, we need to tailor out crucial ecological factors that can ensure cultivation and conservation of medicinal and aromatic herbs. Further, evaluation of phenotypic characters like leaf polymorphism needs to be investigated as leaf morphology remarkably affects active ingredients like in the case of *P. kurroa* exhibiting broader leaf lamina accumulating more active constituents (Kaack et al., 2013; Máthé, 2015). Hence, both genetic and morphological traits should be assessed to conserve superior species. In order to conserve particular species, we need to first know the status of that species, particularly threats to its survival, so as to enlist high-priority taxa and endangered species (Macdonald and Willis, 2013). This can be done by thorough evaluation of species quantitative traits including population structure to population dynamics (Kirkpatrick and Barton, 1997). The prime reason for extinction or vulnerability of Himalayan medicinal herbs is their overexploitation and loss of natural habitat/habitat fragmentation (Hamilton, 2004). Habitat degradation results in 91% loss of plant species at global level (Hundal, 2004). According to IUCN Red List, 620 species were enumerated as threatened, out of which 62 are endemic species and 209 are close at verge of endemism in Indian Himalaya (Canter et al., 2005; Larsen and Olsen, 2007). Threatened species and highly vulnerable plant species are about to extinct due to loss of habitat (Chauhan et al., 2014; Maikhuri et al., 2016). For Indian Himalaya, some important threatened medicinal endemic plants include *Aconitum ferox* (Sikkim Himalaya), *A. deinorrhizum* (Himalaya), *A. falconeri* (H.P.), *Angelica glauca* (Indian Himalaya), *A. nubigena* (Sikkim), *Coptis teeta* and *Panax pseudoginseng* (Eastern Himalaya), *P. kurroa* (Himalaya), and *P. emodi* var. *axillaris* (Sikkim) (Chitale et al., 2014; Singh et al., 2016; Manish et al., 2017). These species are depleted due to overexploitation of wild resources to meet market demands. Himalayan medicinal herbs covered under ranges in HP include 133 medicinal herbs that are enumerated either threatened or endangered species (Sharma and Samant, 2014). The families Apiaceae, Asteraceae, and Ranunculaceae are among the most threatened medicinal plants, for example, *A. glauca, C. asiatica, Inula racemosa, J. dolomiaea,* and *Saussurea costus* (Lone et al., 2014; Malik et al., 2015; Kumar et al., 2017). The endemic taxa, for instance, *A. ferox, A. glauca, Codonopsis affinis, I. racemosa,* and *Saussurea bracteata* owing to their high altitude, require immediate conservation. Medicinal herbs thriving in 3000–4000 m embrace almost 69% threatened taxa; more efforts are required for the assessment of these species (Jee, 2015). For this purpose, we need to evaluate their ecophysiological, endangerment, and adaptive strategies for mass multiplication of endangered species (Breininger et al., 2014; Kuniyal et al., 2015). In order to conserve wild medicinal species, we need to analyze factors that result in endangerment and precise explanation for determining endangerment factors of Indian Himalaya (Patel, 2014; Phondani et al., 2016; Nandi et al., 2016).

1.4 CORRECT IDENTIFICATION AND MANAGEMENT IMPLICATIONS

1.4.1 Correct Identification

Taxonomic classification of living organisms is an imperative step toward advancing research in the field of biology (Gotelli and Colwell, 2001; Carpenter, 2005; Patel, 2014). However, taxonomy as subject always ranked secondary in terms of importance as it seems relatively naive to discover science for naming and classification (Stuessy, 2009; Simpson, 2010). But for conservation of significant species, we need to identify species meticulously for which the subject of taxonomy is prerequisite step. Due to various ambiguity and proclivity of modern science toward advanced research, the discipline of taxonomy has been neglected (Hunt and Handoo, 2009). In the era of climate change, biodiversity science is an indispensable subject for monitoring and assessing of environment and defining methodologies that facilitate conservation strategies (Xu et al., 2009). Taxonomy entails identification, naming, and classification of species and hence emphasizes the fundamentals underlying the study of biodiversity. The illustrative introduction of species and their identification help scientists in unraveling the underlying floristic diversity across the globe (Telwala et al., 2013; Negi et al., 2012). There are consistent characteristic features that differentiate one organism from another, and distinguishing these characteristics is prime in the process of identification (Chawla et al., 2008; Khan et al., 2011, 2013a,b). The process of identification utilizes reference data, for example, text description, illustrations, or DNA sequences from authentic databases to compare with specimen (Shaheen et al., 2011). The specimen that does not match the preexisting data is labeled new species (Khuroo et al., 2007a,b). But this comparative analysis with reference data is quite laborious; hence, taxonomists devised variety of tools known as identification keys to speed up the process of identification (Steele and Pires, 2011). These identification keys include dichotomous computer-aided identification keys and traditional keys. Following species identification, we need to label species with names according to rules established as per international rules (Maxted and Crust, 1995; Belhumeur et al., 2008; Cope et al., 2012). Carl Linnaeus, a Swedish botanist, laid the first foundation of biological nomenclature, and scientists used Latin names for relating with identification. Since local names vary from time to time, this can lead to misidentification; hence, Latin scientific names are preferred due to universality of Latin language (Froese et al., 2000; Qiu, 2008). Species identification is essential for comprehending other disciplines as well; however, there are both applications and limitations of species identification based on conventional taxonomic practices, DNA sequencing, and utility of integrated stratagem to species identification (Armstrong and Ball, 2005; Kress et al., 2005; Group et al., 2009). Ecologists working on species conservation demonstrate the significance of identifying native plants in conserving damaged ecosystems (Hobbs et al., 2009). In the case of wild medicinal plants, ecologists recommend to investigate different phenological stages before actual harvest and effect of invasive species on microflora of aromatic and medicinal herbs (Nautiyal et al., 2009). During recent years, there has been increasing interest toward utilization of varied DNA sequences in deciphering variation in species. However, classical contrivances like taxonomic tools, identification keys for new species. This demonstrates molecular approaches should not override morphological or physical features for comprehending variations in species (Chauhan and Nautiyal, 2005; Semwal et al., 2007) but can be utilized as supplementary evidence for final taxonomic decision. DNA barcoding is another strategy that uses uniform segment of species genome. DNA barcoding has been employed in identifying ferns; however, their utilization for identifying medicinal and aromatic herbs is still lacking (Group et al., 2009; Chen et al., 2010). Plant systematics uses internal transcribed spacers (ITS); region of nuclear ribosomal cistron is the most common sequenced locus at species level (Kress et al., 2005). Schori and Showalter (2011) described numerous chloroplast gene regions; for example, maturase K (mat K) and ribulose-1,5-biphosphate carboxylase/oxygenase large subunit (rbcL) were extensively utilized and known as prime barcodes for plants. The technique of barcoding can be used to refine raw material for procuring authentic herbal products without substitution from other families. Authentic identification of medicinal herbs with respect to purity and quality has enormously become popular due to inclination of modern civilization toward herbal drugs (Han et al., 2016). Aziz et al. (2017) used molecular identification for discriminating medicinal plant species using phylogenetic inference and BLAST method using two genes trnH-psbA and ITS2. Enan et al. (2017) have used DNA barcoding via three barcode loci coding, for genes matK, rbcL, and rpoC1 were amplified for identification of metabolically rich desert herbs in UAE.

1.4.2 Management Implication

Medicinal and aromatic plants are significant in conventional system of medicines and ground raw material for herbal-based advanced drugs. Due to overexploitation, varied ecological pressures, and institutional hindrances, medicinal and aromatic herbs are dwindling very fast. Numerous efforts are being carried out at different levels to tackle this issue via varied strategies. In situ practical training of tribal people to integrate folklore traditional knowledge with government and nongovernment organizations is first step toward conservation and management of Himalayan herbs (Ghimire et al., 2004; Saha and Sundriyal, 2012). Numerous efforts are carried out to elucidate methods and strategies pertaining to conservation, cultivation, and management of marketing in the field of medicinal and aromatic plants (Rana and Samant, 2009).

(a) Integrating tribal knowledge with modern innovations
The quantitative and qualitative data have been generated via integrating traditional knowledge and modern contrivances. However, there is still need to broaden our infrastructure and inculcate further explorations from Indian Himalaya to develop medicinal plant sector (Siwach et al., 2013; Kanwal and Joshi, 2015). There is an urgent need to train local farmers/tribal people and provide them tools/contrivances that encourage them to draw their knowledge and awareness regarding medicinal plants on global level. We need to establish and maintain herbal wealth by regularly monitoring, organizing, and facilitating awareness regarding the know-how about cultivation of medicinal and aromatic plants (Uprety et al., 2011; Laladhas et al., 2015; Vijayan et al., 2017). These efforts need to be mediated through different nongovernment organizations (NGOs) and government organizations. But for effective management, we need to emphasize medicinal plants having greater economic potential. We need to advance your experimentation and introduce hill-specific technologies that include both exploration and evaluation of medicinally important plants (Singh et al., 2010; Krishnan et al., 2011). This could improve identification of quality seeds and enhance productivity of medicinal plants. Identification of potential seed stocks and dissemination of quality plant material should be a priority area for management. In Indian Himalaya, there is an urgent need for advancing cultivation practices and extending

agricultural facilities like polyhouses, polypits, shade nets, and irrigation tools like water harvesting system (Negi et al., 2013; Singh et al., 2014). There should be an integrative effort from research institutes to raise quality propagation of medicinal herbs (Maikhuri et al., 2011; Rana et al., 2013). Mass multiplication of medicinal plants using low-cost propagation techniques was done to accumulate bioactive metabolites under diverse experimental setups.

(b) Conservation and management practices
Collaborations with forest department are to be carried out as per environmental regulations to ensure sustainable management of MAPs. These collaborations are established with tribal community to form Van Panchayats since 1932 (Balooni et al., 2007). Advancing medicinal and aromatic plants for mass cultivation is one of the potential alternatives for augmenting livelihood of tribal communities together with conservation of species in their natural habitat. Consequently, it is more reliable to advance in situ conservation of species (Kala, 2009). For this purpose, we need to first identify superior populations of MAPs for in situ conservation that could provide breeding naturalized population of medicinal plants.

(c) Market investigation of herbal sector
The cost analysis of specific species has to be estimated in market to analyze cost-benefit evaluation of medicinally important species. Among MAPs, *A. heterophyllum* displayed most valuable species with a market rate approx. Rs. 230,833.4 ± 31,258.3/ha, and *Matricaria chamomilla* dispersed least money in a market rate approx. Rs. 36,000 ± 1000/ha. Indian Himalaya yielded maximum monetary benefit Rs. 161,666.7 ± 10,000/ha from *Cymbopogon citratus*, while *S. costus* exhibits maximum income to expenditure ratio following 3 years of cultivation (Kala et al., 2006; Singh, 2011). For expansive marketing of Himalayan herbs, we need to identify major stakeholders implicated in marketing business including primary producers of raw material and then industry for processing of raw material and finally identifying consumers (Olsen, 2005; Aneesh et al., 2009; Kumar et al., 2011; Kandari et al., 2012). There is spectacular rise in consummation of herbs; such proclivity toward herbs appears to be a result of potential novel approach toward economic development together with effective management of valuable Himalayan resources (Nautiyal et al., 2005; Dhiman et al., 2009; Kuniyal et al., 2014). The prime premise is that tribal communities can yield maximum benefit from MAPs and natural resource-based livelihood that could preserve Himalayan herbs and lead to sustainable development of Himalayan resources (Dawa et al., 2013; Bhushan et al., 2016). However, tribal communities need to vigorously grow MAPs to meet market demands so as to respond for increased demands of herbal products (reference). This

could provide tribal people with greater opportunities and more sustainable implications for management of local resources (Kala, 2015). Due to increased market demands, there are higher challenges of combining development of MAP markets and capital raised with conservation of Himalayan resources (Mukherjee et al., 2016). There is urgent need to establish market between primary producers of herbs (growers) and buyers for marketing (consumers). As primary producers incline toward species displaying higher market demands, this results in conservation of selected species (Hussain and Hore, 2007; Khan et al., 2013a,b). Hence, primary producers/growers become the main point of transaction for tribal people, and in majority of cases, they are involved in local business trade of MAPs.

(d) Contribution of various stake holders
Contribution of all stakeholders is indispensable for facilitating growers in leveraging the aid for application of rural technologies in continual way. Majority of primary growers in Northwestern Himalaya are associated with cooperative banks and National Bank for Agriculture and Rural Development (NABARD) for financial support for extensive cultivation of MAPs (Uprety et al., 2011). However, currently, there are no approved guidelines that are available for MAP entrepreneur, but cost-benefit evaluation describes credit necessities that are managed by policy planners and offered to growers (Singh, 2009; Thakur et al., 2013; Phondani et al., 2016).

(e) Current institutional activities
Majority of organizations and institutions executing cultivation and conservation of Himalayan MAPs are primarily arranged along traditional lines while it seems to be indispensable to break boundaries and overcome ecological constraints. In this context, it is necessary to establish associations between diverse departments of both central and state level organizations along nongovernment organizations (Kumar et al., 2014). To establish this strategy, we need to include methods for conservation, cultivation, marketing, and management that could advance monetary incomes to tribal community while managing Himalayan resources as well (Butola and Vashistha, 2013). It is the prime duty of government organizations to adopt bottom-up approach instead of top-down strategy to ensure that each household participates in decision-making policy (Bera, 2010; Joshi and Rao, 2011; Kuniyal et al., 2015). Management implication can furnish an actual effect on advancement of mass cultivation in MAPs; suitable technologies for competence of primary growers and tribal knowledge distribution have to be improved (Ghimire et al., 2008; Choudhary et al., 2013). Contributory research strategies and skill development programs to farmers/growers develop knowledge and general awareness for conservation and management of MAPs. Numerous organizations implicated in

evaluation, advancement, and demonstration of research need to create possible atmosphere that ensures association among farmers and government organizations (Butola, 2011). These associations are supposed to be basic contrivances for providing scientists to assemble suitable data into advanced technology/products to grower's requirements (Chhetri, 2014). Effective partnerships between stakeholders are necessary for mass cultivation of MAPs by advancing information exchange (Singh and Jha, 2008; Mazid et al., 2012). Farming practices in Northwestern Himalaya have observed that training procedures and data brochures can be utilized to encourage information exchange, analyze agrotechnology, and advance the pace of MAP cultivation (Singh et al., 2003; Bisht et al., 2007). The challenge is to integrate tribal information, science, and technological innovations together to meet market demands of the herbal sector. These strategies of MAP sector establish that government approaches pertaining to MAP conservation, cultivation, and management need to be established in milieu of an extensive advancement approaches and definite course of community revolution (Farooquee et al., 2004; Samant et al., 2007a,b; Negi et al., 2010). Medicinal plant cultivation requires effective policies such as marketing as a major component for distributing bulletin in order to collate and spread information regarding cultivation and consumption of MAPs (Palni et al., 2001; Nautiyal et al., 2002). Although in Northwestern Himalaya there has been introduction of strategies to advance and support mass cultivation and conservation together, these policies have not yielded desired results. But we can design new policies for Indian Himalaya that inculcate inputs from all stakeholders that provide comprehensive information on developing their efficient mass cultivation, conservation, and management.

1.5 ALIEN AND INVASIVE FLORA OF KASHMIR

Plant invasions in Indian Himalaya have been an interesting theme during the recent years (Khuroo et al., 2009; Jakobs et al., 2010). There are numerous rationales pertaining to sensitivity of plant invasions in Indian Himalaya due to high endemism (Weber et al., 2008). During the recent years, the biodiversity of Indian Himalaya has been disturbed immensely due to anthropogenic activities, for instance, deforestation and increased urbanization (Khuroo et al., 2008; Sekar, 2012). Due to immense population pressure together with increased transport facilitations, there has been increased incidence of invasive plant species (Khuroo et al., 2010). Though there were alien species already present in mountainous areas but were restricted in lower altitudinal ranges, due to climate change, alien plant species are spreading over high altitudinal ranges (Khuroo et al., 2011). There are increasing impacts on biodiversity due to higher incidence of alien plant species including the rise in endemism as well (Shah and Reshi, 2014). In order to understand plant invasion more clearly, we need to illustrate species richness along altitudinal gradient so as to compare dynamics of alien species with native species (Dar et al., 2015). Altitudinal association among alien and native species along with environmental variables can demonstrate species richness and major threats to biodiversity (Wu et al., 2004; Chandra Sekar et al., 2012). Various studies have illustrated positive association between native species and altitude varying from hump-shaped linkages with maximum species richness at midaltitudes (McDougall et al., 2011) to monotonically declining associations (Shah et al., 2011). Recent research has demonstrated altitudinal biogeographic distributions of the woody flora of Kashmir Himalaya, India. The valley of Kashmir subsists in mountain folds of Indian Himalaya with rich floristic diversity comprising endemics and provides extensive altitudinal ranges (500–4800 m asl). In spite of being at higher risk of plant invasion due to its European colonial past, Indian Himalaya has paucity pertaining to the description of its alien flora (Richardson, 2004). For effective management strategies, corroborated identification, suitable documentation, and description of alien flora are of ultimate significance for promoting studies that would create new approaches for their management (Adhikari et al., 2015). It is more significant to identify native and alien species discreetly so as to expand their application and overcome taxonomic rise in the directories of diversity in alien plant species (Kiehn, 2011; Miller and Schelhas, 2008). Moreover, evaluation of alien flora in phytogeographically distinct regions is an essential strategy for recognizing the discriminating prototypes of plant invasions and indispensable factors of plant invasions (McDougall et al., 2011; Shah et al., 2011). Plant invasion in biodiversity-rich areas needs to be addressed, and for this purpose, we need to assemble alien flora, but no such effort has been made for Indian Himalaya (dan Prospek et al., 2013). Indian Himalaya is more prone to plant invasions due to its proximity with the former Silk route and its endorsement as a global tourist destination resulting in the introduction of diverse flora from varied geographic regions (Khuroo et al., 2012a,b). This dearth of basic information required assemblage of the primary catalog of alien flora from the Kashmir Himalaya. The whole alien flora of Kashmir Valley was characterized by 571 plant species under 352 genera and 104 families. It comprises a comparatively higher (29%) proportion of the whole Kashmir flora. Families with maximum alien representatives are Poaceae (60 species), Asteraceae (54 species), and Brassicaceae (30 species), though families, for instance, Amaranthaceae (83%) and Chenopodiaceae (71%) displayed maximum portion of aliens comparative with their whole Kashmir flora. Among MAPs, maximum number of alien species were reported from *Artemisia* (six spp.), *Hibiscus, Ipomoea, Mentha, Nymphaea, Papaver, Potamogeton,* and *Trifolium* (five spp.). Alien flora was predominantly represented by 181 herbs both annual (32%) and perennial (27%). Among these, 119 (36%) plants were included as ornamentals and 20 (6%) for their medicinal value. Aromatic and medicinal herbs that were introduced include *Lavandula officinalis Chaix, Rosmarinus officinalis* L., *Godetia amoena* Den., *Digitalis grandiflora* Mill., *Nigella sativa* L., *Ranunculus asiaticus* L., and *Petunia hybrid* (Wani and Mushtaq, 2012; Khuroo et al., 2012a, 2012b). In order to identify invasive alien species, we need to follow methodologies pertaining to terminology and concepts of plant invasion that are given previous research reports (Khuroo et al., 2008; Shah et al., 2011). We need to be meticulous while assembling regional floral data, pool it with updated flora for complete explication of likely grounds of phenomenon of invasion, and fortify approaches for effectual prognostic organization, supervising contrivances and management practices. Numerous plant species in various parts across the world led to severe effects on natural ecosystems that are naturalized in Kashmir Himalaya. On contrary, *Aesculus indica* is not invasive anywhere but native in British Isles and New Zealand (Bhat et al., 2013). There is an enormous disparity in the status of alien species among different regions across the world. Indian Himalaya is characterized by severe climatic regimes that describe minimum possibility of invasive species since very few species can inhabit at higher altitudes. But changing climate can facilitate plant invasion through severe effects on native species (Khuroo et al., 2009). We still need illustrative explanation for comparing population dynamics of naturalized alien species at diverse altitudes. Majority of plant invasions are due to historical causes, for example, introductions to mountainous areas from neighboring areas than from overseas. Higher plant species from Asia adapt better in Indian Himalaya than species from abroad. Asian and Eurasian species are common among herbaceous alien species in Indian Himalaya (Khuroo et al., 2007a,b). Alien plant species mask association between native species and ecological factors of habitat. We need to study biogeographic and macroecological evaluation in order to ascertain alien status of species. For this, we need to design inventories containing status of species more elaborately. Alien species occupying majority of Indian Himalaya has substantial effect on plant diversity. Moreover, possible hybridization between native and alien congeneric species is yet another issue. We are oblivious regarding the degree of co-occurrence of these species that could potentially hybridize and that need to be investigated. Since hybridization results in the formation of new invasive taxa and expands the ecological niche of invasive species, we need to investigate plant invasions with regard to their ecological effects, species richness, and hybridization within this Himalayan region.

1.6 CONCLUSIONS

Indian Himalaya is well known for its floristic diversity, and most of the raw material for herbal drugs is exported from India. In order to market demands of herbal drugs, we need to not only conserve but also ensure mass multiplication of MAPs. However, extensive national and international trade has become prime threat to survival of Himalayan herbs. More conservational strategies are required for extensive mass multiplication of MAPs. This chapter reviews biogeographic distribution, correct identification, and management strategies for conservation of MAPs. We have also reviewed the effect of plant invasions on biodiversity of MAPs, and invasive species causes loss of biodiversity including species extinctions and ecosystem function. The competition for nutrients and space could change soil ecology and hamper growth of surviving Himalayan species. For effective conservation and management of Himalayan MAPs, we need to identify and explore biogeographic distribution and assess plant invasion and designing of inventories. We need to integrate phytosociological processes and mapping via GPS to effectively investigate loss of important species and incidences of plant invasions. There is a need to collate information resources from taxonomists, ecologists, and ecophysiologists to monitor and assess biodiversity of Himalayan MAPs. Active contribution of all stakeholders will ensure advancement of MAPs, and the use of modern technological interventions can effectively manage Himalayan resources.

References

Adhikari, D., Tiwary, R., Barik, S.K., 2015. Modelling hotspots for invasive alien plants in India. PLoS One 10 (7), e0134665.

Ahmad, S., Radotra, S., Singh, J.P., Verma, D.K., Sultan, S.M., 2017. Ethnoveterinary uses of some important plants by pastoralists in Kashmir Himalaya. SKUAST J. Res. 19 (1), 121–128.

Alves, R.R.N., Medeiros, M.F.T., Albuquerque, U.P., Rosa, I.L., 2013. From past to present: medicinal animals in a historical perspective. In: Animals in Traditional Folk Medicine. Springer, Berlin Heidelberg, pp. 11–23.

Ameh, S.J., Obodozie, O.O., Inyang, U.S., Abubakar, M.S., Garba, M., 2010. Current phytotherapy—a perspective on the science and regulation of herbal medicine. J. Med. Plant Res. 4 (2), 072–081.

Anand, M., Kumar, K., 2008. Study of research trends in land and water resource management in GB Pant Institute of Himalayan environment and development. Himal. Ecol. 16 (1), 37.

Aneesh, T., Hisham, M., Sekhar, M., Madhu, M., Deepa, T., 2009. International market scenario of traditional Indian herbal drugs-India declining. Int. J. Green Pharm. 3 (3), 184.

Aranda, S.C., Hespanhol, H., Homem, N., Borges, P.A., Lobo, J.M., Gabriel, R., 2015. The iterative process of plant species inventorying for obtaining reliable biodiversity patterns. Bot. J. Linn. Soc. 177 (4), 491–503.

Armstrong, K.F., Ball, S.L., 2005. DNA barcodes for biosecurity: invasive species identification. Philos. Trans. Roy. Soc. Lond. B: Biol. Sci. 360 (1462), 1813–1823.

Aziz, N.A.A., Ahmad, M.I., Naim, D.M., 2017. Species identification based on trnH-psbA and ITS2 genes and analysis of mineral nutrients of selected medicinal plants from Malaysia. Jordan J. Biol. Sci. 10, 117–125.

Badola, H.K., Aitken, S., 2003. The Himalayas of India: a treasury of medicinal plants under siege. Biodiversity 4 (3), 3–13.

Balooni, K., Ballabh, V., Inoue, M., 2007. Declining instituted collective management practices and forest quality in the Central Himalayas. Econ. Polit. Wkly. 1443–1452.

Barboza, G.E., Cantero, J.J., Núñez, C., Pacciaroni, A., Ariza Espinar, L., 2009. Medicinal plants: a general review and a phytochemical and ethnopharmacological screening of the native Argentine Flora. Kurtziana 34 (1-2), 7–365.

Belhumeur, P., Chen, D., Feiner, S., Jacobs, D., Kress, W., Ling, H., Zhang, L., 2008. In: Searching the world's herbaria: a system for visual identification of plant species. Computer Vision—ECCV 2008, pp. 116–129.

Benbrook, C., 2012. Principles governing the long-run risks, benefits, and costs of agricultural biotechnology. In: Biodiversity and the Law: "Intellectual Property, Biotechnology and Traditional Knowledge.". Earthscan, London.

Bera, S.K., 2010. Globalisation and sustainable exports of Indian medicinal and aromatic plants: a protection study, No. 28908. University Library of Munich, Germany.

Bhat, J.A., Kumar, M., Bussmann, R.W., 2013. Ecological status and traditional knowledge of medicinal plants in Kedarnath Wildlife Sanctuary of Garhwal Himalaya, India. J. Ethnobiol. Ethnomed. 9 (1), 1.

Bhat, T.A., Nigam, G., Majaz, M., 2012. Study of some medicinal plants of the Shopian district, Kashmir (India) with emphasis on their traditional use by Gujjar and Bakerwal tribes. Asian J. Pharm. Clin. Res. 5 (Suppl. 2), 94–98.

Bhatt, D., Kumar, R., Tewari, L.M., Joshi, G.C., 2014. *Polygonatum cirrhifolium* Royle and *Polygonatum verticillatum* (L.) Allioni: status assessment and medicinal uses in Uttarakhand, India. J. Med. Plant Res. 8 (5), 253–259.

Bhowmik, D., Kumar, K.P., Pankaj, T., Chiranjib, B., 2009. Traditional herbal medicines: an overview. Arch. Appl. Sci. Res. 1 (2), 165–177.

Bhushan, I., Kumar, A., Patel, J.S., Yadav, R.P., Singh, S., Meena, R., Meena, V.S., 2016. The Indian Himalayan ecosystem as source for survival. In: Conservation Agriculture. Springer, Singapore, pp. 367–380.

Bickford, D., Lohman, D.J., Sodhi, N.S., Ng, P.K., Meier, R., Winker, K., Das, I., 2007. Cryptic species as a window on diversity and conservation. Trends Ecol. Evol. 22 (3), 148–155.

Bisht, I.S., Mehta, P.S., Bhandari, D.C., 2007. Traditional crop diversity and its conservation on-farm for sustainable agricultural production in Kumaon Himalaya of Uttaranchal state: a case study. Genet. Resour. Crop. Evol. 54 (2), 345–357.

Bisht, V.K., Kuniyal, C.P., Nautiyal, B.P., Prasad, P., 2013. Spatial distribution and regeneration of *Quercus semecarpifolia* and *Quercus floribunda* in a subalpine forest of western Himalaya, India. Physiol. Mol. Biol. Plants 19 (3), 443.

Borowitzka, M.A., 2013. High-value products from microalgae—their development and commercialisation. J. Appl. Phycol. 25 (3), 743.

Breininger, D., Duncan, B., Eaton, M., Johnson, F., Nichols, J., 2014. Integrating land cover modeling and adaptive management to conserve endangered species and reduce catastrophic fire risk. Land 3 (3), 874–897.

Butola, J.S., 2011. Post-cultivation evaluation of germplasm in Himalayan threatened medicinal herbs: implication for ex-cultivation and conservation. Natl. Acad. Sci. Lett. 34 (1), 49.

Butola, J.S., Vashistha, R.K., 2013. An overview on conservation and utilization of *Angelica glauca* Edgew. in three Himalayan states of India. Emerg. Med. Plant 5 (3), 171–178.

Canter, P.H., Thomas, H., Ernst, E., 2005. Bringing medicinal plants into cultivation: opportunities and challenges for biotechnology. Trends Biotechnol. 23 (4), 180–185.

Carpenter, C., 2005. The environmental control of plant species density on a Himalayan elevation gradient. J. Biogeogr. 32 (6), 999–1018.

Chakraborty, T., Saha, S., Bisht, N.S., 2017. First Report on the Ethnopharmacological Uses of Medicinal Plants by Monpa Tribe from the Zemithang Region of Arunachal Pradesh, Eastern Himalayas, India. Plants 6 (1), 13.

Chandra Sekar, K., Manikandan, R., Srivastava, S.K., 2012. Invasive alien plants of Uttarakhand Himalaya. Proc. Natl. Acad. Sci. India Sect. B 1–9.

Chaudhuri, A.B., Chaudhuri, A.B., 2003. Megadiversity Conservation: Flora, Fauna and Medicinal Plants of India's Hot Spots. Daya Books, New Delhi.

Chauhan, N.S., 1999. Medicinal and Aromatic Plants of Himachal Pradesh. Indus Publishing, New Delhi.

Chauhan, R.S., Nautiyal, M.C., 2005. Commercial viability of cultivation of an endangered medicinal herb *Nardostachys jatamansi* at three different agroclimatic zones. Curr. Sci. 89 (9), 1481–1488.

Chauhan, R.S., Tiwari, D., Bisht, A.S., Shukla, A., 2014. Ex situ conservation of medicinal and aromatic plants in Bharsar, Uttarakhand, India. Med. Plants Int. J. Phytomed. Relat. Ind. 6 (4), 282–292.

Chaurasia, O.P., Ahmed, Z., Ballabh, B., 2007. Ethnobotany and Plants of Trans-Himalaya. SSPL, New Delhi.

Chawla, A., Rajkumar, S., Singh, K.N., Lal, B., Singh, R.D., Thukral, A.K., 2008. Plant species diversity along an altitudinal gradient of Bhabha Valley in western Himalaya. J. Mt. Sci. 5 (2), 157–177.

Chen, S., Yao, H., Han, J., Liu, C., Song, J., Shi, L., Luo, K., 2010. Validation of the ITS2 region as a novel DNA barcode for identifying medicinal plant species. PLoS One 5 (1), e8613.

Chhetri, D.R., 2014. Medicinal Plants of the Himalaya: Production Technology and Utilization. Agrobios, Jodhpur.

Chitale, V.S., Behera, M.D., Roy, P.S., 2014. Future of endemic flora of biodiversity hotspots in India. PLoS One 9 (12), e115264.

Choudhary, D., Kala, S.P., Todaria, N.P., Rawat, R.B.S., Kunwar, M.S., Kollmair, M., 2013. Upgrading mountain people in medicinal and aromatic plants value chains: lessons for sustainable management and income generation from Uttarakhand, India. Int. J. Sust. Dev. World Ecol. 20 (1), 45–53.

Cope, J.S., Corney, D., Clark, J.Y., Remagnino, P., Wilkin, P., 2012. Plant species identification using digital morphometrics: a review. Expert Syst. Appl. 39 (8), 7562–7573.

Corlett, R.T., 1988. The naturalized flora of Singapore. J. Biogeogr. 15, 657–663.

dan Prospek, M., Dar, G., Khuroo, A., 2013. Floristic diversity in the Kashmir Himalaya: progress, problems and prospects. Sains Malaysiana 42 (10), 1377–1386.

Dar, G.H., Bhagat, R.C., Khan, M.A., 2002. Biodiversity of the Kashmir Himalaya. Valley Book House, Srinagar.

Dar, P.A., Reshi, Z.A., Shah, M.A., 2015. Roads act as corridors for the spread of alien plant species in the mountainous regions: a case study of Kashmir Valley, India. Trop. Ecol. 56 (2), 183–190.

Dawa, S., Hussain, A., Tamchos, T., 2013. Medicinal and aromatic plants as a viable livelihood option for the people of Jammu and Kashmir. Indian Hortic. J. 3 (1&2), 40–42.

Devi, B., Bhardwaj, D.R., Panwar, P., Pal, S., Gupta, N.K., Thakur, C.L., 2013. Carbon allocation, sequestration and carbon dioxide mitigation under plantation forests of north western Himalaya, India. Ann. For. Res. 56 (1), 3.

Dhar, U., 2002. Conservation implications of plant endemism in high-altitude Himalaya. Curr. Sci. 141–148.

Dhar, U., Rawal, R.S., Upreti, J., 2000. Setting priorities for conservation of medicinal plants—a case study in the Indian Himalaya. Biol. Conserv. 95 (1), 57–65.

Dhiman, M., Soumendra, C., Amar, R., Mokthan, M.W., 2009. Differential approach of germplasm conservation of high value of medicinal plants in North-Eastern Himalayan Region. Int. J. Agric. Environ. Biotechnol. 2 (4), 332–340.

Dhyani, A., Nautiyal, B.P., Nautiyal, M.C., 2010. Importance of Astavarga plants in traditional systems of medicine in Garhwal, Indian Himalaya. Int. J. Biodivers. Sci. Ecosyst. Manage. iFirst, 1–7. https://doi.org/10.1080/21513732.2010.521490.

Dubey, N.K., Kumar, R., Tripathi, P., 2004. Global promotion of herbal medicine: India's opportunity. Curr. Sci. 86 (1), 37–41.
Early, R., Bradley, B.A., Dukes, J.S., Lawler, J.J., Olden, J.D., Blumenthal, D.M., Sorte, C.J., 2016. Global threats from invasive alien species in the twenty-first century and national response capacities. Nat. Commun. 7.
Eaton, D., 2015. Variability of climate and riverflow in partially glacierised basins in the Karakoram and western Himalaya. Doctoral dissertation, University of Salford.
Elvin-Lewis, M., 2011. Dereplications can amplify the extent and worth of traditional pharmacopeias. Afr. J. Tradit. Complement. Altern. Med. 8 (5S).
Enan, M.R., Palakkott, A.R., Ksiksi, T.S., 2017. DNA barcoding of selected UAE medicinal plant species: a comparative assessment of herbarium and fresh samples. Physiol. Mol. Biol. Plants 23 (1), 221–227.
Enomoto, T., 1999. Naturalized weeds from foreign countries into Japan. In: Yano, E., Matsuo, K., Shiyomi, M., Andow, D.A. (Eds.), Biological Invasions of Ecosystem Pests and Beneficial Organisms. Yokendo, Tokyo.
Farooquee, N.A., Majila, B.S., Kala, C.P., 2004. Indigenous knowledge systems and sustainable management of natural resources in a high altitude society in Kumaun Himalaya, India. J. Hum. Ecol. 16 (1), 33–42.
Froese, R., Capuli, E., Rañola, M.C., 2000. Challenges to taxonomic information management: how to deal with changes in scientific names. In: Global Environmental Researches on Biological and Ecological Aspects. Center for Global Environmental Research, Tsukuba, Japan, pp. 3–10.
Gairola, S., Sharma, J., Bedi, Y.S., 2014. A cross-cultural analysis of Jammu, Kashmir and Ladakh (India) medicinal plant use. J. Ethnopharmacol. 155 (2), 925–986.
Ghimire, S.K., McKey, D., Aumeeruddy-Thomas, Y., 2004. Heterogeneity in ethnoecological knowledge and management of medicinal plants in the Himalayas of Nepal: implications for conservation. Ecol. Soc. 9 (3), 6.
Ghimire, S.K., Gimenez, O., Pradel, R., McKey, D., Aumeeruddy-Thomas, Y., 2008. Demographic variation and population viability in a threatened Himalayan medicinal and aromatic herb *Nardostachys grandiflora*: matrix modelling of harvesting effects in two contrasting habitats. J. Appl. Ecol. 45 (1), 41–51.
Ghosh, A.K., 2008. A Comprehensive Handbook on Biodiversity. The Energy and Resources Institute (TERI).
Ghosh, S.P., General, F.D.D., 2013. In: Promotion of medicinal and aromatic plants in the Asia-Pacific Region. Expert Consultation on Promotion of Medicinal and Aromatic Plants in the Asia-Pacific Region. Proceedings, p. 33.
Gotelli, N.J., Colwell, R.K., 2001. Quantifying biodiversity: procedures and pitfalls in the measurement and comparison of species richness. Ecol. Lett. 4 (4), 379–391.
Group, C.P.W., Hollingsworth, P.M., Forrest, L.L., Spouge, J.L., Hajibabaei, M., Ratnasingham, S., Fazekas, A.J., 2009. A DNA barcode for land plants. Proc. Natl. Acad. Sci. 106 (31), 12794–12797.
Gupta, M.P., 2011. Perspectives of medicinal and aromatic plants and their products in emerging globalization era. Hortic. Bras. 29 (2), 5831–5855.
Gupta, R.K., 2010. Medicinal and Aromatic Plants. vol. 234. CBS Publishers and Distributors, New Delhi, p. 499.
Gurib-Fakim, A., 2006. Medicinal plants: traditions of yesterday and drugs of tomorrow. Mol. Asp. Med. 27 (1), 1–93.
Hamilton, A.C., 2004. Medicinal plants, conservation and livelihoods. Biodivers. Conserv. 13 (8), 1477–1517.
Han, J., Pang, X., Liao, B., Yao, H., Song, J., Chen, S., 2016. An authenticity survey of herbal medicines from markets in China using DNA barcoding. Sci. Rep. 6, 18723.
Hobbs, R.J., Higgs, E., Harris, J.A., 2009. Novel ecosystems: implications for conservation and restoration. Trends Ecol. Evol. 24 (11), 599–605.
Hui, C., Richardson, D.M., 2017. Invasion Dynamics. Oxford University Press, USA.

Hundal, S.S., 2004. In: Wildlife conservation strategies and management in India: an overview. Proc. Species at Risk 2004 Pathways to Recovery Conference. vol. 2, pp. 2–6.

Hunt, D.J., Handoo, Z.A., 2009. Taxonomy, identification and principal species. In: Root-Knot Nematodes. vol. 1. Formerly Texas A&M University, USA, pp. 55–88.

Hussain, S., Hore, D.K., 2007. Collection and conservation of major medicinal pants of Darjeeling and Sikkim Himalayas. Indian J. Tradit. Knowl. vol. 6 (2), 352–357.

Jakobs, G., Kueffer, C., Daehler, C.C., 2010. Introduced weed richness across altitudinal gradients in Hawai'i: humps, humans and water-energy dynamics. Biol. Invasions 12 (12), 4019–4031.

Jan, S., Kamili, A.N., Parray, J.A., Bedi, Y.S., Ahmad, P., 2016. Microclimatic variation in UV perception and related disparity in tropane and quinolizidine alkaloid composition of *Atropa acuminata, Lupinus polyphyllus* and *Hyoscyamus niger*. J. Photochem. Photobiol. B Biol. 161, 230–235.

Jayakumar, S., Kim, S.S., Heo, J., 2011. Floristic inventory and diversity assessment—a critical review. Proc. Int. Acad. Ecol. Environ. Sci. 1 (3–4), 151.

Jee, V., 2015. Cytogeography of Some Endemic Taxa in Kashmir Himalaya, Ph.D. Thesis, http://hdl.handle.net/10603/32917.

Joshi, D.D., 2012. Herbal Drugs and Fingerprints: Evidence Based Herbal Drugs. Springer Science & Business Media, Heidelberg, New York.

Joshi, P., Rao, N., 2011. Role of indigenous people in conservation of biodiversity of medicinal plants: an Indian case study. Surviv. Sustain. 91–101.

Kaack, K.V., Knudsen, B.F., Rosenberg, L., 2013. In: Guidelines for good agricultural and collection practice for *Sambucus nigra* flowers and berries as active ingredients in traditional herbal medicine and dietary supplements. I International Symposium on Elderberry 1061. pp. 225–235.

Kala, C.P., 2000. Status and conservation of rare and endangered medicinal plants in the Indian trans-Himalaya. Biol. Conserv. 93 (3), 371–379.

Kala, C.P., 2009. Medicinal plants conservation and enterprise development. Med. Plants 1 (2), 79–95.

Kala, C.P., 2015. Medicinal and aromatic plants: boon for enterprise development. J. Appl. Res. Med. Aromat. Plants 2 (4), 134–139.

Kala, C.P., 2005. Ethnomedicinal botany of the Apatani in the Eastern Himalayan region of India. J. Ethnobiol. Ethnomed. 1 (1), 11.

Kala, C.P., Dhyani, P.P., Sajwan, B.S., 2006. Developing the medicinal plants sector in northern India: challenges and opportunities. J. Ethnobiol. Ethnomed. 2 (1), 32.

Kandari, L.S., Phondani, P.C., Payal, K.C., Rao, K.S., Maikhuri, R.K., 2012. Ethnobotanical study towards conservation of medicinal and aromatic plants in upper catchments of Dhauli Ganga in the central Himalaya. J. Mt. Sci. 9 (2), 286–296.

Kanwal, K.S., Joshi, H., 2015. Medicinal plants diversity, indigenous uses and conservation status in Alaknanda Valley of Western Himalaya, Uttarakhand, India. Indian Forester 141 (6), 660–669.

Karki, M., Tiwari, B., Badoni, A., Bhattarai, N., 2003. Creating livelihoods enhancing medicinal and aromatic plants based biodiversity-rich production systems: preliminary lessons from South Asia. In: Oral Paper Presented at The 3rd World Congress on Medicinal and Aromatic Plants for Human Welfare (WOCMAP III), vol. 10. Chiang Mai, Thailand.

Kaul, M.K., 1997. Medicinal Plants of Kashmir and Ladakh: Temperate and Cold Arid Himalaya. Indus Publishing.

Kaul, M.K., 2010. High altitude botanicals in integrative medicine—case studies from Northwest Himalaya. Indian J. Tradit. Knowl. 9 (1), 18–25.

Khan, S.M., Harper, D.M., Page, S., Ahmad, H., 2011. Species and community diversity of vascular flora along environmental gradient in Naran Valley: a multivariate approach through indicator species analysis. Pak. J. Bot. 43 (5), 2337–2346.

Khan, S.M., Page, S.E., Ahmad, H., Harper, D.M., 2013a. Sustainable utilization and conservation of plant biodiversity in montane ecosystems: the western Himalayas as a case study. Ann. Bot. 112 (3), 479–501.

Khan, S.M., Page, S., Ahmad, H., Harper, D., 2013b. Identifying plant species and communities across environmental gradients in the Western Himalayas: method development and conservation use. Eco. Inform. 14, 99–103.

Khare, C.P., 2008. Indian Medicinal Plants: An Illustrated Dictionary. Springer Science & Business Media.

Khuroo, A.A., Rashid, I., Reshi, Z., Dar, G.H., Wafai, B.A., 2007a. The alien flora of Kashmir Himalaya. Biol. Invasions 9, 269–292.

Khuroo, A.A., Dar, G.H., Khan, Z.S., Malik, A.H., 2007b. Exploring an inherent interface between taxonomy and biodiversity: current problems and future challenges. J. Nat. Conserv. 15 (4), 256–261.

Khuroo, A.A., Reshi, Z., Rashid, I., Dar, G.H., Khan, Z.S., 2008. Operational characterization of alien invasive flora and its management implications. Biodivers. Conserv. 17 (13), 3181–3194.

Khuroo, A.A., Weber, E., Malik, A.H., Dar, G.H., Reshi, Z.A., 2010. Taxonomic and biogeographic patterns in the native and alien woody flora of Kashmir Himalaya, India. Nord. J. Bot. 28 (6), 685–696.

Khuroo, A.A., Weber, E., Malik, A.H., Reshi, Z.A., Dar, G.H., 2011. Altitudinal distribution patterns of the native and alien woody flora in Kashmir Himalaya, India. Environ. Res. 111 (7), 967–977.

Khuroo, A.A., Reshi, Z.A., Dar, G.H., 2012a. Kashmir State, India. In: Invasive Alien Plants: An Ecological Appraisal for the Indian Subcontinent. vol. 1. CABI, UK, p. 216.

Khuroo, A.A., Reshi, Z.A., Malik, A.H., Weber, E., Rashid, I., Dar, G.H., 2012b. Alien flora of India: taxonomic composition, invasion status and biogeographic affiliations. Biol. Invasions 14 (1), 99.

Khuroo, A.A., Reshi, Z., Rashid, I., Dar, G.H., Malik, A.H., 2009. Plant invasions in montane ecosystems. Front. Ecol. Environ. 7 (8), 408.

Kiehn, M., 2011. In: Bramwell, D., Caujapé-Castells, J. (Eds.), Invasive Alien Species and Islands. Cambridge University Press, Cambridge, pp. 365–384.

Kirkpatrick, M., Barton, N.H., 1997. Evolution of a species' range. Am. Nat. 150 (1), 1–23.

Koh, K.S., Na, J.G., Suh, M.H., Kil, J.H., Ku, Y.B., Yoon, J.H., Oh, H.K., 2000. The Effects of Alien Plants on Ecosystem and Their Management. The Plant Taxonomic Society of Korea, Korea.

Kress, W.J., Wurdack, K.J., Zimmer, E.A., Weigt, L.A., Janzen, D.H., 2005. Use of DNA barcodes to identify flowering plants. Proc. Natl. Acad. Sci. U. S. A. 102 (23), 8369–8374.

Krishnan, P.N., Decruse, S.W., Radha, R.K., 2011. Conservation of medicinal plants of Western Ghats, India and its sustainable utilization through in vitro technology. In Vitro Cell. Dev. Biol. Plant 47 (1), 110–122.

Kumar, G.P., Kumar, R., Chaurasia, O.P., 2011. Conservation status of medicinal plants in Laclakh: cold arid zone of Trans-Himalayas. Res. J. Med. Plants 5, 685–694.

Kumar, R., Sharma, D., Tiwari, P., 2017. Floristic diversity assessment of dry temperate and alpine region of Kinnaur district of Himachal Pradesh. Int. J. Farm Sci. 7 (1), 57–60.

Kumar, S., 2016. Understanding altered molecular dynamics in the targeted plant species in Western Himalaya in relation to environmental cues: implications under climate change scenario. In: Abiotic Stress Response in Plants. Wiley VCH, Verlag, GmbH & Co, Germany.

Kumar, V., Babu, S., Revale, A.K., Meena, R.K., Ranjan, M.K., Desai, B.S., 2014. Cultivation of medicinal plants in natural ecosystem in Gujarat (India): constraints and conservation need. J. Plant Dev. Sci. 6 (3), 425–435.

Kuniyal, C.P., Bhatt, V.P., Bhatt, V.P., Butola, J.S., Sundriyal, R.C., 2014. Promoting nursery enterprise in high altitude villages: a participatory approach for conservation and commercialization of Himalayan threatened medicinal plants. J. Med. Plant Res. 8 (48), 1399–1407.

Kuniyal, C.P., Bisht, V.K., Negi, J.S., Bhatt, V.P., Bisht, D.S., Butola, J.S., Singh, S.K., 2015. Progress and prospect in the integrated development of medicinal and aromatic plants (MAPs) sector in Uttarakhand, Western Himalaya. Environ. Dev. Sustain. 17 (5), 1141.

Laladhas, K.P., Preetha, N., Baijulal, B., Oommen, V., 2015. Conservation of medicinal plant resources through Community Born Biodiversity Management Committee, Kerala, India. In: Biodiversity Conservation—Challenges for the Future. Bentham Science Publishers, India, p. 27.

Lanker, U., Malik, A.R., Gupta, N.K., Butola, J.S., 2010. Natural regeneration status of the endangered medicinal plant, *Taxus baccata* Hook. F. syn. *T. wallichiana*, in northwest Himalaya. Int. J. Biodivers. Sci. Ecosyst. Serv. Manage. 6 (1-2), 20–27.

Lardos, A., Heinrich, M., 2013. Continuity and change in medicinal plant use: the example of monasteries on Cyprus and historical iatrosophia texts. J. Ethnopharmacol. 150 (1), 202–214.

Larsen, H.O., Olsen, C.S., 2007. Unsustainable collection and unfair trade? Uncovering and assessing assumptions regarding central Himalayan medicinal plant conservation. Biodivers. Conserv. 16 (6), 1679–1697.

Leonti, M., Cabras, S., Weckerle, C.S., Solinas, M.N., Casu, L., 2010. The causal dependence of present plant knowledge on herbals–contemporary medicinal plant use in Campania (Italy) compared to Matthioli (1568). J. Ethnopharmacol. 130, 379–391.

Leppäkoski, E., Gollasch, S., Olenin, S. (Eds.), 2002. Invasive aquatic species of Europe. In Distribution, Impacts and Management. Kluwer Academic Publishers, USA.

Leroux, S.J., Albert, C.H., Lafuite, A.S., Rayfield, B., Wang, S., Gravel, D., 2017. Structural uncertainty in models projecting the consequences of habitat loss and fragmentation on biodiversity. Ecography 40 (1), 36–47.

Lone, P.A., Bhardwaj, A.K., Shah, K.W., Tabasum, S., 2014. Ethnobotanical survey of some threatened medicinal plants of Kashmir Himalaya, India. J. Med. Plant Res. 8 (47), 1362–1373.

Macdonald, D.W., Willis, K.J. (Eds.), 2013. Key Topics in Conservation Biology 2. John Wiley & Sons, West Sussex, UK.

Maikhuri, R.K., Rawat, L.S., Negi, V.S., Purohit, V.K., Rao, K.S., Saxena, K.G., 2011. Managing natural resources through simple and appropriate technological interventions for sustainable mountain development. Curr. Sci. 992–997.

Maikhuri, R.K., Phondani, P.C., Rawat, L.S., Jha, N.K., Maletha, A., Bahuguna, Y.M., Kandari, L.S., 2016. Conservation and management strategies of medicinal plant resources through action research approaches in Indian Himalaya. Iran. J. Sci. Technol. Trans. A: Sci. 1–7.

Malik, S., Sharma, N., Sharma, U.K., Singh, N.P., Bhushan, S., Sharma, M., Ahuja, P.S., 2010. Qualitative and quantitative analysis of anthraquinone derivatives in rhizomes of tissue culture-raised Rheum emodi Wall plants. J. Plant Physiol. 167 (9), 749–756.

Malik, Z.A., Bhat, J.A., Ballabha, R., Bussmann, R.W., Bhatt, A.B., 2015. Ethnomedicinal plants traditionally used in health care practices by inhabitants of Western Himalaya. J. Ethnopharmacol. 172, 133–144.

Mani, M.S. (Ed.), 2012. In: Ecology and Biogeography in India. vol. 23. Springer Science & Business Media, USA.

Manish, K., Pandit, M.K., Telwala, Y., Nautiyal, D.C., Koh, L.P., Tiwari, S., 2017. Elevational plant species richness patterns and their drivers across non-endemics, endemics and growth forms in the Eastern Himalaya. J. Plant Res. 130 (5), 829–844.

Marbuah, G., Gren, I.M., McKie, B., 2014. Economics of harmful invasive species: a review. Diversity 6 (3), 500–523.

Máthé, Á., 2015. Botanical aspects of medicinal and aromatic plants. In: Medicinal and Aromatic Plants of the World. Springer, Netherlands, pp. 13–33.

Maxted, N., Crust, R., 1995. Aids to taxonomic identification. In: Guarino, L., Ramanatha Rao, V., Reid, R. (Eds.), Collecting Plant Genetic Diversity. CAB International, Wallingford, UK.

Mazid, M., Khan, T.A., Mohammad, F., 2012. Medicinal plants of rural India: a review of use by Indian folks. Indo Global J. Pharm. 2 (3), 286–304.

McDougall, K.L., Alexander, J.M., Haider, S., Pauchard, A., Walsh, N.G., Kueffer, C., 2011. Alien flora of mountains: global comparisons for the development of local preventive measures against plant invasions. Divers. Distrib. 17 (1), 103–111.

Measey, G.J., Vimercati, G., Villiers, F.A., Mokhatla, M., Davies, S.J., Thorp, C.J., Kumschick, S., 2016. A global assessment of alien amphibian impacts in a formal framework. Divers. Distrib. 22 (9), 970–981.

Miller, J.H., Schelhas, J., 2008. Adaptive collaborative restoration: a key concept in invasive plant management. In: Invasive Plants and Forest Ecosystems. CRC Press, Taylor and Francis group, Boca Raton, US, p. 251.

Miller, J.R., 2010. Survey of Western Tragopan, Koklass Pheasant, and Himalayan Monal populations in the Great Himalayan National Park, Himachal Pradesh, India. Indian Birds 6 (3), 60–65.

Ministry of Environment Forests (MoEF) of India, 1988. National Forest Policy (1988). Resolution No. 3-1/86-FP. Department of Environment, Forests and Wildlife, New Delhi, India.

Molnar, J.L., Gamboa, R.L., Revenga, C., Spalding, M.D., 2008. Assessing the global threat of invasive species to marine biodiversity. Front. Ecol. Environ. 6 (9), 485–492.

Mukherjee, A., Rakshit, S., Nag, A., Ray, M., Kharbikar, H.L., Shubha, K., Meena, V.S., 2016. Climate change risk perception, adaptation and mitigation strategy: an extension outlook in Mountain Himalaya. In: Conservation Agriculture. Springer, Singapore, pp. 257–292.

Mukherjee, P.K., Wahile, A., 2006. Integrated approaches towards drug development from Ayurveda and other Indian system of medicines. J. Ethnopharmacol. 103 (1), 25–35.

Mukherjee, P.K., Rai, S., Kumar, V., Mukherjee, K., Hylands, P.J., Hider, R.C., 2007. Plants of Indian origin in drug discovery. Expert Opin. Drug Discovery 2 (5), 633–657.

Nandi, S.K., Palni, L.M.S., Pandey, H., Chandra, B., Nadeem, M., 2016. Selection of elites and in vitro propagation of selected high-value Himalayan medicinal herbs for sustainable utilization and conservation. In: Plant Tissue Culture: Propagation, Conservation and Crop Improvement. Springer, Singapore, pp. 15–44.

Nandy, S.N., 2014. Agro-economic indicators—a comparative study of North-Eastern States of India. J. Land Rural Stud. 2 (1), 75–88.

Nandy, S.N., Dhyani, P.P., Samal, P.K., 2006. Resource information database of the Indian Himalaya. In: Environmental Information System on Himalayan Ecology. GP Institute of Himalayan Environment and Development.

Nautiyal, B.P., Nautiyal, M.C., Khanduri, V.P., Rawat, N., 2009. Floral biology of *Aconitum heterophyllum* Wall.: a critically endangered alpine medicinal plant of Himalaya, India. Turk. J. Bot. 33 (1), 13–20.

Nautiyal, S., Maikhuri, R.K., Semwal, R., Saxena, K., 2002. Agroecosystem function around a Himalayan biosphere reserve. J. Environ. Syst. 29 (1).

Nautiyal, S., Rajan, K.S., Shibasaki, R., 2005. Interaction of biodiversity and economic welfare—a case study from the Himalayas of India. J. Environ. Inform. 6 (2), 111–119.

Negi, G.C.S., Samal, P.K., Kuniyal, J.C., Kothyari, B.P., Sharma, R.K., Dhyani, P.P., 2012. Impact of climate change on the western Himalayan mountain ecosystems: an overview. Trop. Ecol. 53 (3), 345–356.

Negi, V.S., Maikhuri, R.K., Phondani, P.C., Rawat, L.S., 2010. An inventory of indigenous knowledge and cultivation practices of medicinal plants in Govind Pashu Vihar Wildlife Sanctuary, Central Himalaya, India. Int. J. Biodivers. Sci. Ecosyst. Serv. Manage. 6 (3–4), 96–105.

Negi, V.S., Maikhuri, R.K., Rawat, L.S., Parshwan, D., 2013. Protected cultivation as an option of livelihood in mountain region of central Himalaya, India. Int. J. Sust. Dev. World Ecol. 20 (5), 416–425.

Norton, B.G., 2014. The Preservation of Species. Princeton University Press, Princeton, New Jersey.

Olsen, C.S., 2005. Valuation of commercial central Himalayan medicinal plants. AMBIO J. Hum. Environ. 34 (8), 607–610.

Palni, L.M.S., Maikhuri, R.K., Rao, K.S., 2001. Conservation of the Himalayan agroecosystems: issues and priorities. In: Economy of Uttaranchal: Profile and Dynamics of Change. Anamika Publishers & Distributors, New Delhi, pp. 135–176.

Pandit, M.K., Manish, K., Koh, L.P., 2014. Dancing on the roof of the world: ecological transformation of the Himalayan landscape. Bioscience 64 (11), 980–992.

Pant, S., Pant, V.S., 2011. Status and conservation management strategies for threatened plants of Jammu and Kashmir. J. Phytol. 3 (7).

Parihaar, R.S., Bargali, K., Bargali, S.S., 2015. Status of an indigenous agroforestry system: a case study in Kumaun Himalaya, India. Indian J. Agric. Sci. 85 (3), 442–447.

Patel, D.K., 2014. Some medicinal and aromatic shrubs introduced in herbal garden for their propagation and ex-situ conservation. J. Med. Plants 2 (6), 9–12.

Phondani, P.C., Bhatt, I.D., Negi, V.S., Kothyari, B.P., Bhatt, A., Maikhuri, R.K., 2016. Promoting medicinal plants cultivation as a tool for biodiversity conservation and livelihood enhancement in Indian Himalaya. J. Asia-Pacific Biodivers. 9 (1), 39–46.

Pyšek, P., Richardson, D.M., 2010. Invasive species, environmental change and management, and health. Annu. Rev. Environ. Resour. 35, 25–55.

Qiu, J., 2008. Scientific publishing: identity crisis. Nat. News 451 (7180), 766–767.

Rai, L.K., Prasad, P., Sharma, E., 2000. Conservation threats to some important medicinal plants of the Sikkim Himalaya. Biol. Conserv. 93 (1), 27–33.

Rana, M.S., Samant, S.S., 2009. Prioritization of habitats and communities for conservation in the Indian Himalayan Region: a state-of-the-art approach from Manali Wildlife Sanctuary. Curr. Sci. 326–335.

Rana, R.S., Bhagat, R.M., Kalia, V., Lal, H., Sen, V., 2013. Indigenous perceptions of Climate change vis-a-vis Mountain Agricultural activities in Himachal Pradesh, India. Indian J. Tradit. Knowl. 12, 596–604.

Rao, M.R., Palada, M.C., Becker, B.N., 2004. Medicinal and aromatic plants in agroforestry systems. In: New Vistas in Agroforestry. Springer, Netherlands, pp. 107–122.

Rawat, S., Andola, H., Giri, L., Dhyani, P., Jugran, A., Bhatt, I.D., Rawal, R.S., 2014. Assessment of nutritional and antioxidant potential of selected vitality strengthening Himalayan medicinal plants. Int. J. Food Prop. 17 (3), 703–712.

Rawat, S., Jugran, A.K., Bahukhandi, A., Bahuguna, A., Bhatt, I.D., Rawal, R.S., Dhar, U., 2016. Anti-oxidant and anti-microbial properties of some ethno-therapeutically important medicinal plants of Indian Himalayan Region. 3 Biotech 6 (2).

Reddy, C.S., 2008. Catalogue of invasive alien flora of India. Life Sci. J. 5 (2), 84–89.

Richardson, D.M., 2004. Plant invasion ecology—dispatches from the front line. Divers. Distrib. 10 (5-6), 315–319.

Rokaya, M.B., Münzbergová, Z., Timsina, B., Bhattarai, K.R., 2012. *Rheum australe* D. Don: a review of its botany, ethnobotany, phytochemistry and pharmacology. J. Ethnopharmacol. 141 (3), 761–774.

Saha, D., Sundriyal, R.C., 2012. Utilization of non-timber forest products in humid tropics: implications for management and livelihood. Forest Policy Econ. 14 (1), 28–40.

Samant, S.S., Pant, S., Singh, M., Lal, M., Singh, A., Sharma, A., Bhandari, S., 2007a. Medicinal plants in Himachal Pradesh, north western Himalaya, India. Int. J. Biodivers. Sci. Manage. 3 (4), 234–251.

Samant, S.S., Dhar, U., 1997. Diversity, endemism and economic potential of wild edible plants of Indian Himalaya. Int. J. Sust. Dev. World Ecol. 4 (3), 179–191.

Samant, S.S., Dhar, U., Palni, L.M.S., 1998. Medicinal Plants of Indian Himalaya. Himavikas Publ.-13, GBPIHED Kosi, Almora, India.

Samant, S.S., Butola, J.S., Sharma, A., 2007b. Assessment of diversity, distribution, conservation status and preparation of management plan for medicinal plants in the catchment area of parbati hydroelectric project stage—III in Northwestern Himalaya. J. Mt. Sci. 4 (1), 034–056.

Sati, V.P., 2013. Cultivation of medicinal plants and its contribution to livelihood enhancement in the Indian Central Himalayan Region. Adv. Med. Plant Res. 1 (2), 17–23.

Schori, M., Showalter, A.M., 2011. DNA barcoding as a means for identifying medicinal plants of Pakistan. Pak. J. Bot. 43, 1–4.

Sekar, K.C., 2012. Invasive alien plants of Indian Himalayan region—diversity and implication. Am. J. Plant Sci. 3 (02), 177.

Semwal, D.P., Saradhi, P.P., Nautiyal, B.P., Bhatt, A.B., 2007. Current status, distribution and conservation of rare and endangered medicinal plants of Kedarnath Wildlife Sanctuary, Central Himalayas, India. Curr. Sci. 1733–1738.

Shah, M.A., Reshi, Z.A., 2014. Characterization of alien aquatic flora of Kashmir Himalaya: implications for invasion management. Trop. Ecol. 55 (2), 143–157.

Shah, M.A., Reshi, Z.A., Lavoie, C., 2011. Predicting plant invasiveness from native range size: clues from the Kashmir Himalaya. J. Plant Ecol. 5 (2), 167–173.

Shaheen, H., Khan, S.M., Harper, D.M., Ullah, Z., Allem Qureshi, R., 2011. Species diversity, community structure, and distribution patterns in western Himalayan alpine pastures of Kashmir, Pakistan. Mt. Res. Dev. 31 (2), 153–159.

Sharma, C.M., Suyal, S., Gairola, S., Ghildiyal, S.K., 2009. Species richness and diversity along an altitudinal gradient in moist temperate forest of Garhwal Himalaya. J. Am. Sci. 5 (5), 119–128.

Sharma, N., Raina, A.K., 2013. Composition, structure and diversity of tree species along an elevational gradient in Jammu province of north-western Himalayas, Jammu and Kashmir, India. J. Biodivers. Environ. Sci. 3 (10), 12–23.

Sharma, P., Samant, S.S., 2014. Diversity, distribution and indigenous uses of medicinal plants in Parbati Valley of Kullu district in Himachal Pradesh, Northwestern Himalaya. Asian J. Adv. Basic Sci. 2, 77–98.

Sharma, P., Samant, S.S., Lal, M., Sharma, A., 2014. Diversity, indigenous uses, threat categorization and conservation prioritization of medicinal plants: a case study from Himachal Pradesh, India. J. Biodivers. Endanger. Species 2, 134.

Sheikh, M.A., 2014. Phytochemical investigation and uses of some of the medicinal plants of Kashmir Himalaya—a review. Int. J. Biol. Pharm. Allied Sci. 3, 462–489.

Simpson, M.G., 2010. Plant Systematics. Academic Press, Burlington, USA.

Singh, G., Rawat, G.S., 2012. Depletion of oak (*Quercus* spp.) forests in the western Himalaya: grazing, fuelwood and fodder collection. In: Global Perspectives on Sustainable Forest Management. InTech, UK.

Singh, G.S., 2011. Commercialization unsustainable to Himalayan environment. Ecol. Quest. 15 (1), 71–75.

Singh, K.M., 2009. Scope of medicinal and aromatic plants farming in Eastern India. Browser Download This Paper.

Singh, K.M., Jha, A.K., 2008. Medicinal and aromatic plants cultivation in Bihar, India: economic potential and condition for adoption. ICAR-RCER, Patna, S.G.I.D.T., Patna 14. February 2008 MPRA Paper No. 47091, posted 20. May 2013 20:34 UTC.

Singh, K.N., Lal, B., Todaria, N.P., 2016. Ecological status and conservation of rare plants in high altitude landscapes of Indian western Himalaya. Glob. J. Res. Med. Plants Indigen. Med. 5 (1), 1.

Singh, M.P., Singh, B.S., Dey, S., 2002. Plant Biodiversity and Taxonomy. Daya Books, New Delhi.

Singh, R.K., Pretty, J., Pilgrim, S., 2010. Traditional knowledge and biocultural diversity: learning from tribal communities for sustainable development in northeast India. J. Environ. Plan. Manag. 53 (4), 511–533.

Singh, R.K., Singh, A., Pandey, C.B., 2014. Agro-biodiversity in rice-wheat-based agroecosystems of eastern Uttar Pradesh, India: implications for conservation and sustainable management. Int. J. Sust. Dev. World Ecol. 21 (1), 46–59.

Singh, V., Singh, B., Kaul, V.K., 2003. Domestication of wild marigold (*Tagetes minuta* L.) as a potential economic crop in western Himalaya and north Indian plains. Econ. Bot. 57 (4), 535–544.

Singh, G.S., Rao, K.S., Saxena, K.G., 1998. Eco-cultural analysis of sacred species and ecosystems in Chhakinal watershed, Himachal Pradesh. Conserving the Sacred for Biodiversity Management. Oxford & IBH, New Delhi.

Siwach, M., Siwach, P., Solanki, P., Gill, A.R., 2013. Biodiversity conservation of Himalayan medicinal plants in India: a retrospective analysis for a better vision. Int. J. Biodivers. Conserv. 5 (9), 529–540.

Steele, P.R., Pires, J.C., 2011. Biodiversity assessment: state-of-the-art techniques in phylogenomics and species identification. Am. J. Bot. 98 (3), 415–425.

Stuessy, T.F., 2009. Plant Taxonomy: The Systematic Evaluation of Comparative Data. Columbia University Press, New York.

Telwala, Y., Brook, B.W., Manish, K., Pandit, M.K., 2013. Climate-induced elevational range shifts and increase in plant species richness in a Himalayan biodiversity epicentre. PLoS One 8 (2), e57103.

Thakur, M.K., Chauhan, R., Pant, K.S., 2013. Bioresources for productivity enhancement in kutki, *Picrorhiza kurooa* Royle ex Benth. Int. J. Bioresour. Stress Manage. 4 (4), 482–486.

Thompson, I.D., Okabe, K., Parrotta, J.A., Brockerhoff, E., Jactel, H., Forrester, D.I., Taki, H., 2014. Biodiversity and ecosystem services: lessons from nature to improve management of planted forests for REDD-plus. Biodivers. Conserv. 23 (10), 2613.

United Nations Office on Drugs and Crime, 2004. World Drug Report 2004. vol. 1. United Nations Publications, Herndon, USA.

Uniyal, S.K., Awasthi, A., Rawat, G.S., 2002. Current status and distribution of commercially exploited medicinal and aromatic plants in upper Gori valley, Kumaon Himalaya, Uttaranchal. Curr. Sci. 1246–1252.

Uprety, Y., Poudel, R.C., Asselin, H., Boon, E.K., Shrestha, K.K., 2011. Stakeholder perspectives on use, trade, and conservation of medicinal plants in the Rasuwa District of Central Nepal. J. Mt. Sci. 8 (1), 75–86.

Vetaas, O.R., Grytnes, J.A., 2002. Distribution of vascular plant species richness and endemic richness along the Himalayan elevation gradient in Nepal. Glob. Ecol. Biogeogr. 11 (4), 291–301.

Vidyarthi, S., Samant, S.S., Sharma, P., 2013. Traditional and indigenous uses of medicinal plants by local residents in Himachal Pradesh, North Western Himalaya, India. Int. J. Biodivers. Sci. Ecosyst. Serv. Manage. 9 (3), 185–200.

Vijayan, N., Swapna, T.S., Kiran, G.S., Abdulhameed, S., 2017. Medicinal plants of India: implications in modern medicine. In: Bioresources and Bioprocess in Biotechnology. Springer, Singapore, pp. 421–443.

Wani, N.R., Mushtaq, S.M., 2012. Status of exotic flora in Kashmir Valley. J. For. Environ. Sci. 28 (1), 8–11.

Weber, E., 2017. Invasive Plant Species of the World: A Reference Guide to Environmental Weeds. CABI.

Weber, E., Sun, S.G., Li, B., 2008. Invasive alien plants in China: diversity and ecological insights. Biol. Invasions 10 (8), 1411–1429.

Wu, S.H., Hsieh, C.F., Chaw, S.M., Rejmánek, M., 2004. Plant invasions in Taiwan: insights from the flora of casual and naturalized alien species. Divers. Distrib. 10 (5-6), 349–362.

Xu, J., Grumbine, R.E., Shrestha, A., Eriksson, M., Yang, X., Wang, Y.U.N., Wilkes, A., 2009. The melting Himalayas: cascading effects of climate change on water, biodiversity, and livelihoods. Conserv. Biol. 23 (3), 520–530.

Zerbe, S., Choi, I.K., Kowarik, I., 2004. Characteristics and habitats of non-native plant species in the city of Chonju, southern Korea. Ecol. Res. 19, 91–98.

Further Reading

Hameed, M., Nawaz, T., Ashraf, M., Ahmad, F., Ahmad, K.S., Ahmad, M.S.A., Ahmad, I., 2012. Floral biodiversity and conservation status of the Himalayan foothill region Punjab. Pak. J. Bot. 44, 143–149.

CHAPTER

2

Utilization and Botanical Significance of Himalayan Herbs

OUTLINE

2.1 INTRODUCTION

With increase in inclination toward herbal-based medicines, medicinal and aromatic herbs gained immense popularity and are currently chief raw resource of conventional medicine system and pharmaceutical (Kunwar et al., 2010; Choudhary and Sekhon, 2011; Phondani et al., 2014). Apart from being chief resource of herbal drugs, MAPs ensure livelihood of diverse tribal communities particularly in Indian Himalaya (Phondani et al., 2014; Yunus, 2015). Diverse microclimatic conditions, high altitudes,

https://doi.org/10.1016/B978-0-08-102227-6.00002-4

and varied habitats were ideal for luxuriant growth of MAPs (Jan et al., 2013, 2016a,b). Due to random harvesting and massive collection of MAPs, there has been alarming threat on availability of these herbs (Phondani et al., 2014; Kuniyal et al., 2015; Bhattacharya et al., 2017). Hence, different strategies for intensive cultivation, optimized harvesting, and security against current threats need to be established to conserve MAPs (Basey et al., 2015). There is an urgent need to convert sociocultural traditions and folklore knowledge into economic possibilities; this will ensure preserving cultural information and practices for further globalization. Numerous novel drugs are plant-based; for example, taxol and its derivatives used as anticancer have incited keen interest in using conventional information to direct drug discovery (Arjmand et al., 2014; Lagunin et al., 2014). There is always strong parallelism between contemporary and conventional medicinal system due to vague information on combinatorial chemistry, functional genomics, proteomics, and conventional system of medicines (Johns et al., 2013; Krumholz, 2014). However, in the past, western medicine is switching toward alternative strategies of health care and exhibits immense interest in reviving conventional medicinal practice (Micozzi, 2014; Langdon et al., 2016). Due to distinct diversity of MAPs and molecules that are produced from these herbs, cornucopia of significant drugs needs to be discovered from these herbs (Bhat et al., 2013a,b; Khan et al., 2014; Tariq et al., 2015). Selection of suitable herb for mass cultivation is cumbersome job especially when huge literature is too oblivious to provide assortment and predilection of economically cultivable medicinal herb (Malik et al., 2015). So, there is a need to produce awareness pertaining to mass cultivation of MAPs among farmer community so as to ensure effective marketing of herbal wealth of Indian Himalaya (Rathore et al., 2015). Significantly, imperative drugs procured from MAPs need to be domesticated and cultivated for establishment of optimum yields of active constituents (Joshi et al., 2016) since the range of active constituents in MAPs differ with respect to ecophysiological features of habitat (Rao et al., 2008; Gairola et al., 2010). Conventional medicine system is the prime resource that fetches easy and rapid route for identification of those MAPs that could prove to be seat for new drug discovery (Katiyar et al., 2012). So in order assimilate this knowledge, we need to assemble all information pertaining to cultivation and utilization of medicinal herbs. In this chapter, we will discuss taxonomic identification and botanical characterization together with ecophysiological and biochemical validation of herbs. We will also discuss pharmacological and medicinal potential of these herbs. Further, we will illustrate different conservation approaches taken into consideration to advance utilization of herbs on mass scale. Since there is huge endemism in Indian Himalaya, this chapter will conclude with note on endemic aromatic herbs of Kashmir Valley.

2.2 BOTANICAL ASPECTS AND TAXONOMICAL EVALUATION

Botany is significant discipline for illustrating knowledge of MAPs in diverse ways. About 40,000 plants collected from both wild and domesticated populations are resources for ethnomedicinal practices (Rao, 2004; Penev et al., 2010). Botany is a wide discipline that entails diverse sciences like plant systematics, morphology, and physiology that can facilitate identification and characterization of MAPs in varied ways (Boyle et al., 2013). Utilization of MAPs commences with authenticated identification of herbs formerly done on basis of morphological characterization (Gupta, 2010). With advancement in technical strategies, chemical evaluation has also been included. Current research trends have opened the possibilities of unraveling DNA and biosynthetic causes of chemodifferentiation that finally provides data of plant metabolome (Máthé, 2010, 2015). Hence, integrating botany with recent technological intervention can open up new possibilities for advent of metabolically rich taxa of MAPs (Leonelli, 2013; Marvin et al., 2016). Inheritance pattern together with intricate economic traits in MAPs is variable issue. Variability together with their ecological plasticity of crops makes it cumbersome to conclude research query and organize inheritable characteristics making MAP breeding even more complex (Macías et al., 2007). The competence among cultivation practices of MAPs is primarily governed by productivity of plant exhibiting highest secondary metabolite accumulation (Maffei et al., 2011; Jaradat, 2015). For utilization, the quantity and composition of secondary metabolite are prerequisite prior to investigation of botanicals to be mass cultivated. Moreover, information and economic mapping of MAP resources is significant for management and utilization of MAPs (Badola and Aitken, 2003; Kumar and Gupta, 2008).

2.2.1 Botanical Aspects of MAPs

Description of MAPs including anatomy and morphology together with classification via systematics particularly with reference to their pharmaceutical properties is significant to commence exploration of MAPs (Ali et al., 2008). Conventionally, botanical science deals with naming and classification of plants. On the contrary, modern botany entails multidisciplinary approaches in dealing with plants and is dependent on diverse areas of plants and their classification (Ramawat and Goyal, 2008; Sen et al., 2011).

(a) Plant morphology
Plant morphology refers to science dealing with structural analysis of plants involving external morphology and internal structure via

anatomy. External morphology describes form, arrangement, and association of varied organs, while as anatomy, it describes finer details of cells (cytology) and tissues (histology). Prior to discovery of theories of inheritance by Charles Darwin, morphological characterization of plant was the basis of plant classification (Dar and Sundarapandian, 2016). Occasionally, pharmaceutical importance was associated with plant form as represented in doctrine of signatures like plant organs having resemblance to some human organs that were used to cure disease of that organ, for example, the liver-shaped *Hepatica* sp. for liver problems and lungwort (*Pulmonaria* sp.) for pulmonary infections (Patil, 2015). Prior to the discovery of theories of inheritance, plant morphology was overrated trait. However, with advent of Darwin's inheritance theory, the progress of first phylogenetic system was developed besides taking plant morphology into consideration for deciphering classification 9 (Pickersgill, 2009; Pascual et al., 2013). However, involvement of plant morphology in classification of MAPs still remains the same since secondary metabolites accumulate in particular plant tissues or cells, for instance, accretion of essential oil in trichomes within the epidermis of family Lamiaceae, while tropane alkaloids are stored in root, leaves, capsules, and berries of members of Solanaceae family. Morphoanatomical characterization of MAPs facilitates authenticated identification of particular traits that associate them with crude drug hence averts any chance of adulteration indispensable for discipline of pharmacognosy (Seigler, 2012; Murthy et al., 2014). Drug identification even in dry samples is facilitated through identification of histological structure via electron microscope, for instance, crystolithic cells, starch grains, polygonal crystalloids, and secretory structures (glandular hairs, schizolysigenous cavities and passage, lactiferous vessels, etc.) (Uggla et al., 2005). Accretion of secondary metabolite in plants is organ-specific, and its utilization is primarily determined by its characteristics of plant. The plant organs exhibiting pronounced accumulation of secondary metabolite are harvested for acquiring its optimized levels in plants. However, there is an extensive heterogeneity pertaining to distribution of secondary metabolite interrelatedness of secondary metabolites with morphoanatomical characteristics of MAPs (Baskaran and Jayabalan, 2005). This heterogeneity is with respect to developmental stages of plants. With development shift from juvenile to adult stage, there is a corresponding metabolic shift generally linked with plant organ (Lombardo et al., 2011). The accretion of secondary metabolite is determined by maturity of organ or tissue associated; for instance, density of glandular trichome varies immensely on plant developmental stage. Moreover, disparity in morphological characters can be determined by ecophysiology of that habitat. Hence, it can be concluded that secondary metabolite accretion is a time- and tissue-specific trait. Hence, information of optimized accretion of secondary metabolites is significant for effective utilization of MAPs.

(b) Plant physiology

Plant physiology is study that deals with functioning of plants. It illustrates the all-inclusive information pertaining to environmental effects on plant metabolism and development (Jones, 2013). Plant physiology demonstrates effect of environmental variables on secondary metabolism. Hence, plant physiology presents significant link for deciphering production of phytochemicals and relates it with plant development. Plant metabolism including both primary and secondary metabolism provides enormous chemical reactions that occur inside plants. Primary metabolites include carbohydrates, proteins, lipids, and nucleic acid, while secondary metabolites are specialized compounds exclusive to crops used for defense and reproduction (Cheng, 2017). Secondary metabolites are derived from primary metabolic pathways in MAPs. In most of cases, secondary metabolites are formed as end products without any relevance with plant metabolism, for example, essential oil secreted by glandular trichome implicated in plant defense. However, secondary metabolites are not ubiquitous among the plant kingdom; most of the time, secondary metabolite accumulation has ecological implications (Metlen et al., 2009). The consumption of MAPs by humans enables ingestion of the same metabolites that proved useful for health. Plant oxidants are prime examples of such secondary metabolites that confer resistance against varied diseases and scavenging of free radicals in humans (Devasagayam et al., 2004). The secondary metabolite accumulated by particular plant taxon demonstrates its speciality as described by its physiological status (Kliebenstein, 2004). The biosynthetic pathway for each class of secondary metabolite is a characteristic in particular plant family. Class of naphthoquinones is a characteristic feature of some of families like plumbagin (Plantaginaceae, Droseraceae, Ebenaceae, and Apocynaceae) and lawsone (Lythraceae, Balsaminaceae, or chimaphillin (Pyrolaceae)). Medicinal and aromatic plants (MAPs) have characteristic feature due to which they undergo ontogenetic changes following several ecological pressure. These changes result in altered biosynthetic pathways, and this ontogenetic change was first observed by Wulff and Nilov in 1934 and Gould (1992). Similar developmental alterations were observed in glycoside and terpenoid biosynthesis occurring in MAPs (Tétényi, 1970). Plant breeders generally use morphological traits in conventional breeding; however, this proves to be misleading.

Determination of harvesting time is another important issue for drug development from MAPs. MAPs are generally harvested at ripening stage as indicated by color development. In most of cases this coincides with enormous content of active metabolites in plant organs to be collected (Croteau et al., 2000). Due to enormous influencing factors, the accurate evaluation of harvesting time is intricate issue that requires to be elucidated further especially for wild-growing MAPs (Ann Lila, 2006). Generally, dry and sunny weather conditions are suitable for wild harvesting of MAPs. Keeping in view seasonal variations that correlate with metabolic processes of plants, underground organs like roots and rhizomes are collected at dormancy stage, while bark is collected following commencement of sap flow in spring (Rousseaux et al., 2009); gemmae and leaves are harvested following bud break and fully opened stage (Glime, 2017).The suitable harvesting time for flowers and inflorescence is full anthesis.

Determination of suitable time for collecting plant material is also necessary for appropriate drug development. Harvest time for raw material and procurement of active ingredient need to be determined. Wild crafting of MAPs remains the main resource of raw material for drug procurement for pharmaceutical industry (Schippmann et al., 2006). Hence, there is an increased risk of damages and losses in production and utilization of MAPs. In this context, selection of suitable collecting practices has gained much significance. The proclivity of species to overcollection differs according to life forms and plant part used. Indiscriminate collection due to wild crafting of root drugs is encountered in the case of *Primula* sp., *Ginseng* sp., etc. Hence, appropriate wild collection of MAPs contributes to mass production, and good quality ensures rise in herbal wealth of Indian Himalaya. As a thumb rule, suitable plant part needs to be dissected/collected at suitable time and in lucid manner so as to avert any damage to plant tissue.

Domestication of wild medicinal plants is quite an intricate process as it requires premeditation of plant from their nativity to new environment. There is an intense connection of MAPs with their ecological niche due to multitude of metabolic fluxes that modulate their rhizosphere accordingly (Nemeth and Bernath, 2008). Diverse researchers have tried to develop structured schemes for thriving plant environment. Domestication of wild MAPs requires modulation of ecophysiological parameters to counterpart with ecological needs for plants to grow and reproduce. The aim of domesticating wild MAPs should be minimization of disparity among wild and cultivated habitats that would detrimentally affect possible production of economically significant natural products (Schippmann et al., 2006). Significant mission of wild MAPs domestication requires adapting crop to new environment and inculcating genetic variation in an established crop so that novel product is generated. Hence, the system of novel crops R + D is primarily composed of five stages, that is, collection, evaluation, enhancement, development, and utilization of germplasm in cultivar development, which is closely implicated with agronomic assessment, and expansion of suitable cultural and management systems and activities connected with complete commercialization.

2.2.2 Plant Systematics and Taxonomy

Classification and naming of MAPs are imperative steps toward authenticated identification of MAPs. MAPs belong to diverse plant families that generally secrete secondary metabolite as result of similarities in biosynthetic pathways; for example, Labiatae (Lamiaceae) family exhibits numerous glandular trichomes that produce essential oil such as lavender, thyme, rosemary, and sage, while other plant families, like the Solanaceae, are characterized by the occurrence of several alkaloid-containing species (belladonna, thorn apple, and tobacco) (Máthé, 2015; Li et al., 2016). Due to immense diversity of MAPs, Bennett and Balick (2014) highlight the significance of accurate and reproducible experiments in medicinal plant research. In this context, they recommended MAP research should commence with identified plant samples within labeled vouchers (Fuller and Murphy, 2014; Meegahakumbura et al., 2016). The respective vouchers should be reserved in herbaria in order to avert any chance of error and ensure correct identification of plant sample. Species as usual remains the basic unit of MAPs that together with other related species constitute genus. In order to differentiate dissimilar populations of wild-growing species, there are categories of subspecies (subsp.), variety (var.), and form (f) (Ghimire et al., 2008). There are infraspecific varieties within both wild and cultivated species, while cultivars are identified according to characteristics acceptable by humans (Schippmann et al., 2006). MAPs are mostly characterized by their specialized morphological, chemical, cytological, and even ecological features utilized for correct identification (reference). Hence, species with similar or little variation constitute taxon, and those with distinctive characteristics constitute variety having either single-gene mutation or particular distinct morphological trait (Rao, 2004; Verma et al., 2010). Single mutation results in new variants exhibiting varied secondary metabolite levels that are not distinct in form termed as chemical races or chemodemes (Tétényi, 1970). Any genetic variation in MAPs results in disparity in secondary metabolism; for example, addition or deletion of chromosome number above or below normal complement results in genetically modified plants (GMPs). Different landraces having diverse species produce several chemical substances such as cyanogenetic glycosides in *Prunus communis*, alkaloids in *Duboisia* species, cardiac glycosides in *Digitalis purpurea*, and essential oils in *Ocimum* spp., *Melissa* spp., and *Thymus* spp. (Evans, 2002; Tiwari et al., 2008). Evolutionary links or hereditary patterns form basis of phylogenetic classification of MAPs, and these systems are extensively based on artificial system of classification given by Linnaeus in 1758. During second century, more significance was associated with the utilization of data pertaining to chemoprofiling resulting in introduction of chemotaxonomy or phytochemical plant systematics (Hegnauer, 1986; Crockett and Robson, 2011; Singh et al., 2016). The chemical substances utilized in chemotaxonomy are not normally amino acids or sugars but major secondary metabolites like alkaloids, isoprenoids, flavonoids, and glycosides (Evans, 2002). Secondary metabolites have ecological implications like defense against abiotic and biotic stress in plants (Reigosa et al., 2006; Chen et al., 2009). However, chemotaxonomy has taken back seat in plant systematics due to advent of advanced molecular technologies (Denford, 1984; Crawford, 1990). This results in more intricate study that reveals quantitative analysis of molecular structure and mechanisms underlying plant metabolism (Thorne, 1992). Hence, with advancing molecular biology, there is a significant progress toward implication of molecular tools in botanical classification termed as molecular systematics (Simpson, 2010). Molecular biology reveals central dogma underlying secondary metabolism, thereby exposing genetic diversity among MAPs. Integrating cytology with molecular biology has led to the discovery of efficient tool for botanical classification of MAPs and rendered new amendments in already established phylogenetic system (Rudall et al., 1998; Sytsma and Pires, 2001; Goldman et al., 2004). Thus, modern botany has now emerged as multidisciplinary subject integrating varied areas of science and technology (Stopes, 1919; Copeland and McDonald, 2012).

Following authenticated identification and naming of MAPs, we need to monitor quality and purity of crude drug that has significance in botanical sciences (Gupta, 2010). Even though diversity of MAPs and secondary metabolite produced is too enormous; however, there is a disparity with regard to production process of botanicals (Wink, 2003; Kroymann, 2011). Among production processes, main steps involved include collection practices, proper processing, drying, packaging, and correct storage (Yadav and Dixit, 2008; Folashade et al., 2012; Nafiu et al., 2017). The prime factor that facilitates species identification even among relatively similar species is knowledge about plant habitat, for example, *Tussilago farfara* and *Petasites hybridus* (Caballero and Aranda, 2010). This can provide us information pertaining to environmental conditions of plant habitat such as heavy-metal contamination and pesticide/herbicide residues. Following the characteristics of plant habitat, we need to determine plant part and developmental stage that yields maximum secondary metabolite and is suitable for drug development (Verpoorte, 1998; Hussain et al., 2012). In herbal drugs, where floral parts are collected for active substance that is in relatively lower concentration in stem, hence, stem diameter and length are restricted. The flower stalks are also limited in the case of *Chamomillae flos* and *Sambuci flos* where quality of flower drug can be declined by number of flower stalks (Schulz and Blumenthal, 2004; Máthé, 2015; González et al., 2017). Similarly, in the case of drugs procured from roots, stem parts need to be minimized (e.g., *Valerianae radix* and *Primulae radix*). While in the case of fruit drugs, significant trait to be considered is fruit color; for example, in *Rosa dumalis* and *R. rubiginosa*, color changes determine ripening of rose fruits and hue related to variation in sugar levels that is possible indicator of suitable harvesting time (Uggla et al., 2005; Upton et al., 2016). These data and expertise in standardization of drug development and stakeholders (farmers) who collect MAPs need to be trained and educated (Balunas and Kinghorn, 2005; Rawal et al., 2010; Chawla et al., 2013).

2.3 ECOPHYSIOLOGY AND BIOCHEMICAL FEATURES

Due to ecological imbalance and indiscriminate use of natural resources, there has been a devastating effect on plant biodiversity in Indian Himalaya. In central and northwestern Himalaya, there is a huge pressure on plant diversity due to irrational, naive extraction, and overgrazing pressure (Ghimire et al., 2006). Though domestication of wild MAPs has been emphasized, there is dearth of such practices across Indian Himalaya. However, rural settlers of central Himalayas have established rainfed cultivation of rare, endangered, and threatened MAPs as documented in Red Data Book (Amujoyegbe et al., 2012; Cotter et al., 2014). Diverse strategies have been devised to conserve and cultivate Himalayan herbs at low altitudinal clines (Rawat and Chandrabani, 2007; Singh et al., 2010a,b; Kumar and Vats, 2017). However, the information pertaining to ecophysiological and biochemical characterization is prerequisite for any applied research whose aim is to expand the confines of alpine medicinal vegetation across Himalaya and to evaluate the possible potential of choosing and breeding valuable high-yielding MAPs in the mountain topography of the Himalaya (Chauhan, 1999; Palni et al., 2001; Kala et al., 2006). The ecological plasticity of species is directly/indirectly linked with the competence to take up the energy from microclimate and modulate biochemical and physiological status of the leaf and subsequently the function in production of the entire plant (Smith-Gill, 1983; Lüttge, 2007; Charrier et al., 2015).

Large-scale distribution of MAPs or plant on the whole is determined by climatic conditions; alterations of climate inevitably lead to shifts of distributional limits (Pearson and Dawson, 2003; Mooney et al., 2009; Jackson et al., 2009). These shifts are closely associated with variation in plant ecophysiology, growth, and production (Rout et al., 2000; Cavaliere, 2009). In the past, global warming has more consequent effect on high mountain ecosystem than plain lowlands (Pauli et al., 1996; Körner, 2003, 2007; Chen et al., 2011). These effects have been recorded for European and American mountains, while the largest and highest mountainous Indian Himalayas are not illustrated yet. Long-term observations have elucidated marked variations in global biodiversity in response to climate change (Parmesan, 2006). Mountain regions were found to be more liable to biodiversity loss because of fragmented ecosystem (Poiani et al., 2000; Steffen et al., 2006; Luo et al., 2011). In Indian Himalayas, upward shifting of vegetation zones, range expansion, and contraction together with shifts in phenology and ecophysiological characters marked by climate variation have been recorded in these ecologically delicate areas (Larcher, 2003). Regional climate changes and physiological process are required to discover climate-induced species shift (Walther et al., 2002; Thuiller, 2007; Chown et al., 2010). It has been supposed that with 0.5°C decline in temperature per 100 m rise in altitude, climate change is determined to result in an ascendant shift of vegetation belts at the rate of 8–10 m per decade (Dobhal and Mehta, 2010), with danger of extinction of plants limited to mountain pinnacle (Vetaas and Grytnes, 2002; Oommen and Shanker, 2005; Telwala et al., 2013).

The upward shift of vegetation will move upper vegetation cover toward much higher altitudes in mountains. In mountain ecosystems, ecophysiology and structural changes in plants modulate metabolite production (Nautiyal et al., 2001; Vashistha et al., 2009; Janatková et al., 2013). Among diverse physiological parameters, CO_2 has most detrimental effect on physiological status of plants, for instance, stomatal parameters, water relations, and secondary metabolism. Stomatal characteristics are entirely under genetic control, and their density is not influenced by ecological factors. However, stomatal movements are influenced by increased CO_2 emission rate (Chaturvedi et al., 2013; Agnihotri et al., 2017). Likewise, accumulation of secondary metabolite is influenced by environmental variables that are more distinctive at highest altitudes (Figueiredo et al., 2008). Elevated levels of CO_2 concentration provide surplus of carbon required for growth, and carbon partitioning could result in biosynthesis of secondary metabolites like phenolics, sterols, flavonoids, and sterol compounds (Xu et al., 2009a,b; Berli et al., 2010; Dong et al., 2011). The Himalaya has experienced drastic climatic change in the past resulting in consequent effect on biodiversity, ecological balance, and anthropogenic activities that is more pronounced on mountain ecosystem than others (Bennett, 2004; Schickhoff, 2011). Due to increased urbanization in India, natural environment of Himalaya has been significantly declined. The climate changes at regional level are quite different than at global level. Human development along with increased interventions of modern technologies led to accelerated deforestation and increased soil erosion and contamination of soil, air, and water resources (Wackernagel and Rees, 1998; Melnick, 2005). Himalaya is rapidly losing its ecological plasticity in order to avert climatic fluctuation (Bolch et al., 2012). Himalayan flora exhibits hermetic response to climate change; there are meager research reports on systematic research documenting impacts of climate change on endemic flora of Himalaya (Lall, 1981; Shrestha and Aryal,

2011). There is scarcity of research studies demonstrating climate-induced shifting of higher altitudes for Himalayan flora (reference). Recent research reports demonstrate effects of climate change on MAPs, like *Geranium nepalense* Sweet (Geraniaceae), *Inula racemosa* Hook. f. (Asteraceae), and *Lavatera kashmiriana* Camb. (Malvaceae) (Kaul, 1997; Bhattarai and Vetaas, 2003; Chawla et al., 2008). These studies compared old herbaria (1900–60) with recently generated herbaria (1961–2010) containing information on stomatal density and metabolite concentration of herbarium specimen. This study evaluated impacts of climate change on elevational distribution pattern and ecophysiological characteristics of plants thriving in northwestern Himalaya during the last 110 years and recommends that rising temperature and higher CO_2 rate alter altitudinal distribution pattern, stomatal movements, and secondary metabolite levels with passage of time compared within herbaceous species (De Lillis et al., 2004; Jan et al., 2013). Significant altitudinal shifts was demonstrated in all herbaceous species based on records of old herbaria. The mean elevation of these medicinal species exhibited an average rise of 55.2 m/decade between 1900 and 2010, higher than the average global rate of 6.1 m/decade (Patiño et al., 2016; Agnihotri et al., 2017). More pronounced shifts were demonstrated in highest elevation for *G. nepalense* and *I. racemosa* than for *L. kashmiriana*, which is a high-altitude species (Pörtner et al., 2009; Sharafzadeh and Ordookhani, 2011; Jump et al., 2012; Dolezal et al., 2016). In Indian Himalaya particularly in the northwestern region during 1900–2010, there has been temperature rise by 0.31°C/100 years, and the mean annual average minimum temperature increased by 0.79°C/100 years (Yadav et al., 2012, 2013). This clearly indicates upward shifting and alteration in ecophysiological parameters; in the case of northwestern Himalaya, climate has warmed based on its minimum temperature limits (Maikhuri et al., 2000; Shrestha et al., 2014; Linares, 2011). In addition to rise in temperature, there has been rise in precipitation levels by 8.15 mm during the past six decades (Iqbal et al., 2009; Oguntunde et al., 2012; Camuffo, 2013). Northwestern Himalaya represents longest bioclimatic elevation gradient having richest plant biodiversity. The MAPs collected from this belt provide valuable resource for studying climate change on plant metabolism and development. The plants, *G. nepalense*, are significant in evaluating soil erosion and water loss; *L. kashmiriana* and *I. racemosa* are endemic to this Himalayan region (Samant et al., 2007). Climate change activates distribution of *I. racemosa* that was normally distributed within elevation levels (1700–1800 m amsl). Marked upward altitudinal shifts were more pronounced in herbaceous species than in woody plants because of their shorter life cycle, faster maturation, and limited adaptability to sustain drastic climatic conditions (Klimes, 2003). Indian Himalaya exhibited 72.94 ppm rise in CO_2 concentration (316.90–389.84 ppm) as recorded in the last 50 years resulting in significant effect on ecophysiology of herbs (Pörtner et al., 2009; Rana et al., 2011; Barua et al., 2014). Higher CO_2 levels and increased temperature can affect stomatal density, thereby affecting cell differentiation and physiological metabolism (Teng et al., 2006). Montane flora experiences lesser partial pressure of CO_2 and is more susceptible to variation in CO_2 partial pressure than flora thriving in elevated CO_2 levels of environment (Casson and Gray, 2008; Fitter and Hay, 2012). This indicates that higher CO_2 levels will lead to a decline in stomatal density and decreased carbon isotope disparity at elevated altitude (Larcher, 2003). There was inverse correlation between carbon dioxide levels and stomatal density as CO_2 levels rise and consequent decline in stomatal density occurs that helps plants to avert excessive water loss and induce increased stomatal conductance at lesser CO_2 levels (Long et al., 2004; Ainsworth and Rogers, 2007). Metabolic pathways implicated in secondary metabolism generate huge spectrum of phytochemicals majority of which are pharmaceutically active such as gallic acid and quercetin isolated from *G. nepalense* and inulin from *I. racemosa*. Alterations in secondary metabolites influenced environmental variables like temperature, water availability, and CO_2 levels together with increasing altitude (Caldwell et al., 2007). During the last 100 years, there is a significant decline in fatty acids with increase in altitude. Concentration of phytosterols in some Himalayan herbs is immensely affected by increased temperature, higher CO_2 levels, and higher incidences of precipitation in Himalaya (Kurian and Sankar, 2007; Agarwal, 2011). Climate changes alter ecophysiology leading to upward shift of Himalayan herbs at much faster pace (55.2 m/decade) than 8–10 m/decade in the worldwide mountain ecosystems (Manish et al., 2016). Thus, upward shifting of mountainous herbs depends on levels of climatic variation including anthropogenic pressures that act discretely in each mountain system across world. High-altitude temperate herbs exhibit significant variation about 13-fold increment in antioxidants including phenols and flavonoids in *Hypericum perforatum* L., *Matricaria chamomilla* L., *Thymus vulgaris* L., *Cynara cardunculus* L., and *Echinacea purpurea* L. (Kunwar et al., 2010: Tangjang et al., 2011). Among these plants, *H. perforatum* and *M. chamomilla* have revealed increased accumulation of phenols and xanthophylls at lower elevations and in Jammu (305 m) indicating prospective for raising their geographic area (Sultana et al., 2012; Kaur et al., 2016). *Roscoea procera* is a significant medicinal herb with high antioxidant potential and high total phenolics and flavonoid content in the rhizomes (Nawchoo, 2012). The antioxidant potential of these wild herbs was determined by RP-HPLC evaluating significant variation in gallic acid, catechin, and *p*-coumaric acid content. These data revealed that wild population of herbs growing in open grass fields accumulates higher levels of antioxidants (Kalia et al., 2008; Shabir et al., 2011: Kumar et al., 2016).

2.4 PHARMACOLOGICAL SIGNIFICANCE

Standardized herbal medicines with high quality and quantity of active ingredients can be procured from raw and pure raw herbal material (Schultz et al., 2001; Bandaranayake, 2006). The authenticated and properly characterized quality raw material can generate reliable herbal formulation based on clinical trials derived from efficacy tests and safety measures (Shukla et al., 2009). Proper validation and exhaustive analysis of herbal formulations are prerequisites prior to commercialization of any herbal products so as to avert any assumption stating "herbal products are natural" (Yeung et al., 2008; Joshi, 2012). Safety and efficacy reports pertaining to particular herbal drugs need to be handed over to doctors and other health professionals to prevent any impropriate drug consumption practice. Numerous herbs have been used to treat gastrointestinal tract diseases across the world based on conventional practice of treatment. Earlier, health professionals used to treat human disease via natural medicine following exhaustive evaluation and trail methods (WHO, 2000; Pope et al., 2007) to generate information regarding herbal formulation. This knowledge was transferred to next generations and updated regularly. Currently, there is a shift to conventional medicine system from modern medicine in order to avert toxic effects and inaccessibility of contemporary medicines (Tabi et al., 2006; Grover and Vats, 2001; Grover et al., 2007). Present scientific research demonstrates superiority of traditional system of medicine over mainstream allopathic medicines (Pal and Shukla, 2003; Patwardhan et al., 2008). In the recent years, numerous research reports demonstrate herbal medicine as preferred alternative for combating various diseases. The discipline of pharmacognosy depends on morphoanatomical features of crude drugs so as to prevent adulteration (Máthé, 2015). Pharmacognosy utilizes histological characters like reserve starch grains and crystolithic cells and morphological features like glandular trichome, schizolysigenous cavities, and lactiferous vessels for drug identification even in dried crude drug (Khandelwal, 2008; Pandey and Tripathi, 2014). Secondary metabolite accumulation in plants is organ-/tissue-specific, and utilization of these organs/tissues containing highest secondary metabolite content is a prime step toward drug formulation (Shah, 2009). However, there is an immense diversity regarding content of secondary metabolites and morphoanatomical features of MAPs. The variability is with respect to morphological characteristics and optimization of harvesting time. Hence in pharmacognosy, there is an urgent need for meticulous observation of herbal drugs especially botanical and taxonomic evaluation of crude herbs (Yaniv and Bachrach, 2005; Prakashan, 2009). Validation of 48 medicinal herbs from central Himalaya was subjected to pharmacological analysis and validated for their folk uses such as *Acacia catechu* used for cold and cough, *Aconitum spicatum* as an analgesic, *Aesculus indica* for joint pain, *Andrographis paniculata* for fever, *Anisomeles indica* for urinary affections, *Azadirachta indica* for fever, *Euphorbia hirta* for asthma, *Taxus wallichiana* for tumor control, and *Tinospora sinensis* for diabetes. These observations were found consistent with current pharmacological properties together with ayurvedic and previous uses. Pharmacological assessment is validated with correct identification and uses active ingredients for treatment of diseases (Kaul, 2010; Shaheen et al., 2012; Baig et al., 2013). The perennial alpine Himalayan herb *Nardostachys jatamansi* has been used in cure of numerous chronic diseases. Pharmacological studies validated anticonvulsant, antiparkinson, tranquilizing, hepatoprotective, neuroprotective, hypotensive, and antidiabetic activities (Phillipson, 2007; Dhami, 2013). High-altitude herb, *Potentilla fulgens Wall. ex Hook.*, from Rosaceae family was pharmacologically evaluated for its antihyperglycemic, hypoglycemic, antihyperlipidemic, antitumor, antioxidant, anti-inflammatory, and antiulcerogenic properties, thereby validating its ethnotherapeutic use (Laloo et al., 2013a, b). The pharmacological significance of some MAPs is summarized in Table 2.1.

TABLE 2.1 Description of Diverse Plant Species With Their Pharmacological Significance

Plant species	Pharmacological significance
Abies pindrow (Royle ex D. Don) Royle (Pinaceae)	Leaves are used for curing bronchitis and asthma and inner bark for constipation, and the cones are used as a diuretic and purgative
Aegle marmelos L. *Corrêa* (Rutaceae)	Decoction of leaves used for treatment of dysentery, diarrhea, respiratory tract infections, heart ailments, and digestive disorders
Achillea millefolium L. (Asteraceae)	Tribal people of Jammu and Kashmir use plant infusion as diuretic, and decoction is used for treatment of cold and fever
Ageratum conyzoides L. (Asteraceae)	Leaf extract can be used to stop bleeding and to treat skin diseases
Acorus calamus L. (Araceae)	Baitadi and Darchula districts of far-western Nepal rhizome juice used as antihelmintic and for stomachache. Western Nepal rhizome used to treat coughs, colds, and sore throat and as wound poultice
Ajuga parviflora Benth. (Lamiaceae)	Tribal people in the Mornaula Reserve Forest of Kumaun, west Himalaya, use its leaves as an antihelmintic (*Ascaris*)
Callistemon citrinus (Curtis) Skeels (Myrtaceae)	Used as antifungal and antibacterial
Cannabis sativa L. (Cannabaceae)	Juice of leaves used as wound poultice and for the cure of tumors

TABLE 2.1 Description of Diverse Plant Species With Their Pharmacological Significance—cont'd

Plant species	Pharmacological significance
Carum carvi L. (Apiaceae)	In western Nepal, fruits are applied to treat swelling of the breast and testicles, and fruits are used for stomachic problems
Cedrus deodara (Roxb. ex D.Don) G.Don (Pinaceae)	In Jammu and Kashmir, India, the bark is used as a diuretic, carminative, and antiflatulent and for urinary disorders. People in Kumaun, Uttarakhand, use the fumes from the bark and wood as a snake repellent. People living in the Nanda Devi National Park (Uttarakhand), India, use bark decoction to treat fever and dysentery
Boenninghausenia albiflora (Hook.) Rchb. ex Meisn. (Rutaceae)	In Jammu and Kashmir, India, shoot is used as insect repellent and for relieving toothache
Ficus religiosa L. (Moraceae)	In Nepal, the bark juice is applied for paralysis, and in Jammu and Kashmir, India, leaves and young shoots are used as a purgative
Juniperus recurva Buch.-Ham. ex D. Don (Cupressaceae)	In Nepal, leaf decoction is used to treat fever, headache, coughs, and colds; paste of the leaves and berries is used to treat skin diseases. In the Nubra Valley (Jammu and Kashmir), a leaf decoction is used to reduce fever
Mentha arvensis L. (Lamiaceae)	In Kashmir, leave decoction is used to treat gastroenteritis; aerial parts of plants are used to treat cough, sore throat, indigestion, and constipation. In Maccheguan, Nepal, the leaves (mixed with *Ocimum sanctum*) are applied externally and taken orally to treat fever, cold, cough
Mentha longifolia L. *Huds.* (Lamiaceae)	In Jammu and Kashmir, India, the essential oil is used for flavoring confectionery; decoction of leaves is taken as a cooling medicine. In Uttarakhand, India, the herb is used for gastrointestinal disorders and cough
Lantana camara L. (Verbenaceae)	Decoction of aerial parts used for rheumatism, diuretic, snakebite, fever, and colds; crushed leaves used externally on wounds, ulcers, skin sores
Juniperus macropoda Boiss. (syn. J. excelsa M. Bieb.) (Cupressaceae)	In Kashmir Valley, its aerial parts are used as incense
Filipendula vestita (Wall. ex G. Don) Maxim. (Rosaceae)	In Kashmir, a leaf paste is applied to wound
Blumea lacera (Burm. f.) DC. (Asteraceae)	In Jammu and Kashmir, leaves are used as an antipyretic, febrifuge, diuretic, and antihelmintic

(Continued)

TABLE 2.1 Description of Diverse Plant Species With Their Pharmacological Significance—cont'd

Plant species	**Pharmacological significance**
Artemisia vulgaris L. (Asteraceae)	In Nepal, crushed leaves are used to stop nosebleeds and for mouth ulcers. The leaf extract is used for fever
Artemisia maritima L. (Asteraceae)	In India, Himachal Pradesh, root juice is used for treatment of boils. Decoction of the leaves is used as febrifuge
Artemisia nilagirica (C.B. Clarke) *Pamp.* (Asteraceae)	Leaves are used as poultice for wounds and cuts and used for treatment of oral ulcers
Artemisia roxburghiana Besser (Asteraceae)	The whole plant extract is used as a tonic and to relieve fever and as febrifuge
Artemisia scoparia Waldst. and Kit. (Asteraceae)	Tribal community of Jammu and Kashmir used it as febrifuge
Artemisia parviflora Buch.-Ham. ex D. Don (Asteraceae)	Used as diuretic and in prevention of skin diseases
Artemisia japonica Thunb. (Asteraceae)	Leaves used as incense and insecticide. Leaves are also used in cure of malarial diseases
A. indica Willd. (Asteraceae)	Whole plant used as antileech
Artemisia gmelinii Weber ex Stechm. (Asteraceae)	Paste leaves used for pimples and boil
Artemisia dubia Wall. ex Besser (Asteraceae)	Leaf juices used for treatment of cuts and wounds
Carum carvi L. (Apiaceae)	Used for prevention of swelling in breasts and testicles. The decoction of aerial shoot used in the curing cough and cold
Filipendula vestita (Wall. ex G. Don) Maxim. (Rosaceae)	In Kashmir, a leaf paste is applied to wounds
Artemisia dracunculus L. (Asteraceae)	*Artemisia dracunculus* (tarragon) is used throughout the world for flavoring food; in Nubra (Kashmir) Valley, Kibber Wildlife Sanctuary (Himachal Pradesh), and Lahual Valley (Himachal Pradesh), its extract of the plant is used to relieve toothache and reduce fever and as a treatment for gastrointestinal problems
Aristolochia indica L. (Aristolochiaceae)	Whole plant was given for fever, as an antihelmintic, and to treat snakebites
Blumea lacera (Burm. f.) DC. (Asteraceae)	In Jammu and Kashmir, plants are used as an antipyretic, febrifuge, diuretic, and antihelmintic
Boenninghausenia albiflora (Hook.) Rchb. ex Meisn. (Rutaceae)	Used for eradication ticks, mites, and human lice

TABLE 2.1 Description of Diverse Plant Species With Their Pharmacological Significance—cont'd

Plant species	Pharmacological significance
Aralia cachemirica Decne. (Araliaceae)	Root used for gastric intestines
Nepeta erecta (Royle ex Benth.) Benth. (Lamiaceae)	Kashmiri uses the leaves of *N. erecta* to cure cough, cold, fever
Nepeta juncea Benth. (Lamiaceae)	Essential oil procured from aerial parts of plant used as antifungal (*Aspergillus umigatus, Trichophyton mentagrophytes, T. rubrum*)
Nepeta laevigata (D. Don) Hand.-Mazz. (Lamiaceae)	Used seeds of plants for cure of dysentery
Nepeta leucophylla Benth. (Lamiaceae)	Used for cure of fever and cough
Nepeta longibracteata Benth. (Lamiaceae)	The whole plant is used for stomach disorders
Nepeta govaniana (Wall. ex Benth.) Benth. (Lamiaceae)	Decoction of whole plant used for treatment of colds, influenza, diarrhea, colic, insomnia, and menstrual cramp, and decoction of whole plant taken for sore throat and as a cardiac tonic
Nepeta floccosa Benth. (Lamiaceae)	People in the cold desert of Ladakh, Kashmir, prepare a decoction of the leaves as a remedy for colds, coughs, and fever
Ocimum basilicum L. (Lamiaceae)	Decoction of seeds used to treat urinary disorders; a leaf paste is used externally to treat skin diseases and fungal infections
Origanum vulgare L. (Lamiaceae)	The aromatic oil of *O. vulgare* is used as stimulant, rubefacient, and tonic. Women in the Gurez Valley of Kashmir take a warm decoction of the plant to lighten menstrual discomfort
Anisomeles indica (L.) *Kuntze* (Lamiaceae)	In western Himalaya, the people use the whole plant as an antidote to poisonous bites, and the leaf extract is taken for urinary complaints
Perovskia abrotanoides Kar. (Lamiaceae)	The plant extract is used by people in the Nubra valley, Jammu, and Kashmir, to treat coughs and headache
Stachys sericea Wall. ex Benth. (Lamiaceae)	In Kashmir, the whole plant is taken internally to treat epilepsy
Tanacetum gracile Hook. f. and Thomson (Asteraceae)	In Kashmir, the leaves are used as an antihelmintic (intestinal worms)
Tanacetum longifolium Wall. ex DC. (Asteraceae)	In Kashmir, the root powder is taken with tea to relieve stomach pain. Local inhabitants of the Kedarnath Wildlife Sanctuary, Uttarakhand, use the leaves to treat stomachache and indigestion

(Continued)

TABLE 2.1 Description of Diverse Plant Species With Their Pharmacological Significance—cont'd

Plant species	Pharmacological significance
Pinus roxburghii Sarg. (Pinaceae)	In Kashmir, the bark resin is used as expectorant for bronchitis. In far-western Nepal, a paste made from the bark is used to treat burns and scalds; the bark resin is applied to boils. In Uttarakhand, the bark resin is used to treat snakebite and scorpion sting
Vitex negundo L. (Verbenaceae)	Use the aromatic leaves as a tonic and vermifuge and leaf juice for stomachache
Thymus serpyllum L. (Lamiaceae)	Ethnic people of Almora district of Uttarakhand use the juice of the whole plant orally to treat cough and asthma; the paste of whole plant is used externally to treat arthritis
Thymus linearis Benth. (Lamiaceae)	Tribal people of Jammu and Kashmir used oil from herb to the gums for toothache and used febrifuge as well. Decoction of the aerial parts is drunk to treat cough, cold, stomachache, gastritis, diarrhea, indigestion. Powdered leaf (with honey) is used by people living in the Nanda Devi National Park (Uttarakhand, India) to treat eczema and psoriasis
Lantana camara L. (Verbenaceae)	Decoction taken for rheumatism, diuretic, snakebite, fever, colds; crushed leaves used externally on wounds, ulcers, skin sores. The tribal people of Jammu and Kashmir use decoction of the plant to treat tetanus, rheumatism, and malaria
Mentha longifolia (L.) *Huds.* (Lamiaceae)	Essential oil used for flavoring confectionery and as coolant. Leaf extract used for treatment of cough, cold, and gastrointestinal infections. Paste of powdered leaf applied to wounds and burns
Nepeta elliptica Royle ex Benth. (Lamiaceae)	In Uttarakhand and Jammu and Kashmir, an infusion of the seeds is used for digestive disorders
Mentha arvensis L. (Lamiaceae)	Local people in the Mornaula Reserve Forest of Kumaun, India, use the plant to releave stomachache, vomiting. In Kashmir, decoction of the leaves is taken to treat gastroenteritis; powder from the aerial parts is taken to treat cough, sore throat, indigestion, and constipation
Juniperus recurva Buch.-Ham. ex D. Don (Cupressaceae)	Used for the treatment of fever, headache, coughs, and colds, and a paste of the leaves and berries is applied externally to treat skin diseases
Amomum subulatum Roxb. (Zingiberaceae)	In Ayurveda, the plant is used to treat indigestion, vomiting, biliousness, abdominal pains, rectal diseases; throat trouble, lung congestion, pulmonary tuberculosis

2.5 MEDICINAL AND THERAPEUTIC POTENTIAL

Indian Himalaya embraces about 8000 vascular plants out of which 1748 are known for their healing properties. In earlier times, medicinal herbs were utilized as both medicine and food by tribal communities in Himalaya (Samal et al., 2003, 2010). In Indian Himalaya, a wide range of MAPs from different areas are scattered sparsely across different bioclimatic belts of the Himalaya. The major fraction of MAPs was collected from Bhutan, Nepal, and the Indian Himalaya of Uttarakhand, Himachal Pradesh, and Jammu and Kashmir (the Himalaya). Indian Himalaya has distinct and varied microclimatic conditions that functions as ideal environment for accumulation of secondary metabolites in MAPs. These MAPs exhibiting highest secondary metabolite are usually endemic to Himalaya (Kuniyal et al., 2005; Khan et al., 2009; Song et al., 2013). These MAPs function as major source of income for tribal communities for they used them as conventional herbal medicines. However, there is scarcity of research reports that provide inclusive information pertaining to biological activities of herbs and their ethnobotanical uses (Singh and Lal, 2008; Kumar et al., 2009; Bhattarai et al., 2010). However, some of reports document biological activities of crude extracts and essential oil procured from MAPs (Mukherjee, 2001; Singh and Panda, 2005; Kunwar et al., 2010). These biological activities are due to nonvolatile fraction of plant extract. But researchers overlook bioactivity screening associated with herbs. Some of MAPs have commercial importance due to their potential of being used as flavoring agents, perfumes, cosmetics, and pesticides (Thomas et al., 2001; Kumar, 2006; Venkateshappa and Sreenath, 2013). In cultivation and domestication of wild-growing MAPs as in Bhutan, three species of MAPs, namely, *Carum carvi*, *Saussurea lappa*, and *Zanthoxylum armatum*, are being cultivated in 30 acres of private land (Dhar et al., 2000; Rao et al., 2004; Kala, 2007). In Nepal, special training was given for suitable gathering of MAPs like *Z. armatum* and *Sapindus mukorossi* (Gaire and Subedi, 2011). Several tons of MAPs procured from forests are collected from Indian Himalaya, and the government has encouraged development of herbal enterprise for livelihood of Himalayan tribes (Kala et al., 2006; Kala, 2009). The target species that need to be cultivated and domesticated include *S. mukorossi* (ritha), *Valeriana jatamansi* (sugandhawal), *Swertia chirata* (chiraito), *Z. armatum* (timur), *S. costus* (kuth), *C. carvi* (zeera), *Z. armatum* (thingye), and *Rauvolfia serpentina* (sarpagandha). The most used herbs include *Taraxacum officinale* (hand), *Artemesia absinthium* (tethwen), *Malva sylvestris* (sonchal), *Bergenia ciliata* (Zakhmi-Hayat), *Rheum emodi* (pumb hak), and *Cedrus deodara* (Rai et al., 2000; Rinzin et al., 2009; Sharma et al., 2010; Mir et al., 2014). Among MAPs, dicots are most dominant (89%) followed by gymnosperms (7%) and monocots (4%). Majority of these plants were herbs (75%), most likely because they are more abundant followed by tress (15%), shrubs (8%), and climbers (2%) (Bhattacharya et al., 2008; Kalaichelvi and Swaminathan, 2009). Due to an increasing demand of essential oil and inclination of modern medicine toward herbal resources, there is need to explore, document, and collect information pertaining to medicinal and therapeutic potential of MAPs thriving in Himalayan herbs (Table 2.1). Moreover, we need to monitor environmental degradation, plant invasion, and impacts of climate change that threaten Himalayan flora. There is an urgent need to preserve traditional knowledge of Himalayan herbs, and more steps should be undertaken to maintain Himalayan ecology.

2.6 ADVANCEMENT IN UTILIZATION AND CONSERVATION STRATEGIES

Indian Himalaya is hot spot for plant conservation as it embraces huge plant biodiversity as communities display upward shift along altitudinal and climatic gradient counting higher proportion of endemic species (Rawat and Chandrabani, 2007; Zurick, 2006). Conservation of Himalayan flora requires transdisciplinary strategies to establish domestication and cultivation of MAPs (Rai et al., 2000; Kala, 2005). We need to integrate geographers, conservationists, ecologists, and social scientists. Due to indiscriminate harvesting, erroneous collection practices and rapid deforestation followed by habitat loss have led to drastic decline in biodiversity (Pandit et al., 2007; Xu et al., 2009a,b). The idea of conserving natural resources was first discovered by the Chiang Mai Declaration (1988) that revealed immense concern over loss of plant biodiversity. This declaration highlighted "the urgent need for international cooperation and coordination to establish programs for the conservation of medicinal plants to ensure that adequate quantities are available for future generations" (Akerele, 1993; Morris, 1998). It has also demonstrated need for including both in situ and ex situ conservation strategies. There are also numerous declaration followed by this statement and collections of recommendations identifying the conservation and sustainable utilization of MAPs biodiversity (Dhar et al., 2002; Naidoo et al., 2008; Krishnan et al., 2011). Among these, Convention on Biological Diversity (CBD), an international, legally binding treaty reached at the Earth Summit, in Rio de Janeiro (1992), established objectives for conservation of biodiversity, sustainable use of its components, and the fair and equitable sharing of the benefits from the use of genetic resources (Cunningham, 2001). Domestic and international demand is met primarily via indefensible in situ harvesting that has become a main risk to the survival of numerous MAPs (Rishi and Gauli, 2005; Kartikasari, 2008). Further management programs are immediately requisite in addition to more research on cultivation of wild Himalayan taxa. Some of MAPs are at the verge of extinction due to the lack of inclusive policies and regulations for sustainable utilization, trade, and management (Larsen and Olsen, 2007; Uprety et al., 2010; Pant and Pant, 2011). The conservation strategies for preservation of MAPs in their natural habitats will be extensively discussed in next chapter. In this chapter, we will discuss advanced contrivances for conserving MAPs in the Himalaya. For ecological sustainability of Himalayan flora, besides fulfilling knowledge gaps and mass multiplication of MAPs, conversationalists are trying to include tribal communities as developmental strategy for adding economic value to forest products (Uprety et al., 2012; Bhat et al., 2013a, b; Dhyani and Dhyani, 2016). There has been increasing inclination toward development of local enterprises that ensure ecological sustainability of Himalayan MAPs (Nepal and Weber, 1995; Rangan, 2004). Extensive research is being conducted to identify most significant MAPs having promising potential in health care and trade (Gupta, 2010; Kumar et al., 2011a,b). With advent of advanced technologies, conservation strategies are upgraded with novel and updated versions of integrated methods. Molecular tools are employed for recording effects on genetic diversity (Husband et al., 2002; Agarwal et al., 2008). We are also employing bioinformatics for investigating threats to biodiversity and monitoring extinction risk analysis (Frankham, 2010). The development of software and applications ensures storing of conventional knowledge of Himalayan flora (Sharma et al., 2009a,b; Singh et al., 2011; Phondani et al., 2016). Further sustainable management of Himalayan flora includes four management tasks (modified after Peters (1994)): (i) most exploited species to be first selected; (ii) baseline information regarding the present resource base, their abundance (density and productivity) and distribution, and local use and management practices are then collected through inventory; (iii) demographic studies; and (iv) monitoring harvesting process.

2.7 COMMERCIAL HERB UTILIZATION

Following the conservation strategies, we need to promote Himalayan herbs as alternative cash crops that could improve economy of tribal communities (Kaul et al., 2011). We need to encourage tribal communities for protection and preservation of wild MAPs and traditional knowledge concerned MAPs. This could be achieved through practical blending of in situ and ex situ involvement (Chen et al., 2016). This could lead to development of herbal enterprise that serves as alternative option for livelihood of several landless laborers, marginal farmers, and tribal communities (Silori and Badola, 2000; Bawa et al., 2007; Phondani et al., 2011). Tribal people and marginal farmers involved in wild crafting of MAPs can function as grassroot-level intermediaries (Chauhan, 2010; Sati, 2013; Pangriya, 2015).

MAP-based schemes allocate the expansion of area- and product-focused involvement directly aimed at the producers based on their knowledge and skills. At higher altitudinal mountain terrain, there is lesser scope for agricultural productivity; hence, MAP-based enterprise provides novel sustainability option (Chhetri, 2014; Sundriyal and Sharma, 2016). The conventional practices of resource utilization facilitate rural economy with equity and social justice. Overcommercialization of biological resources due to indiscriminate harvesting by untrained laborers especially in remote areas had led to overexploitation of natural resources at alarming rate (Rawal, 1996; Siddique and Jeelani, 2015). In addition, ambiguous government forest policy and overlooking of farmer's participation in decision-making together with climatic change are determinant factors for shifting resource management (IPCC, 2014; Pant et al., 2017). Previous research reports from northwestern Himalaya demonstrate conventional knowledge from ethnobotanical angle (Anand and Kumar, 2008; Das, 2013). In present scenario, no measures are taken to analyze factors that led to depilation of conventional practices and exhaustive extraction of natural resources.

Commercialization of MAPs is hampered due to dramatic decline in quality and quantity of herbs collected by local farmers during premature stage (Bhattacharya et al., 2008; Subramoniam, 2014). There are various other reasons that describe prime cause of decline in commercial trade of MAPs like overharvesting of underground parts like rhizome *Jurinea macrocephala*. Indiscriminate and premature harvesting of undergrounds parts in herbs leads to reduction in size of rhizomes in *J. macrocephala* (Singh, 2011; Uniyal et al., 2011; Dutt et al., 2015). Similarly, rhizomes and tubers from *S. lappa*, *Picrorrhiza kurrooa*, *J. macrocephala*, and *Aconitum heterophyllum* harvested after every year lead to declining effect on root regeneration (Kala, 2005, 2006; Gairola et al., 2014). On the contrary, conventional harvesting of these herbs was done following 2–3 year gap so as to get maximum yield without hampering their regeneration potential (Chandra et al., 2006; Vashistha et al., 2010; Ticktin and Shackleton, 2011). In the past, species richness of *Morchella esculenta* and *Dioscorea deltoidea* was enormous, but now, their population has become too sparse to be found (Sharma, 1998; Bhat et al., 2014; Sharma and Sagar, 2015). Similarly, *Dactylorhiza hatagirea*, *Ainsliaea aptera*, and *A. heterophyllum* are sparsely distributed, while *S. lappa* is listed under threatened category (Bisht et al., 2006; Rana and Samant, 2011; Devi et al., 2014). For commercialization of medicinal herbs, introduction of high-yielding varieties will have negative impacts on medicinal values (Farooquee and Saxena, 1996; Olsen, 2005). Conversion of forest to agricultural farms leads to destruction of habitat loss in various herbs (Ranjan and Upadhyay, 1999; Afreen et al., 2011). Naturally occurring herbs that grow near farm fields are disappearing due to intensive and extensive cultivation of cash crops (Palni et al., 2001; Banskota, 2000). Commercialization vis-à-vis sustainability ensures management and utilization of MAPs by increasing local utilization plant resources for subsistence that geared forward conventional management (Rashid et al., 2014). Increased market demand of herbal drugs and conventional farming practices have improved economic growth of local people at the cost of ecological imbalance (Tandon, 2003; Singh et al., 2013). Commercialization of herbs has not only changed crop cultivation, cropping patterns, land cover, socioeconomic, and sociocultural background but also concurrently lead to soil erosion, climatic change resulting in loss of biodiversity, limiting forest cover, and natural habitat destruction (Dovie et al., 2007; Xu et al., 2009a,b). These variations are inducing unsustainability of Himalayan ecology. Conservation and supervision strategy are always focused toward management of timber producing tree species for trade and industrial benefits. However, we can procure minor forest products without disturbing regeneration potential of natural ecosystems based on conventional methods (Samal et al., 2003). Data pertaining to several stages on trade like value addition, processing, and marketing should be amended, compiled, and distributed in local farmers so that new system of regeneration, cultivation, and domestication can be established. In the past years, medicinal herbs like *S. lappa*, *I. racemosa*, and *C. carvi* were cultivated and domesticated in Lahaul and Spiti district making it highest producer of MAPs (Chauhan, 2004; Singh et al., 2009). However, forest policy is against is cultivation of such medicinal herbs at commercial level (Rao, 2004). However, these herbs can be cultivated in farmer's field and community lands. For fortification of farm fields, there is an urgent need to redirect conventional information, empower local institutions, establish scientific methods of collection processers, and ascertain association with government agencies and nongovernment organizations (Bisht et al., 2017). Moreover, government support for cultivation MAPs and revival of conventional knowledge and implication of the entire community in decision-making could be an alternative for extenuating commercial utilization of MAPs. Waste land and degraded land can be used for establishing plantation of MAPs so as to complement conservation of Himalayan flora (Rao and Saxena, 1994; Sharma et al., 2004).

2.8 ENDEMIC AROMATIC HERBS OF KASHMIR VALLEY

Endemism is prime criterion for identifying hot spot since endemic species are totally dependent on only area for their survival and are restricted to certain areas making them vulnerable (Dhar, 2002; Dar et al., 2012). Hence, species confined to restricted areas of Himalayan ecosystem will be at higher risk of extinction therefore need instant and effective conservation approaches (Dar and Naqshi, 2001; Dar et al., 2007). Plant endemism and degree of threat establish any area as hot spot. Planet Earth comprises only 2% of hot-spot area though there are greater than 50% species diversity (Heywood, 1995). In the recent years, endemism has been marked as distinct feature of biodiversity and is imperative to ecologists, conservation biologists and phylogenetic biogeographers as represented in Fig. 2.1. The endemic species are significant for assessing, monitoring, and conservating biodiversity (Dar and Dar, 2017). Endemic species are source of enthrallment that makes them floristically exclusive (Raju et al., 2010; Dar et al., 2012). Endemic species highlight biogeography of particular ecoregion, speciation, and adaptive evolution of its flora and fauna (Pant and Pant, 2011; Jee et al., 1989). Endemic plants consist significant part of any flora and have received global attention with current rise in comprehending value of biodiversity (Dhar and Samant, 1993; Raju et al., 2010). Hence, areas whether small or large need to be identified.

Kashmir is known for its rich biodiversity parallel with an extensive assortment of habitats. The flora of Kashmir has experienced an immense transition from tropical through subtropical to temperate climate zones throughout the glacial phase of the Pleistocene (Khuroo et al., 2007a,b, 2010, 2011). Kashmir is now combination of microclimatic zones (reference). Endemism has been surveyed by several researchers, and various endemic plant taxa were procured from Kashmir Himalaya (Kachroo, 1993, 1995; Dar et al., 2002). However, there are erroneous reports that cover taxa from Afghanistan eastward to Burma, thus presenting significantly higher figures (up to 58%) for flowering plant endemism in

FIG. 2.1 Schematic representation of important localities in Kashmir Valley. https://books.google.co.in/books?id=PCC7DQAAQBAJ&redir_esc=y.

Kashmir than known for the whole of the Himalayan range (reference). Kashmir besides having rich floristic diversity is composed of several endemic taxa restricted to this region. Dhar and Kachroo (1983) observed 589 (36.58%) taxa as endemic in the alpine and subalpine region of Kashmir Himalaya (Kashmir and Ladakh), 39% dicots and 19% monocots endemics. Asteraceae comprises highest number of endemic species about 21 infraspecific taxa followed by Rosaceae (11), Fabaceae and Scrophulariaceae (9), Gentianaceae (8), and Poaceae (6). *Gentiana* has the largest number of endemic taxa (6), followed by *Astragalus* (5), *Berberis, Corydalis, Festuca, Potentilla,* and *Pseudomertensia* (4). The ecological regions containing highest number of endemics are enumerated as endemic centers of the country. There are about 28 endemic centers in India; Kashmir is one of the endemic centers. Maximum endemism is reported in northeast India, followed by southern parts of Peninsular India and northwestern Himalaya (Dar and Dar, 2017).

Endemic aromatic herbs having medicinal, ornamental, and food properties from Kashmir are included in Table 2.2. About 40% of endemic

TABLE 2.2 Description of Medicinal, Ornamental, and Food Endemic MAPs in Indian Himalaya

Medicinal endemic MAPs	Ornamental endemic MAPs	Food endemic MAPs
Aconitum heterophyllum	*Aquilegia nivalis*	*Allium kachrooi*
Artemisia amygdalin	*Chelonopsis cashmerica*	*Berberis* spp.
Artemisis benghalensis	*Corydalis* spp.	*Crataegus clarkei*
Artemesia cashmerica	*Delphinium cashmerianum*	*Fagopyrum kashmirianium*
Berberis kashmiriana	*Dianthus cachemericus*	*Lychnis cashmeriana*
Berberis pseudumbelata subsp. *pseudumbelata*	*Epipogium tuberosum, Euphorbia* spp.	*Megacarpaea bifida*
Dolomiaea baltalensis	*Gentiana cachemirica*	*Megacarpaea polyandra*
Inula royleana	*Geranium pratense* subsp. *stewartianum*	*Ribes himalense*
Iris kashmiriana	*Iris kashmiriana*	*Rubus antennifer*
Jurinea ceratocarpa	*Meconopsis latifolia*	*Sorbus cashmeriana*
Lagotis cashmeriana	*Primula* spp.	*Elymus jacquemontii*
Saussurea costus	*Saxifraga lilacina*	*Lepidium virginicum* var. *kashmiricum*
Senecio acquemontianus	*Swertia thomsonii*	*Trigonella podperae*
Valeriana jaeschkei		

MAPs in Kashmir Valley are endangered because of marked changes in climate in which alpine endemics grow and their utilization for medicinal, ornamental, and fodder uses (reference). Currently, an immense floristic diversity exhibits about 8% endemic taxa as valley is relatively younger geologically. Moreover, northwestern Himalaya is not thoroughly surveyed; there is an immense need for in-depth analysis of new taxa. The endemic flora of Kashmir Valley (Table 2.2) is imperative for phytogeographical distribution of MAPs. The endemism in Kashmir is threatened by biotic interference as well, and this requires developing practical policy for their sustainable utilization and conservation. Endemism in overall Indian Himalaya is also documented; we have assembled data from different resources that enclose list of endemic species across the Indian Himalaya as represented in Table 2.3.

TABLE 2.3 Endemic MAPs From Indian Himalaya

Endemic MAPs from Indian Himalaya		
Abies densa	*Didymocarpus pulchera*	*Hedychium venustum*
Acanthus leucostachyus	*Dioscorea belophylla*	*Himalayacalamus falconeri*
Acer caesium	*Dioscorea bulbifera*	*Hoya lobbii*
Acer campbellii	*Dioscorea deltoidea*	*Hypericum oblongifolium*
Acer oblongum	*Diospyros montana*	*Ilex dipyrena*
Achyranthes aspera	*Diplazium esculentum*	*Ilex embelioides*
Aconitum assamicum	*Diploknema butyracea*	*Ilex insignis*
Aconitum ferox	*Dracaena petiolata*	*Ilex khasiana*
Aconitum nagarum	*Dysoxylum reticulatum*	*Ilex venulosa*
Actephila excelsa	*Elaeocarpus glandulosus*	*Impatiens discolor*
Aesculus indica	*Elaeocarpus sikkimensis*	*Impatiens jurpia*
Agapetes griffithii	*Elatostema decipiens*	*Impatiens minor*
Agapetes incurvata	*Elatostema griffithii*	*Impatiens scabrida*
Alseodaphne khasyana	*Elatostema lineolatum*	*Impatiens stenantha*
Alysicarpus vaginalis	*Elatostema papillosum*	*Indoneesiella echioides*
Amomum subulatum	*Ellertonia rheedei*	*Ixora finlaysoniana*
Ampelocissus latifolia	*Enkianthus himalaicus*	*Ixora subsessilis*
Androsace lanuginosa	*Epilobium latifolium*	*Jasminum multiflorum*
Anemone elongata	*Eragrostis nutans*	*Jasminum scandens*
Anemone griffithii	*Eriobotrya anguistissima*	*Justicia vasculosa*

TABLE 2.3 Endemic MAPs From Indian Himalaya—cont'd

Endemic MAPs from Indian Himalaya		
Antidesma bunius	*Eulalia mollis*	*Khasiaclunea oligocephala*
Aquilaria malaccensis	*Euonymus attenuatus*	*Knema cinerea*
Aquilegia pubiflora	*Euonymus echinatus*	*Lagerstroemia microcarpa*
Archidendron bigeminum	*Euonymus indicus*	*Lasianthus biermanii*
Ardisia griffithii	*Euphorbia royleana*	*Lasianthus hookeri*
Ardisia pedunculosa	*Evodia lunu-ankenda*	*Lasianthus tubiferus*
Areca nagensis	*Flacourtia indica*	*Leptodermis lanceolata*
Arenaria glanduligera	*Garuga pinnata*	*Lindelofia longiflora*
Arisaema intermedium	*Gaultheria fragrantissima*	*Lindera latifolia*
Aristolochia cathcartii	*Gaultheria nummularioides*	*Litsea meissneri*
Artemisia roxburghiana	*Gentiana speciosa*	*Livistona jenkinsiana*
Arundinaria callosa	*Geranium wallichianum*	*Macaranga indica*
Arundinaria falcata	*Geum elatum*	*Magnolia pterocarpa*
Arundinaria maling	*Globba multiflora*	*Malaxis muscifera*
Aster molliusculus	*Globba pauciflora*	*Mallotus tetracoccus*
Aster peduncularis	*Globba racemosa*	*Maytenus heyneana*
Aster thomsonii	*Glochidion calocarpum*	*Mesua ferrea*
Astragalus candolleanus	*Glochidion khasicum*	*Michelia baillonii*
Astragalus himalayanus	*Glochidion malabaricum*	*Miliusa macrocarpa*
Atalantia wightii	*Grewia optiva*	*Morina longifolia*
Athyrium mackinnoniorum	*Pachylarnax pleiocarpa*	*Morus serrata*
Bambusa arundinacea	*Parthenocissus semicordata*	*Musa velutina*
Bambusa balcooa	*Pennisetum flaccidum*	*Nepeta hindostana*
Bambusa tulda	*Persea gamblei*	*Nepeta leucophylla*
Bauhinia semla	*Persea gammieana*	*Ochna obtusata*
Bauhinia vahlii	*Persea globularia*	*Olea glandulifera*
Begonia rex	*Persicaria amplexicaulis*	*Ophiorrhiza fasciculata*
Begonia sikkimensis	*Peucedanum nagpurense*	*Ophiorrhiza treutleri*
Beilschmiedia brandisii	*Phlogacanthus tubiflorus*	*Oxyspora vagans*
Beilschmiedia pseudomicrocarpa	*Phoebe cooperiana*	*Oxytropis cachemirica*

(Continued)

TABLE 2.3 Endemic MAPs From Indian Himalaya—cont'd

Endemic MAPs from Indian Himalaya		
Deutzia staminea	*Pholidota articulata*	*Pieris formosa*
Picea smithiana	*Pilea insolens*	
Pinus roxburghii	*Themeda onathera*	*Uvaria eucinata*
Pogostemon benghalense	*Trachelospermum lucidum*	*Vaccinium retusum*
Polyalthia fragrans	*Trichodesma khasianum*	*Vaccinium vacciniaccum*
Polygala tricholopha	*Tsuga dumosa*	*Vaccinium venosum*
Polygonum rude	*Ventilago denticulata*	*Ventilago maderaspatana*
Potentilla argyrophylla	*Viburnum cotinifolium*	*Waldheimia tomentosa*
Potentilla atrisanguinea	*Viola pilosa*	*Wallichia densiflora*
Potentilla microphylla	*Ribes griffithii*	*Quercus ferox*
Potentilla nepalensis	*Rosa macrophylla*	*Quercus floribunda*
Premna barbata	*Rubia manjith*	*Reinwardtiodendron anamalaiense*
Premna bengalensis	*Rubus birmanicus*	*Senecio graciliflorus*
Premna milleflora	*Rubus hamiltonii*	*Senecio pachycarpus*
Prunus arborea	*Rubus lucens*	*Senecio quinquelobus*
Prunus punctata	*Rubus paniculatus*	*Shorea assamica*
Pseudophegopteris levingei	*Rubus rosaefolius*	*Shorea robusta*
Psychotria adenophylla	*Salix sikkimensis*	*Sloanea sterculiacea*
Psychotria anamallayana	*Salvia lanata*	*Smilax glaucophylla*
Psychotria monticola	*Sapium eugeniaefolium*	*Smilax rigida*
Pyrus pashia	*Sarcosperma arboreum*	*Spondias pinnata*
Saurauia roxburghii	*Rhamnus procumbens*	*Stellaria monosperma*
Schima wallichii	*Rhamnus triqueter*	*Sterculia villosa*
Schisandra grandiflora	*Rhaphidophora calophyllum*	*Strobilanthes angustifrons*
Scrophularia elatior	*Rhaphidophora glauca*	*Strobilanthes atropurpureus*
Scutellaria discolor	*Rhaphidophora hookeri*	*Strobilanthes dalhousianus*
Scutellaria repens	*Rheum australe*	*Strobilanthes discolor*
Selaginella adunca	*Rheum spiciforme*	*Strobilanthes extensus*
Selinum vaginatum	*Rheum webbianum*	*Strobilanthes secunda*
Selinum wallichianum	*Rhododendron anthopogon*	*Sumbaviopsis albicans*

TABLE 2.3 Endemic MAPs From Indian Himalaya—cont'd

Endemic MAPs from Indian Himalaya		
Rhododendron dalhousiae	*Rhododendron arboreum*	*Swertia ciliata*
Rhododendron edgeworthii	*Rhododendron barbatum*	*Symplocos cochinchinensis*
Polygonatum brevistylum	*Rhododendron campanulatum*	*Tylophora himalaica*
Polygonum affine	*Rhododendron campylocarpum*	*Ulmus wallichiana*
Polygonum recumbens	*Symplocos laurina*	*Tetrastigma dubium*
Rhododendron lepidotum	*Syzygium balsameum*	*Tetrastigma planicaule*
Rhododendron megeratum	*Syzygium kurzii*	*Thalictrum radiatum*

Now, research is being carried out for analyzing effect of reproductive biology, low fruit setting, poor pollen viability, commercial utilization, overgrazing, and predation on MAPs distribution in Kashmir Valley (Dar et al., 2008, 2009, 2010). There is an immense threat to endemism in Kashmir especially to the Transhimalaya rich in endemics because Kashmir covers only 0.48% area of India (reference). Moreover, increasing threat to endemics is due to their restriction to alpine and subalpine regions of Himalaya. Several instant conservation strategies for endemic MAPs have been mentioned in numerous research reports. These are following conservation strategies for conserving plant endemism in Kashmir:

(i) Establishment of Himalayan flora
Majority of endemic Himalayan MAPs are distributed in high altitudes, alpine, and subalpine zones of Kashmir Valley. Conservation of endemics in their natural habitat can prove valuable genetic resource.

(ii) Development of protected area network
Expansion of protected area to conserve endemic MAPs should be initiated so as to ensure mass cultivation and preservation of MAPs.

(iii) Fortification of ex situ conservation measures
Micropropagation and mass cultivation of endemics in preserved areas will ensure rehabilitation of MAPs. This could be achieved by various micropropagation techniques like tissue culture.

(iv) Identification and rehabilitation of degraded habitats
Natural habitats that are suitable for vegetative growth of MAPs are destroyed or urbanized by humans. These natural habitats need to be identified and established.

(v) Public awareness and education
Endemic species needs protection from indiscriminate harvesting and overgrazing. In this context, we need to educate public and organize sustainability programs for educating the people about

significance of bio-resources that could help in implication of conservation programs at grassroot level.

(vi) Research needs
There is an urgent need to explore and investigate diverse ecosystems that require to be undertaken for evaluating determining factors for their threat.

(vii) Prohibition of commercial exploitation
Overexploitation of endemic species of MAPs needs to checked, and development of strict legislation should be established for proper marketing. This will improve regeneration and multiplication of MAPs.

References

Afreen, S., Sharma, N., Chaturvedi, R.K., Gopalakrishnan, R., Ravindranath, N.H., 2011. Forest policies and programs affecting vulnerability and adaptation to climate change. Mitig. Adapt. Strateg. Glob. Chang. 16 (2), 177–197.

Agarwal, M., Shrivastava, N., Padh, H., 2008. Advances in molecular marker techniques and their applications in plant sciences. Plant Cell Rep. 27 (4), 617–631.

Agarwal, R., 2011. Phytochemical analysis of some plants of family Lamiaceae and Amaranthaceae and screening of their biological activities. Doctoral dissertation GB Pant University of Agriculture and Technology, Pantnagar-263145 (Uttarakhand).

Agnihotri, P., Husain, T., Shirke, P.A., Sidhu, O.P., Singh, H., Dixit, V., Nautiyal, C.S., 2017. Climate change-driven shifts in elevation and ecophysiological traits of Himalayan plants during the past century. Curr. Sci. *112* (3), 00113891.

Ainsworth, E.A., Rogers, A., 2007. The response of photosynthesis and stomatal conductance to rising [CO_2]: mechanisms and environmental interactions. Plant Cell Environ. 30 (3), 258–270.

Akerele, O., 1993. Nature's medicinal bounty: don't throw it away. World Health Forum 14 (4), 390–395.

Ali, S.S., Kasoju, N., Luthra, A., Singh, A., Sharanabasava, H., Sahu, A., Bora, U., 2008. Indian medicinal herbs as sources of antioxidants. Food Res. Int. 41 (1), 1–15.

Amujoyegbe, B.J., Agbedahunsi, J.M., Amujoyegbe, O.O., 2012. Cultivation of medicinal plants in developing nations: means of conservation and poverty alleviation. Int. J. Med. Aromat. Plants 2, 345–353.

Anand, M., Kumar, K., 2008. Study of research trends in land and water resource management in GB Pant Institute of Himalayan Environment and Development. Himal. Ecol. 16 (1), 37.

Ann Lila, M., 2006. The nature-versus-nurture debate on bioactive phytochemicals: the genome versus terroir. J. Sci. Food Agric. 86 (15), 2510–2515.

Arjmand, F., Parveen, S., Tabassum, S., Pettinari, C., 2014. Organo-tin antitumor compounds: their present status in drug development and future perspectives. Inorg. Chim. Acta 423, 26–37.

Badola, H.K., Aitken, S., 2003. The Himalayas of India: a treasury of medicinal plants under siege. Biodiversity 4 (3), 3–13.

Baig, B.A., Ramamoorthy, D., Bhat, T.A., 2013. Threatened medicinal plants of Menwarsar Pahalgam, Kashmir Himalayas: distribution pattern and current conservation status. Proc. Int. Acad. Ecol. Environ. Sci. 3 (1), 25.

Balunas, M.J., Kinghorn, A.D., 2005. Drug discovery from medicinal plants. Life Sci. 78 (5), 431–441.

Bandaranayake, W.M., 2006. Quality control, screening, toxicity, and regulation of herbal drugs. In: Modern Phytomedicine: Turning Medicinal Plants Into Drugs, pp. 25–57.

Banskota, M., 2000. The Hindu Kush-Himalayas: searching for viable socioeconomic and environmental options. In: Growth, Poverty Alleviation, and Sustainable Resource Management in the Mountain Areas of South Asia, pp. 57–106.

Barua, A., Katyaini, S., Mili, B., Gooch, P., 2014. Climate change and poverty: building resilience of rural mountain communities in South Sikkim, Eastern Himalaya, India. Reg. Environ. Chang. 14 (1), 267–280.

Basey, A.C., Fant, J.B., Kramer, A.T., 2015. Producing native plant materials for restoration: 10 rules to collect and maintain genetic diversity. Native Plants J. 16 (1), 37–53.

Baskaran, P., Jayabalan, N., 2005. An efficient micropropagation system for Eclipta alba–a valuable medicinal herb. In Vitro Cell. Dev. Biol. Plant 41 (4), 532–539.

Bawa, K.S., Joseph, G., Setty, S., 2007. Poverty, biodiversity and institutions in forest-agriculture ecotones in the Western Ghats and Eastern Himalaya ranges of India. Agric. Ecosyst. Environ. 121 (3), 287–295.

Bennett, G., 2004. Integrating Biodiversity Conservation and Sustainable Use: Lessons Learned From Ecological Networks. IUCN.

Bennett, B.C., Balick, M.J., 2014. Does the name really matter? The importance of botanical nomenclature and plant taxonomy in biomedical research. J. Ethnopharmacol 152 (3), 387–392.

Berli, F.J., Moreno, D., Piccoli, P., Hespanhol-Viana, L., Silva, M.F., Bressan-Smith, R., Bottini, R., 2010. Abscisic acid is involved in the response of grape (*Vitis vinifera* L.) cv. Malbec leaf tissues to ultraviolet-B radiation by enhancing ultraviolet-absorbing compounds, antioxidant enzymes and membrane sterols. Plant Cell Environ. 33 (1), 1–10.

Bhat, F., Mahajan, D.M., Sayyed, M.R.G., Bhat, A., 2014. Ethno-medicinal survey of north Kashmir Himalaya—a case study of Lolab vally (J & K), India. Ecol. Environ. Conserv. 20 (1), 59–71.

Bhat, J.A., Kumar, M., Bussmann, R.W., 2013a. Ecological status and traditional knowledge of medicinal plants in Kedarnath Wildlife Sanctuary of Garhwal Himalaya, India. J. Ethnobiol. Ethnomed. 9 (1), 1.

Bhat, J.A., Kumar, M., Negi, A.K., Todaria, N.P., 2013b. Informants consensus on ethnomedicinal plants in Kedarnath Wildlife Sanctuary of Indian Himalayas. J. Med. Plant Res. 7 (4), 148–154.

Bhattacharya, P., Prasad, R., Bhattacharyya, R., Asokan, A., 2008. Towards certification of wild medicinal and aromatic plants in four Indian states. Unasylva 230 (59), 35–44.

Bhattacharya, R.D., Parmar, K.M., Itankar, P.R., Prasad, S.K., 2017. Phytochemical and pharmacological evaluation of organic and non-organic cultivated nutritional *Centella asiatica* collected after different time intervals of harvesting. S. Afr. J. Bot. 112, 237–245.

Bhattarai, K.R., Vetaas, O.R., 2003. Variation in plant species richness of different life forms along a subtropical elevation gradient in the Himalayas, east Nepal. Glob. Ecol. Biogeogr. 12 (4), 327–340.

Bhattarai, S., Chaudhary, R.P., Quave, C.L., Taylor, R.S., 2010. The use of medicinal plants in the trans-Himalayan arid zone of Mustang district, Nepal. J. Ethnobiol. Ethnomed. 6 (1), 14.

Bisht, A.K., Bhatt, A., Rawal, R.S., Dhar, U., 2006. Prioritization and conservation of Himalayan medicinal plants: *Angelica glauca* Edgew. as a case study. Ethnobot. Res. Appl. 4, 011–024.

Bisht, I.S., Mehta, P.S., Negi, K.S., Verma, S.K., Tyagi, R.K., Garkoti, S.C., 2017. Farmer's rights, local food systems and sustainable household dietary diversification: a case of Uttarakhand Himalaya in north-western India. Agroecol. Sustain. Food Syst. 42, 77–113.

Bolch, T., Kulkarni, A., Kääb, A., Huggel, C., Paul, F., Cogley, J.G., Bajracharya, S., 2012. The state and fate of Himalayan glaciers. Science 336 (6079), 310–314.

Boyle, B., Hopkins, N., Lu, Z., Garay, J.A.R., Mozzherin, D., Rees, T., Lowry, S., 2013. The taxonomic name resolution service: an online tool for automated standardization of plant names. BMC Bioinform. 14 (1), 16.

Caballero, C., Aranda, M.C., 2010. In: Plant species identification using leaf image retrieval. Proceedings of the ACM International Conference on Image and Video Retrieval. ACM, pp. 327–334.

Caldwell, M.M., Bornman, J.F., Ballaré, C.L., Flint, S.D., Kulandaivelu, G., 2007. Terrestrial ecosystems, increased solar ultraviolet radiation, and interactions with other climate change factors. Photochem. Photobiol. Sci. 6 (3), 252–266.

Camuffo, D., 2013. Microclimate for Cultural Heritage: Conservation, Restoration, and Maintenance of Indoor and Outdoor Monuments. Elsevier, USA.

Casson, S., Gray, J.E., 2008. Influence of environmental factors on stomatal development. New Phytol. 178 (1), 9–23.

Cavaliere, C., 2009. The effects of climate change on medicinal and aromatic plants. Herbal Gram 81, 44–57.

Chandra, B., Palni, L.M.S., Nandi, S.K., 2006. Propagation and conservation of *Picrorhiza kurrooa* Royle ex Benth.: an endangered Himalayan medicinal herb of high commercial value. Biodivers. Conserv. 15 (7), 2325.

Charrier, G., Ngao, J., Saudreau, M., Améglio, T., 2015. Effects of environmental factors and management practices on microclimate, winter physiology, and frost resistance in trees. Front. Plant Sci. 6, 259. https://doi.org/10.3389/fpls.2015.00259.

Chaturvedi, A.K., Prasad, P., Nautiyal, M.C., 2013. Impact of elevated CO_2 on growth, morphology and dry matter partitioning in alpine growth forms of north western Himalayas. Indian J. Plant Physiol. 18 (2), 118–124.

Chauhan, N.S., 1999. Medicinal and Aromatic Plants of Himachal Pradesh. Indus Publishing, New Delhi.

Chauhan, N.S., 2004. Present status and future strategies for popularising medicinal and aromatic plants in Himachal Pradesh. Adv. Hortic. 60, 213–217.

Chauhan, R.S., 2010. Socioeconomic improvement through medicinal and aromatic plants (MAPs) cultivation in Uttarakhand, India. J. Sustain. Agric. 34 (6), 647–658.

Chawla, A., Rajkumar, S., Singh, K.N., Lal, B., Singh, R.D., Thukral, A.K., 2008. Plant species diversity along an altitudinal gradient of Bhabha Valley in western Himalaya. J. Mt. Sci. 5 (2), 157–177.

Chawla, R., Thakur, P., Chowdhry, A., Jaiswal, S., Sharma, A., Goel, R., Arora, R., 2013. Evidence based herbal drug standardization approach in coping with challenges of holistic management of diabetes: a dreadful lifestyle disorder of 21st century. J. Diabetes Metab. Disord. 12 (1), 35.

Chen, F., Liu, C.J., Tschaplinski, T.J., Zhao, N., 2009. Genomics of secondary metabolism in Populus: interactions with biotic and abiotic environments. Crit. Rev. Plant Sci. 28 (5), 375–392.

Chen, I.C., Hill, J.K., Ohlemüller, R., Roy, D.B., Thomas, C.D., 2011. Rapid range shifts of species associated with high levels of climate warming. Science 333 (6045), 1024–1026.

Chen, S.L., Yu, H., Luo, H.M., Wu, Q., Li, C.F., Steinmetz, A., 2016. Conservation and sustainable use of medicinal plants: problems, progress, and prospects. Chin. Med. 11 (1), 37.

Cheng, D.K., 2017. Primary and secondary metabolites. Ethnobot. Phytochem. Perspect. 111.

Chhetri, D.R., 2014. Medicinal Plants of the Himalaya: Production Technology and Utilization. Agrobios, Bangalore, India.

Choudhary, N., Sekhon, B.S., 2011. An overview of advances in the standardization of herbal drugs. J. Pharm. Educ. Res. 2 (2), 55.

Chown, S.L., Hoffmann, A.A., Kristensen, T.N., Angilletta Jr., M.J., Stenseth, N.C., Pertoldi, C., 2010. Adapting to climate change: a perspective from evolutionary physiology. Clim. Res. 43 (1/2), 3–15.

Copeland, L.O., McDonald, M., 2012. Principles of Seed Science and Technology. Springer Science & Business Media, Springer-Verlag, USA.

Cotter, M., Berkhoff, K., Gibreel, T., Ghorbani, A., Golbon, R., Nuppenau, E.A., Sauerborn, J., 2014. Designing a sustainable land use scenario based on a combination of ecological assessments and economic optimization. Ecol. Indic. 36, 779–787.

Crawford, D.J., 1990. Plant Molecular Systematics: Macromolecular Approaches. John Wiley & Sons, Inc., New York, USA.

Crockett, S.L., Robson, N.K., 2011. Taxonomy and chemotaxonomy of the genus Hypericum. Med. Aromat. Plant Sci. Biotechnol. 5 (Special Issue 1), 1.

Croteau, R., Kutchan, T.M., Lewis, N.G., 2000. Natural products (secondary metabolites). Biochem. Mol. Biol. Plants 24, 1250–1319.

Cunningham, A.B., 2001. Applied Ethnobotany: People, Wild Plant Use and Conservation. Earthscan, UK.

Dar, A.R., Dar, G.H., 2017. Conservation of Kashmir Himalayan Endemic Plants. Anchor Academic Publishing, Hamburg, Germany.

Dar, A.R., Dar, G.H., Reshi, Z., Khuroo, A.A., 2007. An assessment of endangered plant species in the Kashmir Himalaya. In: Souvenir & Abstracts of XVII Annual Conference of Indian Association for Angiosperm Taxonomy and International Seminar on Changing Scenario in Angiosperm Systematics, 19–21 November, 2007. Department of Botany, Shivaji University, Kolhapur, India, pp. 141. (Abstract).

Dar, A.R., Dar, G.H., Reshi, Z., 2008. Narrow endemic angiosperms of the Kashmir Himalaya: threat assessment and conservation. In: Chisti, M.Z., Ahmad, F. (Eds.), Science for Better Tomorrow. Universal Printers, Khanyar, Srinagar, pp. 31–39.

Dar, A.R., Reshi, Z., Dar, G.H., 2009. Germination studies on three critically endangered endemic angiosperm species of the Kashmir Himalaya, India. Plant Ecol. 200, 105–115.

Dar, A.R., Reshi, Z.A., Dar, G.H., 2010. Factors contributing to critically endangered status of Aquilegia nivalis Falc. ex Jackson - an alpine endemic angiospenn in the Kashmir Himalaya, India. (In press).

Dar, G.H., Naqshi, A.R., 2001. Threatened flowering plants of the Kashmir Himalaya—a checklist. Oriental Sci. 6 (1), 23–53.

Dar, G.H., Bhagat, R.C., Khan, M.A., 2002. Biodiversity of the Kashmir Himalaya. Valley Book House, Srinagar.

Dar, G.H., Khuroo, A.A., Nasreen, A., 2012. In: Mukherjee, S.K., Maiti, G.G. (Eds.), Endemism in the angiosperm flora of Kashmir Valley, India: Stocktaking. Proceedings of International Seminar on Multidisciplinary Approaches in Angiosperm Systematics, Kolkata.

Dar, J.A., Sundarapandian, S., 2016. Patterns of plant diversity in seven temperate forest types of Western Himalaya, India. J. Asia-Pacific Biodivers. 9 (3), 280–292.

Das, N., 2013. Assessment of ecotourism resources: an applied methodology to Nameri National Park of Assam-India. J. Geogr. Region. Plan. 6 (6), 218.

De Lillis, M., Matteucci, G., Valentini, R., 2004. Carbon assimilation, nitrogen, and photochemical efficiency of different Himalayan tree species along an altitudinal gradient. Photosynthetica 42 (4), 597–605.

Denford, K.E., 1984. Phytochemical approaches to biosystematics. In: Plant Biosystematics. Academic Press, Canada.

Devasagayam, T.P.A., Tilak, J.C., Boloor, K.K., Sane, K.S., Ghaskadbi, S.S., Lele, R.D., 2004. Free radicals and antioxidants in human health: current status and future prospects. Japi 52 (10), 794–804.

Devi, U., Sharma, P., Rana, J.C., Sharma, A., 2014. Phytodiversity assessment in Sangla valley, Northwest Himalaya, India. Check List 10 (4), 740–760.

Dhami, N., 2013. Trends in Pharmacognosy: a modern science of natural medicines. J. Herb. Med. 3 (4), 123–131.

Dhar, U., 2002. Conservation implications of plant endemism in high-altitude Himalaya. Curr. Sci. 141–148.

Dhar, U., Kachroo, P., 1983. Alpine Flora of Kashmir Himalaya. Scientific Publishers, Jodhpur, India.

Dhar, U., Samant, S.S., 1993. Endemic plant diversity in the Indian Himalaya I. Ranunculaceae and Paeoniaceae. J. Biogeogr. 659–668.

Dhar, U., Rawal, R.S., Upreti, J., 2000. Setting priorities for conservation of medicinal plants—a case study in the Indian Himalaya. Biol. Conserv. 95 (1), 57–65.

Dhar, U., Manjkhola, S., Joshi, M., Bhatt, A., Bisht, A.K., Joshi, M., 2002. Current status and future strategy for development of medicinal plants sector in Uttaranchal, India. Curr. Sci. 83 (8), 956–964.

Dhyani, S., Dhyani, D., 2016. Strategies for reducing deforestation and disaster risk: lessons from Garhwal Himalaya, India. In: Ecosystem-Based Disaster Risk Reduction and Adaptation in Practice. Springer International Publishing, Switzerland, pp. 507–528.

Dobhal, D.P., Mehta, M., 2010. Surface morphology, elevation changes and terminus retreat of Dokriani Glacier, Garhwal Himalaya: implication for climate change. Himal. Geol. 31 (1), 71–78.

Dolezal, J., Dvorsky, M., Kopecky, M., Liancourt, P., Hiiesalu, I., Macek, M., Borovec, J., 2016. Vegetation dynamics at the upper elevational limit of vascular plants in Himalaya. Sci. Rep. 6, 24881. https://doi.org/10.1038/srep24881.

Dong, J., Ma, X., Wei, Q., Peng, S., Zhang, S., 2011. Effects of growing location on the contents of secondary metabolites in the leaves of four selected superior clones of Eucommia ulmoides. Ind. Crop. Prod. 34 (3), 1607–1614.

Dovie, D.B., Shackleton, C.M., Witkowski, E.T.F., 2007. Conceptualizing the human use of wild edible herbs for conservation in South African communal areas. J. Environ. Manag. 84 (2), 146–156.

Dutt, H.C., Bhagat, N., Pandita, S., 2015. Oral traditional knowledge on medicinal plants in jeopardy among Gaddi shepherds in hills of northwestern Himalaya, J&K, India. J. Ethnopharmacol. 168, 337–348.

Evans, W.C., 2002. Trease and Evans Pharmacognosy, 15th ed. WB Saunders Co. Ltd., Singapore, pp. 357–375.

Farooquee, N.A., Saxena, K.G., 1996. Conservation and utilization of medicinal plants in high hills of the central Himalayas. Environ. Conserv. 23 (1), 75–80.

Figueiredo, A.C., Barroso, J.G., Pedro, L.G., Scheffer, J.J., 2008. Factors affecting secondary metabolite production in plants: volatile components and essential oils. Flavour Fragr. J. 23 (4), 213–226.

Fitter, A.H., Hay, R.K., 2012. Environmental Physiology of Plants. Academic Press, New York.

Folashade, O., Omoregie, H., Ochogu, P., 2012. Standardization of herbal medicines—a review. Int. J. Biodivers. Conserv. 4 (3), 101–112.

Frankham, R., 2010. Challenges and opportunities of genetic approaches to biological conservation. Biol. Conserv. 143 (9), 1919–1927.

Fuller, D.Q., Murphy, C., 2014. Overlooked but not forgotten: India as a center for agricultural domestication. Gen. Anthropol. 21 (2), 1–8.

Gaire, B.P., Subedi, L., 2011. Medicinal plant diversity and their pharmacological aspects of Nepal Himalayas. Pharm. J. 3 (25), 6–17.

Gairola, S., Shariff, N.M., Bhatt, A., 2010. Influence of climate change on production of secondary chemicals in high altitude medicinal plants: issues needs immediate attention. J. Med. Plant Res. 4 (18), 1825–1829.

Gairola, S., Sharma, J., Bedi, Y.S., 2014. A cross-cultural analysis of Jammu, Kashmir and Ladakh (India) medicinal plant use. J. Ethnopharmacol. 155 (2), 925–986.

Ghimire, S.K., Mckey, D., Aumeeruddy-Thomas, Y., 2006. Himalayan medicinal plant diversity in an ecologically complex high altitude anthropogenic landscape, Dolpo, Nepal. Environ. Conserv. 33 (2), 128–140.

Ghimire, S.K., Gimenez, O., Pradel, R., McKey, D., Aumeeruddy-Thomas, Y., 2008. Demographic variation and population viability in a threatened Himalayan medicinal and aromatic herb *Nardostachys grandiflora*: matrix modelling of harvesting effects in two contrasting habitats. J. Appl. Ecol. 45 (1), 41–51.

Glime, J.M., 2017. Ecophysiology of Development: Spore Germination. Chapt. 5-2. In: Glime, J.M. (Ed.), Bryophyte Ecology, vol. 1. 5-2 Physiological Ecology. Ebook sponsored by Michigan Technological University and the International Association of Bryologists.

Goldman, D.H., Jansen, R.K., van den Berg, C., Leitch, I.J., Fay, M.F., Chase, M.W., 2004. Molecular and cytological examination of Calopogon (Orchidaceae, Epidendroideae): circumscription, phylogeny, polyploidy, and possible hybrid speciation. Am. J. Bot. 91 (5), 707–723.

González, J.A., Carvalho, A.M., Vallejo, J.R., Amich, F., 2017. Plant-based remedies for wolf bites and rituals against wolves in the Iberian Peninsula: therapeutic opportunities and cultural values for the conservation of biocultural diversity. J. Ethnopharmacol. 209, 124–139.

Gould, S.J., 1992. Roots: ontogeny and phylogeny—revisited and reunited. BioEssays 14 (4), 275–279.

Grover, J.K., Vats, V., 2001. Shifting paradigm: from conventional to alternative medicines—an introduction on traditional Indian medicines. Asia-Pacific Biotech News 5 (01), 28–32.

Grover, J.K., Vats, V., Rathi, S.S., 2007. Nefazodone—a new anti-depressant. J. Inst. Med. 22 (1).

Gupta, R.K., 2010. Medicinal and Aromatic Plants. vol. 234. CBS Publishers and Distributors499.

Hegnauer, R., 1986. Phytochemistry and plant taxonomy—an essay on the chemotaxonomy of higher plants. Phytochemistry 25 (7), 1519–1535.

Heywood, V.H., 1995. Global Biodiversity Assessment. Cambridge University Press, Cambridge.

Husband, R., Herre, E.A., Turner, S.L., Gallery, R., Young, J.P.W., 2002. Molecular diversity of arbuscular mycorrhizal fungi and patterns of host association over time and space in a tropical forest. Mol. Ecol. 11 (12), 2669–2678.

Hussain, M.S., Fareed, S., Saba Ansari, M., Rahman, A., Ahmad, I.Z., Saeed, M., 2012. Current approaches toward production of secondary plant metabolites. J. Pharm. Bioallied Sci. 4 (1), 10.

Intergovernmental Panel on Climate Change, 2014. Climate Change 2014—Impacts, Adaptation and Vulnerability: Regional Aspects. Cambridge University Press.

Iqbal, M.M., Hussain, S.S., Goheer, M.A., Sultana, H., Salik, K.M., Mudasser, M., Khan, A.M., 2009. Climate Change and Wheat Production in Pakistan: Calibration, Validation and Application of CERES-Wheat Model. Global Change Impact Studies Centre, Islamabad.

Jackson, S.T., Betancourt, J.L., Booth, R.K., Gray, S.T., 2009. Ecology and the ratchet of events: climate variability, niche dimensions, and species distributions. Proc. Natl. Acad. Sci. 106 (Suppl. 2), 19685–19692.

Jan, S., Kamili, A.N., Hamid, R., Parray, J.A., 2013. Variation in adaptation mechanisms of medicinal herbs to the extreme winter conditions across the North Western Himalaya. Israel J. Plant Sci. 61 (1-4), 1–11.

Jan, S., Kamili, A.N., Parray, J.A., Bedi, Y.S., Ahmad, P., 2016a. Microclimatic variation in UV perception and related disparity in tropane and quinolizidine alkaloid composition of *Atropa acuminata, Lupinus polyphyllus* and *Hyoscyamus niger*. J. Photochem. Photobiol. B Biol. 161, 230–235.

Jan, S., Kamili, A.N., Parray, J.A., Bedi, Y.S., 2016b. Differential response of terpenes and anthraquinones derivatives in *Rumex dentatus* and *Lavandula officinalis* to harsh winters across north-western Himalaya. Nat. Prod. Res. 30 (5), 608–612.

Janatková, K., Řeháková, K., Doležal, J., Šimek, M., Chlumská, Z., Dvorský, M., Kopecký, M., 2013. Community structure of soil phototrophs along environmental gradients in arid Himalaya. Environ. Microbiol. 15 (9), 2505–2516.

Jaradat, A.A., 2015. Beyond biodiversity: ecosystem services of crop wild relatives. In: Crop Wild Relatives and Climate Change, pp. 336–349.

Jee, V., Dhar, U., Kachroo, P., 1989. Cytogeography of some endemic taxa of Kashmir Himalaya. Proc. Ind. Natl. Sci. Acad. (Bio.Sci.) B 55, 177–184.

Johns, T., Powell, B., Maundu, P., Eyzaguirre, P.B., 2013. Agricultural biodiversity as a link between traditional food systems and contemporary development, social integrity and ecological health. J. Sci. Food Agric. 93 (14), 3433–3442.

Jones, H.G., 2013. Plants and Microclimate: A Quantitative Approach to Environmental Plant Physiology. Cambridge University Press, UK.

Joshi, D.D., 2012. Herbal Drugs and Fingerprints: Evidence Based Herbal Drugs. Springer Science & Business Media, CBS Publishers, New Delhi, India.

Joshi, R.K., Satyal, P., Setzer, W.N., 2016. Himalayan aromatic medicinal plants: a review of their ethnopharmacology, volatile phytochemistry, and biological activities. Medicines 3 (1), 6–65.

Jump, A.S., Huang, T.J., Chou, C.H., 2012. Rapid altitudinal migration of mountain plants in Taiwan and its implications for high altitude biodiversity. Ecography 35 (3), 204–210.

Kachroo, P., 1993. Plant diversity in Northwest Himalaya—a preliminary survey. In: Dhar, U. (Ed.), Himalayan Biodiversity: Conservation Strategies. G B. Pant Institute of Himalayan Environment and Development, Almora, India, pp. 111–132.

Kachroo, P., 1995. Central Asia and Kashmir Himalaya—Archaeobotany and Floristics. Scientific Publishers, Jodhpur, India.

Kala, C.P., 2005. Indigenous uses, population density, and conservation of threatened medicinal plants in protected areas of the Indian Himalayas. Conserv. Biol. 19 (2), 368–378.

Kala, C.P., 2006. Medicinal plants of the high altitude cold desert in India: diversity, distribution and traditional uses. Int. J. Biodivers. Sci. Manage. 2 (1), 43–56.

Kala, C.P., 2007. Prioritization of cultivated and wild edibles by local people in the Uttaranchal hills of Indian Himalaya. Indian J. Tradit. Knowl. 6, 239–243.

Kala, C.P., 2009. Medicinal plants conservation and enterprise development. Med. Plants 1 (2), 79–95.

Kala, C.P., Dhyani, P.P., Sajwan, B.S., 2006. Developing the medicinal plants sector in northern India: challenges and opportunities. J. Ethnobiol. Ethnomed. 2 (1), 32.

Kalaichelvi, K., Swaminathan, A.A., 2009. Alternate land use through cultivation of medicinal and aromatic plants—a review. Agric. Rev. 30 (3), 176–183.

Kalia, K., Sharma, K., Singh, H.P., Singh, B., 2008. Effects of extraction methods on phenolic contents and antioxidant activity in aerial parts of *Potentilla atrosanguinea* Lodd. and quantification of its phenolic constituents by RP-HPLC. J. Agric. Food Chem. 56 (21), 10129–10134.

Kartikasari, S.N., 2008. Your biodiversity in my backyard: key local stakeholders' perceptions of biodiversity conservation in Gorontalo, Indonesia. Doctoral dissertation, Lincoln University.

Katiyar, C., Gupta, A., Kanjilal, S., Katiyar, S., 2012. Drug discovery from plant sources: an integrated approach. Ayu 33 (1), 10.

Kaul, K., Jaitak, V., Kaul, V.K., 2011. Review on pharmaceutical properties and conservation measures of *Potentilla fulgens* Wall. ex Hook.—a medicinal endangered herb of higher Himalaya. Indian J. Nat. Prod. Resour. 2 (3), 298–306.

Kaul, M.K., 1997. Medicinal plants of Kashmir and Ladakh: temperate and cold arid Himalaya. Indus Publishing, New Delhi, India.

Kaul, M.K., 2010. High altitude botanicals in integrative medicine—case studies from Northwest Himalaya. Indian J. Tradit. Knowl. 9 (1), 18–25.

Kaur, T., Bhat, R., Vyas, D., 2016. Effect of contrasting climates on antioxidant and bioactive constituents in five medicinal herbs in Western Himalayas. J. Mt. Sci. 13 (3), 484.

Khan, A.L., Hussain, J., Hamayun, M., Shinwari, Z.K., Khan, H., Kang, Y.H., Lee, I.J., 2009. Inorganic profile and allelopathic effect of endemic Inula koelzii from Himalaya Pakistan. Pak. J. Bot. 41 (5), 2517–2527.

Khan, I., AbdElsalam, N.M., Fouad, H., Tariq, A., Ullah, R., Adnan, M., 2014. Application of ethnobotanical indices on the use of traditional medicines against common diseases. Evid. Based Complement. Alternat. Med. 2014, 635371. 21 pp.

Khandelwal, K., 2008. Practical Pharmacognosy. Pragati Books Pvt. Ltd., Maharashtra, India.

Khuroo, A.A., Rashid, I., Reshi, Z., Dar, G.H., Wafai, B.A., 2007a. The alien flora of Kashmir Himalaya. Biol. Invasions 9, 269–292.

Khuroo, A.A., Dar, G.H., Khan, Z.S., Malik, A.H., 2007b. Exploring an inherent interface between taxonomy and biodiversity: current problems and future challenges. J. Nat. Conserv. 15 (4), 256–261.

Khuroo, A.A., Weber, E., Malik, A.H., Dar, G.H., Reshi, Z.A., 2010. Taxonomic and biogeographic patterns in the native and alien woody flora of Kashmir Himalaya, India. Nord. J. Bot. 28 (6), 685–696.

Khuroo, A.A., Weber, E., Malik, A.H., Reshi, Z.A., Dar, G.H., 2011. Altitudinal distribution patterns of the native and alien woody flora in Kashmir Himalaya, India. Environ. Res. 111 (7), 967–977.

Kliebenstein, D.J., 2004. Secondary metabolites and plant/environment interactions: a view through *Arabidopsis thaliana* tinged glasses. Plant Cell Environ. 27 (6), 675–684.

Klimes, L., 2003. Life-forms and clonality of vascular plants along an altitudinal gradient in E Ladakh (NW Himalayas). Basic Appl. Ecol. 4 (4), 317–328.

Körner, C., 2003. Alpine Plant Life: Functional Plant Ecology of High Mountain Ecosystems; With 47 Tables. Springer Science & Business Media, Springer-Verlag, Berlin, Heidelberg.

Körner, C., 2007. The use of "altitude" in ecological research. Trends Ecol. Evol. 22 (11), 569–574.

Krishnan, P.N., Decruse, S.W., Radha, R.K., 2011. Conservation of medicinal plants of Western Ghats, India and its sustainable utilization through in vitro technology. In Vitro Cell. Dev. Biol. Plant 47 (1), 110–122.

Kroymann, J., 2011. Natural diversity and adaptation in plant secondary metabolism. Curr. Opin. Plant Biol. 14 (3), 246–251.

Krumholz, H.M., 2014. Big data and new knowledge in medicine: the thinking, training, and tools needed for a learning health system. Health Aff. 33 (7), 1163–1170.

Kumar, A., Kumar, P., Koundal, R., Agnihotri, V.K., 2016. Antioxidant properties and UPLC-MS/MS profiling of phenolics in jacquemont's hazelnut kernels (*Corylus jacquemontii*) and its byproducts from western Himalaya. J. Food Sci. Technol. 53 (9), 3522–3531.

Kumar, G., Gupta, S., Murugan, M., Bala Singh, S., 2009. Ethnobotanical studies of Nubra Valley—a cold arid zone of Himalaya. Ethnobot. Leaflets 2009 (6), 9.

Kumar, G.P., Kumar, R., Chaurasia, O.P., 2011a. Conservation status of medicinal plants in Laclakh: cold arid zone of Trans-Himalayas. Res. J. Med. Plants 5, 685–694.

Kumar, J., Gupta, P.K., 2008. Molecular approaches for improvement of medicinal and aromatic plants. Plant Biotechnol. Rep. 2 (2), 93.

Kumar, N.J.B.M., 2006. Introduction to Spices, Plantation Crops, Medicinal and Aromatic Plants. Oxford and IBH Publishing, New Delhi, India.

Kumar, S., Vats, S.K., 2017. Plant Adaptation in Mountain Ecosystem. In: Plant Biotechnology: Principles and Applications. Springer, Singapore, pp. 249–271.

Kumar, S., Kumar, R., Khan, A., 2011b. Medicinal plant resources: manifestation and prospects of life-sustaining healthcare system. Cont. J. Biol. Sci. 4 (1), 19–29.

Kuniyal, C.P., Rawat, Y.S., Oinam, S.S., Kuniyal, J.C., Vishvakarma, S.C., 2005. Kuth (*Saussurea lappa*) cultivation in the cold desert environment of the Lahaul valley, northwestern Himalaya, India: arising threats and need to revive socio-economic values. Biodivers. Conserv. 14 (5), 1035–1045.

Kuniyal, C.P., Bisht, V.K., Negi, J.S., Bhatt, V.P., Bisht, D.S., Butola, J.S., Singh, S.K., 2015. Progress and prospect in the integrated development of medicinal and aromatic plants (MAPs) sector in Uttarakhand, Western Himalaya. Environ. Dev. Sustain. 17 (5), 1141.

Kunwar, R.M., Shrestha, K.P., Bussmann, R.W., 2010. Traditional herbal medicine in Far-west Nepal: a pharmacological appraisal. J. Ethnobiol. Ethnomed. 6 (1), 35.

Kurian, A., Sankar, M.A., 2007. Medicinal Plants. vol. 2. New India Publishing, New Delhi, India.

Lagunin, A.A., Goel, R.K., Gawande, D.Y., Pahwa, P., Gloriozova, T.A., Dmitriev, A.V., Druzhilovsky, D.S., 2014. Chemo- and bioinformatics resources for in silico drug discovery from medicinal plants beyond their traditional use: a critical review. Nat. Prod. Rep. 31 (11), 1585–1611.

Lall, J.S., 1981. The Himalaya Aspects of Change. Oxford University Press, Delhi.

Laloo, D., Kumar, M., Prasad, S.K., Hemalatha, S., 2013a. Quality control standardization of the roots of *Potentilla fulgens* Wall.: a potent medicinal plant of the Western Himalayas and North-eastern India. Pharm. J. 5 (3), 97–103.

Laloo, D., Prasad, S.K., Krishnamurthy, S., Hemalatha, S., 2013b. Gastroprotective activity of ethanolic root extract of *Potentilla fulgens* Wall. ex Hook. J. Ethnopharmacol. 146 (2), 505–514.

Langdon, A., Crook, N., Dantas, G., 2016. The effects of antibiotics on the microbiome throughout development and alternative approaches for therapeutic modulation. Genome Med. 8 (1), 39.

Larcher, W., 2003. Physiological Plant Ecology: Ecophysiology and Stress Physiology of Functional Groups. Springer Science & Business Media, Springer-Verlag, Berlin, Heidelberg.

Larsen, H.O., Olsen, C.S., 2007. Unsustainable collection and unfair trade? Uncovering and assessing assumptions regarding central Himalayan medicinal plant conservation. Biodivers. Conserv. 16 (6), 1679–1697.

Leonelli, S., 2013. Integrating data to acquire new knowledge: three modes of integration in plant science. Stud. Hist. Philos. Sci. Part C: Stud. Hist. Philos. Biol. Biomed. Sci. 44 (4), 503–514.

Li, F., Wei, C., Qiao, C., Chen, Z., Wang, P., Wei, P., Wang, R., Jin, L., Yang, J., Lin, F., Luo, Z., 2016. Three duplication events and variable molecular evolution characteristics involved in multiple GGPS. J. Genet. 95 (2), 453–457.

Linares, J.C., 2011. Biogeography and evolution of Abies (Pinaceae) in the Mediterranean basin: the roles of long-term climatic change and glacial refugia. J. Biogeogr. 38 (4), 619–630.

Lombardo, V.A., Osorio, S., Borsani, J., Lauxmann, M.A., Bustamante, C.A., Budde, C.O., Drincovich, M.F., 2011. Metabolic profiling during peach fruit development and ripening reveals the metabolic networks that underpin each developmental stage. Plant Physiol. 157 (4), 1696–1710.

Long, S.P., Ainsworth, E.A., Rogers, A., Ort, D.R., 2004. Rising atmospheric carbon dioxide: plants FACE the future. Annu. Rev. Plant Biol. 55, 591–628.

Luo, Y., Melillo, J., Niu, S., Beier, C., Clark, J.S., Classen, A.T., Czimczik, C.I., 2011. Coordinated approaches to quantify long-term ecosystem dynamics in response to global change. Glob. Chang. Biol. 17 (2), 843–854.

Lüttge, U., 2007. Physiological Ecology of Tropical Plants. Springer Science & Business Media, Springer-Verlag, Berlin, Heidelberg.

Macías, F.A., Galindo, J.L., Galindo, J.C., 2007. Evolution and current status of ecological phytochemistry. Phytochemistry 68 (22), 2917–2936.

Maffei, M.E., Gertsch, J., Appendino, G., 2011. Plant volatiles: production, function and pharmacology. Nat. Prod. Rep. 28 (8), 1359–1380.

Maikhuri, R.K., Semwal, R.L., Rao, K.S., Singh, K., Saxena, K.G., 2000. Growth and ecological impacts of traditional agroforestry tree species in Central Himalaya, India. Agrofor. Syst. 48 (3), 257–271.

Malik, Z.A., Bhat, J.A., Ballabha, R., Bussmann, R.W., Bhatt, A.B., 2015. Ethnomedicinal plants traditionally used in health care practices by inhabitants of Western Himalaya. J. Ethnopharmacol. 172, 133–144.

Manish, K., Telwala, Y., Nautiyal, D.C., Pandit, M.K., 2016. Modelling the impacts of future climate change on plant communities in the Himalaya: a case study from Eastern Himalaya, India. Model. Earth Syst. Environ. 2 (2), 1–12.

Marvin, D.C., Koh, L.P., Lynam, A.J., Wich, S., Davies, A.B., Krishnamurthy, R., Asner, G.P., 2016. Integrating technologies for scalable ecology and conservation. Glob. Ecol. Conserv. 7, 262–275.

Máthé, A., 2010. In: A new look at medicinal and aromatic plants. XXVIII International Horticultural Congress on Science and Horticulture for People (IHC2010): A New Look at Medicinal and 925, August, pp. 13–20.

Máthé, Á., 2015. Botanical aspects of medicinal and aromatic plants. In: Medicinal and Aromatic Plants of the World. Springer, Netherlands, pp. 13–33.

Meegahakumbura, M.K., Wambulwa, M.C., Thapa, K.K., Li, M.M., Möller, M., Xu, J.C., Yang, J.B., Liu, B.Y., Ranjitkar, S., Liu, J., Li, D.Z., Gao, L.M., 2016. Indications for three independent domestication events for the tea plant (*Camellia sinensis* (L.) O. Kuntze) and new insights into the origin of tea germplasm in China and India revealed by nuclear microsatellites. PLoS One 11 (5), e0155369.

Melnick, D., 2005. Environment and Human Well-Being: A Practical Strategy. Earthscan, UK.

Metlen, K.L., Aschehoug, E.T., Callaway, R.M., 2009. Plant behavioural ecology: dynamic plasticity in secondary metabolites. Plant Cell Environ. 32 (6), 641–653.

Micozzi, M.S., 2014. Fundamentals of Complementary and Alternative Medicine-E-Book. Elsevier Health Sciences, Elsevier Saunders, Missouri., US.

Mir, A.H., Upadhaya, K., Choudhury, H., 2014. Diversity of endemic and threatened ethnomedicinal plant species in Meghalaya, North-East India. Int. Res. J. Environ. Sci. 3, 64–78.

Mooney, H., Larigauderie, A., Cesario, M., Elmquist, T., Hoegh-Guldberg, O., Lavorel, S., Yahara, T., 2009. Biodiversity, climate change, and ecosystem services. Curr. Opin. Environ. Sustain. 1 (1), 46–54.

Morris, R.C., 1998. Surviving pleasure at the periphery: Chiang Mai and the photographies of political. Publ. Cult. 10 (2), 341–370.

Mukherjee, P.K., 2001. Evaluation of Indian traditional medicine. Drug Inf. J. 35 (2), 623–632.

Murthy, H.N., Lee, E.J., Paek, K.Y., 2014. Production of secondary metabolites from cell and organ cultures: strategies and approaches for biomass improvement and metabolite accumulation. Plant Cell Tiss. Org. Cult. 118 (1), 1–16.

Nafiu, M.O., Hamid, A.A., Muritala, H.F., Adeyemi, S.B., 2017. Preparation, standardization, and quality control of medicinal plants in Africa. In: Medicinal Spices and Vegetables From Africa: Therapeutic Potential Against Metabolic, Inflammatory, Infectious and Systemic Diseases. Academic Press, USA, pp. 171.

Naidoo, R., Balmford, A., Costanza, R., Fisher, B., Green, R.E., Lehner, B., Ricketts, T.H., 2008. Global mapping of ecosystem services and conservation priorities. Proc. Natl. Acad. Sci. 105 (28), 9495–9500.

Nautiyal, M.C., Nautiyal, B.P., Prakash, V., 2001. Phenology and growth form distribution in an alpine pasture at Tungnath, Garhwal, Himalaya. Mt. Res. Dev. 21 (2), 168–174.

Nawchoo, I.A., 2012. In vitro antibacterial activity and phytochemical studies of methanolic extract of leaves of *Hypericum perforatum* L. growing wild in Kashmir Himalaya. Asian J. Plant Sci. Res. 2 (4), 414–420.

Nemeth, E., Bernath, J., 2008. Biological activities of yarrow species (Achillea spp.). Curr. Pharm. Des. 14 (29), 3151–3167.

Nepal, S.K., Weber, K.W., 1995. Managing resources and resolving conflicts: national parks and local people. Int. J. Sustain. Dev. World Ecol. 2 (1), 11–25.

Oguntunde, P.G., Abiodun, B.J., Olukunle, O.J., Olufayo, A.A., 2012. Trends and variability in pan evaporation and other climatic variables at Ibadan, Nigeria, 1973–2008. Meteorol. Appl. 19 (4), 464–472.

Olsen, C.S., 2005. Valuation of commercial central Himalayan medicinal plants. AMBIO J. Hum. Environ. 34 (8), 607–610.

Oommen, M.A., Shanker, K., 2005. Elevational species richness patterns emerge from multiple local mechanisms in Himalayan woody plants. Ecology 86 (11), 3039–3047.

Pal, S.K., Shukla, Y., 2003. Herbal medicine: current status and the future. Asian Pac. J. Cancer Prev. 4 (4), 281–288.

Palni, L.M.S., Maikhuri, R.K., Rao, K.S., 2001. Conservation of the Himalayan agroecosystems: issues and priorities. In: Economy of Uttaranchal: Profile and Dynamics of Change. Anamika Publishers & Distributors, New Delhi, India, pp. 135–176.

Pandey, A., Tripathi, S., 2014. Concept of standardization, extraction and pre phytochemical screening strategies for herbal drug. J. Pharmacog. and Phytochem. 2 (5), 115–119.

Pandit, M.K., Sodhi, N.S., Koh, L.P., Bhaskar, A., Brook, B.W., 2007. Unreported yet massive deforestation driving loss of endemic biodiversity in Indian Himalaya. Biodivers. Conserv. 16 (1), 153–163.

Pangriya, R., 2015. Study of aromatic and medicated plants in Uttrakhand, India: with focus on role in employment generation and supply chain management. Int. J. Soc. Sci. Manage. 2 (2), 148–156.

Pant, B., Rai, R., Wallrapp, C., Ghate, R., Shrestha, U., Ram, A., 2017. Horizontal integration of multiple institutions: solutions for Yarshagumba related conflict in the Himalayan region of Nepal? Int. J. Commons 11 (1), 464–486.

Pant, S., Pant, V.S., 2011. Status and conservation management strategies for threatened plants of Jammu and Kashmir. J. Phytology 3 (7), 50–56.

Parmesan, C., 2006. Ecological and evolutionary responses to recent climate change. Annu. Rev. Ecol. Evol. Syst. 37, 637–669.

Pascual, L., Xu, J., Biais, B., Maucourt, M., Ballias, P., Bernillon, S., Bouchet, J.P., 2013. Deciphering genetic diversity and inheritance of tomato fruit weight and composition through a systems biology approach. J. Exp. Bot. 64 (18), 5737–5752.

Patil, D., 2015. The doctrine of signatures: a warfare between myth and reality. Life Sci. Leaflets 59, 32–50.

Patiño, J., Mateo, R.G., Zanatta, F., Marquet, A., Aranda, S.C., Borges, P.A., Muñoz, J., 2016. Climate threat on the Macaronesian endemic bryophyte flora. Sci. Rep. 6, 29156.

Patwardhan, B., Vaidya, A.D., Chorghade, M., Joshi, S.P., 2008. Reverse pharmacology and systems approaches for drug discovery and development. Curr. Bioact. Compd. 4 (4), 201–212.

Pauli, H., Gottfried, M., Grabherr, G., 1996. Effects of climate change on mountain ecosystems—upward shifting of alpine plants. World Resour. Rev. 8 (3), 382–390.

Pearson, R.G., Dawson, T.P., 2003. Predicting the impacts of climate change on the distribution of species: are bioclimate envelope models useful? Glob. Ecol. Biogeogr. 12 (5), 361–371.

Penev, L., Roberts, D., Smith, V., Agosti, D., Erwin, T., 2010. Taxonomy shifts up a gear: new publishing tools to accelerate biodiversity research. ZooKeys 50, 1–4.

Peters, C.M., 1994. Sustainable Harvest of Nontimber Forest Plant Resources in Tropical Moist Forest: An Ecological Primer. Biodiversity Support Program, Washington, DC, USA.

Phillipson, J.D., 2007. Phytochemistry and pharmacognosy. Phytochemistry 68 (22), 2960–2972.

Phondani, P.C., Negi, V.S., Bhatt, I.D., Maikhuri, R.K., Kothyari, B.P., 2011. Promotion of medicinal and aromatic plants cultivation for improving livelihood security: a case study from West Himalaya, India. Int. J. Med. Aromat. Plants 1 (3), 245–252.

Phondani, P.C., Maikhuri, R.K., Saxena, K.G., 2014. The efficacy of herbal system of medicine in the context of allopathic system in Indian Central Himalaya. J. Herb. Med. 4 (3), 147–158.

Phondani, P.C., Bhatt, I.D., Negi, V.S., Kothyari, B.P., Bhatt, A., Maikhuri, R.K., 2016. Promoting medicinal plants cultivation as a tool for biodiversity conservation and livelihood enhancement in Indian Himalaya. J. Asia-Pacific Biodivers. 9 (1), 39–46.

Pickersgill, B., 2009. Domestication of plants revisited—Darwin to the present day. Bot. J. Linn. Soc. 161 (3), 203–212.

Poiani, K.A., Richter, B.D., Anderson, M.G., Richter, H.E., 2000. Biodiversity conservation at multiple scales: functional sites, landscapes, and networks. Bioscience 50 (2), 133–146.

Pope, C., Mays, N., Popay, J., 2007. Synthesising Qualitative and Quantitative Health Evidence: A Guide to Methods: A Guide to Methods. McGraw-Hill Education, UK.

Pörtner, H.O., Farrell, A.P., Knust, R., Lannig, G., Mark, F.C., Storch, D., 2009. Adapting to climate change-response. Science 323, 876–877.

Prakashan, N., 2009. Pharmacognosy. Nirali Prakashan.

Rai, L.K., Prasad, P., Sharma, E., 2000. Conservation threats to some important medicinal plants of the Sikkim Himalaya. Biol. Conserv. 93 (1), 27–33.

Raju, V.S., Reddy, C.S., Suthari, S., 2010. Flowering plant diversity and endemism in India: an overview. Algae 2 (40000), 6–25.

Ramawat, K.G., Goyal, S., 2008. The Indian herbal drugs scenario in global perspectives. Bioact. Mol. Med. Plants, 325–347.

Rana, C.S., Rana, V., Bisht, M.P.S., 2011. New distributional record of *Gentiana tetrasepala* Biswas (Gentianales: Gentianaceae) from the Valley of Flowers National Park, Garhwal Himalaya. J. Threatened Taxa 3 (9), 2100–2103.

Rana, M.S., Samant, S.S., 2011. Diversity, indigenous uses and conservation status of medicinal plants in Manali wildlife sanctuary, North western Himalaya. Indian J. Tradit. Knowl. 10, 439–459.

Rangan, H., 2004. The environment of protest and development in the Indian Himalaya. In: Liberation Ecologies: Environment, Development and Social Movements, Routledge, USA, pp. 338.

Ranjan, R., Upadhyay, V.P., 1999. Ecological problems due to shifting cultivation. Curr. Sci. 77 (10), 1246–1250.

Rao, K.S., Saxena, K.G., 1994. Sustainable Development and Rehabilitation of Degraded Village Lands in Himalaya. Bishen Singh Mahendra Pal Singh, Dehradun, India.

Rao, K.S., Maikhuri, R.K., Chauhan, K., 2008. Effect of pre-sowing, temperature and light on the seed germination of *Arnebia benthamii* (Wall. ex G. Don): an endangered medicinal plant of Central Himalaya, India. African J. Plant Sci. 2 (1), 005–011.

Rao, M.R., Palada, M.C., Becker, B.N., 2004. Medicinal and aromatic plants in agroforestry systems. In: New Vistas in Agroforestry. Springer, Netherlands, pp. 107–122.

Rao, M.V., 2004. The importance of botanical nomenclature and synonymy in taxonomy and biodiversity. Curr. Sci. 87 (5), 602–606.

Rashid, A.M., Tunon, H., Khan, N.A., Mukul, S.A., 2014. Commercial cultivation by farmers of medicinal plants in northern Bangladesh. Eur. J. Environ. Sci. 4 (1), 60–68.

Rathore, S., Tiwari, J.K., Malik, Z.A., 2015. Ethnomedicinal survey of herbaceous flora traditionally used in health care practices by inhabitants of Dhundsir Gad watershed of Garhwal Himalaya, India. Glob. J. Res. Med. Plants Indigenous Med. 4 (4), 65–78.

Rawal, R.B., 1996. Status of commercialization of medicinal and aromatic plants in Nepal. In: The Role of Bamboo, Rattan and Medicinal Plants in Mountain Development. 174.

Rawal, R.S., Bhatt, I.D., Garhwal, S., 2010. Chromatographic and spectral fingerprinting standardization of traditional medicines: an overview as modern tools. Res. J. Phytochem. 4 (4), 234–241.

Rawat, G.S., Chandrabani, D.D., 2007. Alpine vegetation of the western Himalaya: species diversity, community structure, dynamics and aspects of conservation. (Doctoral dissertation, D.Sc. Thesis) Kumaun University, Nainital. 239 pp.

Reigosa, M.J., Pedrol, N., González, L., 2006. Allelopathy: A Physiological Process With Ecological Implications. Springer Science & Business Media, Springer, Netherlands.

Rinzin, C., Vermeulen, W.J., Wassen, M.J., Glasbergen, P., 2009. Nature conservation and human well-being in Bhutan: an assessment of local community perceptions. J. Environ. Dev. 18 (2), 177–202.

Rishi, P., Gauli, K., 2005. Who should lead us?: the voice of the western Terai region of Nepal. Int. For. Rev. 7 (3), 235–240.

Rousseaux, M.C., Figuerola, P.I., Correa-Tedesco, G., Searles, P.S., 2009. Seasonal variations in sap flow and soil evaporation in an olive (*Olea europaea* L.) grove under two irrigation regimes in an arid region of Argentina. Agric. Water Manag. 96 (6), 1037–1044.

Rout, G.R., Samantaray, S., Das, P., 2000. In vitro manipulation and propagation of medicinal plants. Biotechnol. Adv. 18 (2), 91–120.

Rudall, P.J., Engleman, E.M., Hanson, L., Chase, M.W., 1998. Embryology, cytology and systematics of *Hemiphylacus, Asparagus* and *Anemarrhena* (*Asparagales*). Plant Syst. Evol. 211 (3-4), 181–199.

Samal, P.K., Palni, L.M.S., Agrawal, D.K., 2003. Ecology, ecological poverty and sustainable development in Central Himalayan region of India. Int. J. Sustain. Dev. World Ecol. 10 (2), 157–168.

Samal, P.K., Dhyani, P.P., Dollo, M., 2010. Indigenous medicinal practices of Bhotia tribal community in Indian Central Himalaya. Indian J. Tradit. Knowl. 9 (1), 140–144.

Samant, S.S., Butola, J.S., Sharma, A., 2007. Assessment of diversity, distribution, conservation status and preparation of management plan for medicinal plants in the catchment area of Parbati hydroelectric project stage—III in Northwestern Himalaya. J. Mt. Sci. 4 (1), 034–056.

Sati, V.P., 2013. Cultivation of medicinal plants and its contribution to livelihood enhancement in the Indian Central Himalayan Region. Adv. Med. Plant Res. 1 (2), 17–23.

Schickhoff, U., 2011. Dynamics of mountain ecosystems. In: Millington, A., Blumler, M., Schickhoff, U. (Eds.), Handbook of Biogeography. SAGE Publications, London, pp. 313–337.

Schippmann, U.W.E., Leaman, D., Cunningham, A.B., 2006. A comparison of cultivation and wild collection of medicinal and aromatic plants under sustainability aspects. Frontis 17, 75–95.

Schultz, V., Hänsel, R., Tyler, V.E., 2001. Rational Phytotherapy: A Physician's Guide to Herbal Medicine. Psychology Press, Springer-Verlag, Berlin, Heidelberg.

Schulz, V., Blumenthal, M., 2004. Medicinal plants, phytomedicines, and phytotherapy. In: Rational Phytotherapy. Springer, Berlin Heidelberg, pp. 1–42.

Seigler, D.S., 2012. Plant Secondary Metabolism. Springer Science & Business Media, New York.

Sen, S., Chakraborty, R., De, B., 2011. Challenges and opportunities in the advancement of herbal medicine: India's position and role in a global context. J. Herb. Med. 1 (3), 67–75.

Shabir, G., Anwar, F., Sultana, B., Khalid, Z.M., Afzal, M., Khan, Q.M., Ashrafuzzaman, M., 2011. Antioxidant and antimicrobial attributes and phenolics of different solvent extracts from leaves, flowers and bark of Gold Mohar [*Delonix regia* (Bojer ex Hook.) Raf]. Molecules 16 (9), 7302–7319.

Shah, B.N., 2009. Textbook of Pharmacognosy and Phytochemistry. Elsevier, India.

Shaheen, H., Shinwari, Z.K., Qureshi, R.A., Ullah, Z., 2012. Indigenous plant resources and their utilization practices in village populations of Kashmir Himalayas. Pak. J. Bot. 44 (2), 739–745.

Sharafzadeh, S., Ordookhani, K., 2011. Influence of carbon dioxide enrichment on accumulation of secondary metabolites in plants. Aust. J. Basic Appl. Sci. 5 (11), 1681–1686.

Sharma, C.M., Suyal, S., Gairola, S., Ghildiyal, S.K., 2009a. Species richness and diversity along an altitudinal gradient in moist temperate forest of Garhwal Himalaya. J. Am. Sci. 5 (5), 119–128.

Sharma, P., Rai, S.C., Sharma, R., Sharma, E., 2004. Effects of land-use change on soil microbial C, N and P in a Himalayan watershed. Pedobiologia 48 (1), 83–92.

Sharma, R.S., Sagar, M.P., 2015. Impacts of hydropower construction projects on economically important plants in the Upper Beas Valley, Himachal Pradesh. Transactions 37 (1), 120–132.

Sharma, S., Bajracharya, R., Sitaula, B., 2009b. Indigenous Technology Knowledge in Nepal—a review. Indian J. Tradit. Knowl. 8 (4), 569–576.

Sharma, S., Rathi, N., Kamal, B., Pundir, D., Kaur, B., Arya, S., 2010. Conservation of biodiversity of highly important medicinal plants of India through tissue culture technology—a review. Agric. Biol. J. N. Am. 1 (5), 827–833.

Sharma, V., 1998. Assessment of herb and mushroom collection in Great Himalayan National Park Conservation Area, Kullu District, HP. Forest research education and extension project—Great Himalayan National Park (FREEPGHNP), Research Project. Dehradun: Wildlife Inst. India 2 (8), 1–71.

Shrestha, A.B., Aryal, R., 2011. Climate change in Nepal and its impact on Himalayan glaciers. Reg. Environ. Chang. 11 (1), 65–77.

Shrestha, K.B., Hofgaard, A., Vandvik, V., 2014. Recent treeline dynamics are similar between dry and mesic areas of Nepal, central Himalaya. J. Plant Ecol. 8 (4), 347–358.

Shukla, S.S., Saraf, S., Saraf, S., 2009. Approaches towards standardization and quality assessment of herbals. J. Res. Educ. Indian Med. 15 (1), 25–32.

Siddique, M.A.A., Jeelani, S.M., 2015. Production technology for endangered medicinal plants of Kashmir Himalayas II. Cultivation profile of *Inula racemosa* Hook. f. Indian J. Nat. Prod. Resour. 6 (3), 231–236.

Silori, C.S., Badola, R., 2000. Medicinal plant cultivation and sustainable development: a case study in the buffer zone of the Nanda Devi Biosphere Reserve, Western Himalaya, India. Mt. Res. Dev. 20 (3), 272–279.

Simpson, M.G., 2010. Plant Systematics. Academic Press, USA.

Singh, A., Lal, M., Samant, S.S., 2009. Diversity, indigenous uses and conservation prioritization of medicinal plants in Lahaul valley, proposed Cold Desert Biosphere Reserve, India. Int. J. Biodivers. Sci. Manag. 5 (3), 132–154.

Singh, G.S., 2011. Commercialization unsustainable to Himalayan environment. Ecol. Quest. 15 (1), 71–75.

Singh, K.M., Kumar, A., Singh, R.K.P., Kumar, U., 2013. Medicinal and Aromatic Plants for Enhancing Farm Income: The Case of Bihar. http://sci-hub.bz/http://dx.doi.org/10.2139/ssrn.2341506.

Singh, K.N., Lal, B., 2008. Ethnomedicines used against four common ailments by the tribal communities of Lahaul-Spiti in western Himalaya. J. Ethnopharmacol. 115 (1), 147–159.

Singh, K.N., Lal, B., Todaria, N.P., 2016. Ecological status and conservation of rare plants in high altitude landscapes of Indian western Himalaya. Glob. J. Res. Med. Plants Indigenous Med. 5 (1), 1.

Singh, M.P., Panda, H., 2005. Medicinal Herbs With Their Formulations. Daya Books, New Delhi, India.

Singh, R.K., Pretty, J., Pilgrim, S., 2010a. Traditional knowledge and biocultural diversity: learning from tribal communities for sustainable development in northeast India. J. Environ. Plan. Manag. 53 (4), 511–533.

Singh, S.P., Singh, V., Skutsch, M., 2010b. Rapid warming in the Himalayas: ecosystem responses and development options. Clim. Dev. 2 (3), 221–232.

Singh, V., Sharma, S., Singh, N.K., Shukla, K.P., Kumar, V., 2011. Tapping the potential of traditional knowledge associated with medicinal plants of tribal communities in Central India: perspective and avenues. J. Phytology 3 (6), 42–50.

Smith-Gill, S.J., 1983. Developmental plasticity: developmental conversion versus phenotypic modulation. Am. Zool. 23 (1), 47–55.

Song, B., Zhang, Z.Q., Stöcklin, J., Yang, Y., Niu, Y., Chen, J.G., Sun, H., 2013. Multifunctional bracts enhance plant fitness during flowering and seed development in *Rheum nobile* (Polygonaceae), a giant herb endemic to the high Himalayas. Oecologia 172 (2), 359–370.

Steffen, W., Sanderson, R.A., Tyson, P.D., Jäger, J., Matson, P.A., Moore III, B., Wasson, R.J., 2006. Global Change and the Earth System: A Planet Under Pressure. Springer Science & Business Media, Springer-Verlag, Berlin, Heidelberg.

Stopes, M.C., 1919. Botany or, the Modern Study of Plants. vol. 5. Routledge, London and Edinburgh.

Subramoniam, A., 2014. Present scenario, challenges and future perspectives in plant based medicine development. Ann. Phytomed. 3 (1), 31–36.

Sultana, M., Verma, P.K., Raina, R., Prawez, S., Dar, M.A., 2012. Quantitative analysis of total phenolic, flavonoids and tannin contents in acetone and n-hexane extracts of *Ageratum conyzoides*. Int. J. ChemTech Res. 3, 996–999.

Sundriyal, M., Sharma, B., 2016. Status of biodiversity in Central Himalaya. Appl. Ecol. Environ. Sci. 4 (2), 37–43.

Sytsma, K.J., Pires, J.C., 2001. Plant systematics in the next 50 years re-mapping the new frontier. Taxon, 713–732.

Tabi, M.M., Powell, M., Hodnicki, D., 2006. Use of traditional healers and modern medicine in Ghana. Int. Nurs. Rev. 53 (1), 52–58.

Tandon, P., 2003. Biodiversity—A Scientific Approach: Agenda for the 21st Century. Sci. Lett. 26 (5), 111–118.

Tangjang, S., Namsa, N.D., Aran, C., Litin, A., 2011. An ethnobotanical survey of medicinal plants in the Eastern Himalayan zone of Arunachal Pradesh, India. J. Ethnopharmacol. 134 (1), 18–25.

Tariq, A., Mussarat, S., Adnan, M., 2015. Review on ethnomedicinal, phytochemical and pharmacological evidence of Himalayan anticancer plants. J. Ethnopharmacol. 164, 96–119.

Telwala, Y., Brook, B.W., Manish, K., Pandit, M.K., 2013. Climate-induced elevational range shifts and increase in plant species richness in a Himalayan biodiversity epicentre. PLoS One 8 (2), e57103.

Teng, N., Wang, J., Chen, T., Wu, X., Wang, Y., Lin, J., 2006. Elevated CO_2 induces physiological, biochemical and structural changes in leaves of *Arabidopsis thaliana*. New Phytol. 172 (1), 92–103.

Tétényi, P., 1970. Infraspeciflc chemical taxa of medicinal plants. Feddes Repert, 83, Akadémiai Kiadó, Budapest, pp. 125–126.

Thorne, R.F., 1992. Classification and geography of the flowering plants. Bot. Rev. 58 (3), 225–327.

Thuiller, W., 2007. Biodiversity: climate change and the ecologist. Nat. Rep. Clim. Change, 60–62.

Ticktin, T., Shackleton, C., 2011. Harvesting non-timber forest products sustainably: opportunities and challenges. In: Non-timber Forest Products in the Global Context. Springer, Berlin, Heidelberg, pp. 149–169.

Uggla, M., Gustavsson, K.E., Olsson, M.E., Nybom, H., 2005. Changes in colour and sugar content in rose hips (*Rosa dumalis* L. and *R. rubiginosa* L.) during ripening. J. Hortic. Sci. Biotechnol. 80 (2), 204–208.

Uniyal, A., Uniyal, S.K., Rawat, G.S., 2011. Commercial extraction of *Picrorhiza kurrooa* Royle ex Benth. in the Western Himalaya. Mt. Res. Dev. 31 (3), 201–208.

Uprety, Y., Asselin, H., Boon, E.K., Yadav, S., Shrestha, K.K., 2010. Indigenous use and bio-efficacy of medicinal plants in the Rasuwa District, Central Nepal. J. Ethnobiol. Ethnomed. 6 (1), 3.

Uprety, Y., Poudel, R.C., Shrestha, K.K., Rajbhandary, S., Tiwari, N.N., Shrestha, U.B., Asselin, H., 2012. Diversity of use and local knowledge of wild edible plant resources in Nepal. J. Ethnobiol. Ethnomed. 8 (1), 16.

Upton, R., Graff, A., Jolliffe, G., Länger, R., Williamson, E., 2016. American Herbal Pharmacopoeia: Botanical Pharmacognosy-Microscopic Characterization of Botanical Medicines. CRC Press, Boca Raton.
Vashistha, R.K., Rawat, N., Chaturvedi, A.K., Nautiyal, B.P., Prasad, P., Nautiyal, M.C., 2009. An exploration on the phenology of different growth forms of an alpine expanse of North-West Himalaya, India. N. Y. Sci. J. 2 (6), 29–41.
Vashistha, R.K., Butola, J.S., Nautiyal, B.P., Nautiyal, M.C., 2010. Phenological attributes of *Angelica glauca* and *A. archangelica* expressed at two different climatic zones in Western Himalaya. Open Access J. Med. Aromatic Plants 1 (1), 7.
Venkateshappa, S.M., Sreenath, K.P., 2013. Potential medicinal plants of Lamiaceae. Am. Int. J. Res. Formal, Appl. Nat. Sci. 3 (1), 82–87.
Verpoorte, R., 1998. Exploration of nature's chemodiversity: the role of secondary metabolites as leads in drug development. Drug Discov. Today 3 (5), 232–238.
Vetaas, O.R., Grytnes, J.A., 2002. Distribution of vascular plant species richness and endemic richness along the Himalayan elevation gradient in Nepal. Glob. Ecol. Biogeogr. 11 (4), 291–301.
Wackernagel, M., Rees, W., 1998. Our Ecological Footprint: Reducing Human Impact on the Earth (No. 9). New Society Publishers, Canada.
Walther, G.R., Post, E., Convey, P., Menzel, A., Parmesan, C., Beebee, T.J., Bairlein, F., 2002. Ecological responses to recent climate change. Nature 416 (6879), 389–395.
Wink, M., 2003. Evolution of secondary metabolites from an ecological and molecular phylogenetic perspective. Phytochemistry 64 (1), 3–19.
World Health Organization, 2000. General Guidelines for Methodologies on Research and Evaluation of Traditional Medicine.
Xu, J., Grumbine, R.E., Shrestha, A., Eriksson, M., Yang, X., Wang, Y.U.N., Wilkes, A., 2009a. The melting Himalayas: cascading effects of climate change on water, biodiversity, and livelihoods. Conserv. Biol. 23 (3), 520–530.
Xu, J.G., Tian, C.R., Hu, Q.P., Luo, J.Y., Wang, X.D., Tian, X.D., 2009b. Dynamic changes in phenolic compounds and antioxidant activity in oats (*Avena nuda* L.) during steeping and germination. J. Agric. Food Chem. 57 (21), 10392–10398.
Yadav, R.B.S., Bayrak, Y., Tripathi, J.N., Chopra, S., Bayrak, E., 2012. Regional variation of the ω-upper bound magnitude of GIII distribution in Hindukush-Pamir Himalaya and the adjacent regions: a perspective on earthquake hazard. Tectonophysics 544, 1–12.
Yadav, R.B.S., Tsapanos, T.M., Koravos, G.C., Bayrak, Y., Devlioti, K.D., 2013. Spatial mapping of earthquake hazard parameters in the Hindukush-Pamir Himalaya and adjacent regions: implication for future seismic hazard. J. Asian Earth Sci. 70, 115–124.
Yaniv, Z., Bachrach, U., 2005. Handbook of Medicinal Plants. CRC Press, Boca Raton.
Yeung, K.S., Gubili, J., Cassileth, B., 2008. Evidence-based botanical research: applications and challenges. Hematol. Oncol. Clin. North Am. 22 (4), 661–670.
Yunus, M., 2015. Herbal medicine in India. Ann. Phytomed. 4 (2), 1–3.
Zurick, D., 2006. Illustrated Atlas of the Himalaya. India Research Press, New Delhi, India.

Further Reading

Adhikari, D., Tiwary, R., Barik, S.K., 2015. Modelling hotspots for invasive alien plants in India. PLoS One 10 (7), e0134665.
Alves, R.R.N., Medeiros, M.F.T., Albuquerque, U.P., Rosa, I.L., 2013. From past to present: medicinal animals in a historical perspective. In: Animals in Traditional Folk Medicine. Springer, Berlin Heidelberg, pp. 11–23.
Ameh, S.J., Obodozie, O.O., Inyang, U.S., Abubakar, M.S., Garba, M., 2010. Current phytotherapy—a perspective on the science and regulation of herbal medicine. J. Med. Plant Res. 4 (2), 072–081.

Aneesh, T., Hisham, M., Sekhar, M., Madhu, M., Deepa, T., 2009. International market scenario of traditional Indian herbal drugs-India declining. Int. J. Green Pharm. *3* (3), 184.

Aranda, S.C., Hespanhol, H., Homem, N., Borges, P.A., Lobo, J.M., Gabriel, R., 2015. The iterative process of plant species inventorying for obtaining reliable biodiversity patterns. Bot. J. Linn. Soc. 177 (4), 491–503.

Armstrong, K.F., Ball, S.L., 2005. DNA barcodes for biosecurity: invasive species identification. Philos. Trans. R. Soc. Lond. B: Biol. Sci. 360 (1462), 1813–1823.

Aziz, N.A.A., Ahmad, M.I., Naim, D.M., 2017. Species identification based on trnH-psbA and ITS2 genes and analysis of mineral nutrients of selected medicinal plants from Malaysia. Jordan J. Biol. Sci. 10, 117–125.

Balooni, K., Ballabh, V., Inoue, M., 2007. Declining instituted collective management practices and forest quality in the Central Himalayas. Econ. Polit. Wkly., 1443–1452.

Barboza, G.E., Cantero, J.J., Núñez, C., Pacciaroni, A., Ariza Espinar, L., 2009. Medicinal plants: a general review and a phytochemical and ethnopharmacological screening of the native Argentine Flora. Kurtziana 34 (1–2), 7–365.

Belhumeur, P., Chen, D., Feiner, S., Jacobs, D., Kress, W., Ling, H., Zhang, L., 2008. Searching the world's herbaria: a system for visual identification of plant species. Comput. Vis. ECCV 2008, 116–129.

Benbrook, C., 2012. Principles governing the long-run risks, benefits, and costs of agricultural biotechnology. In: Biodiversity and the Law: "Intellectual Property, Biotechnology and Traditional Knowledge". Routledge, USA.

Bera, S.K., 2010. Globalisation and Sustainable Exports of Indian Medicinal and Aromatic Plants: A Protection Study. University Library of Munich, Germany.

Bhat, T.A., Nigam, G., Majaz, M.A., 2012. Study of some medicinal plants of the Shopian district, Kashmir (India) with emphasis on their traditional use by Gujjar and Bakerwal tribes. Asian J. Pharm. Clin. Res. 5 (Suppl. 2), 94–98.

Bhatt, D., Kumar, R., Tewari, L.M., Joshi, G.C., 2014. *Polygonatum cirrhifolium* Royle and *Polygonatum verticillatum* (L.) Allioni: status assessment and medicinal uses in Uttarakhand, India. J. Med. Plant Res. 8 (5), 253–259.

Bhowmik, D., Kumar, K.P., Pankaj, T., Chiranjib, B., 2009. Traditional herbal medicines: an overview. Arch. Appl. Sci. Res. 1 (2), 165–177.

Bhushan, I., Kumar, A., Patel, J.S., Yadav, R.P., Singh, S., Meena, R., Meena, V.S., 2016. The Indian Himalayan ecosystem as source for survival. In: Conservation Agriculture. Springer, Singapore, pp. 367–380.

Bickford, D., Lohman, D.J., Sodhi, N.S., Ng, P.K., Meier, R., Winker, K., Das, I., 2007. Cryptic species as a window on diversity and conservation. Trends Ecol. Evol. 22 (3), 148–155.

Bisht, I.S., Mehta, P.S., Bhandari, D.C., 2007. Traditional crop diversity and its conservation on-farm for sustainable agricultural production in Kumaon Himalaya of Uttaranchal state: a case study. Genet. Resour. Crop. Evol. 54 (2), 345–357.

Bisht, V.K., Kuniyal, C.P., Nautiyal, B.P., Prasad, P., 2013. Spatial distribution and regeneration of *Quercus semecarpifolia* and *Quercus floribunda* in a subalpine forest of western Himalaya, India. Physiol. Mol. Biol. Plants 19 (3), 443.

Borowitzka, M.A., 2013. High-value products from microalgae—their development and commercialisation. J. Appl. Phycol. 25 (3), 743.

Breininger, D., Duncan, B., Eaton, M., Johnson, F., Nichols, J., 2014. Integrating land cover modeling and adaptive management to conserve endangered species and reduce catastrophic fire risk. Land 3 (3), 874–897.

Butola, J.S., 2011. Post-cultivation evaluation of germplasm in Himalayan threatened medicinal herbs: implication for ex-cultivation and conservation. Natl. Acad. Sci. Lett. 34 (1), 49.

Butola, J.S., Vashistha, R.K., 2013. An overview on conservation and utilization of *Angelica glauca* Edgew. in three Himalayan states of India. Emerg. Med. Plant 5 (3), 171–178.

C. P. W, G., Hollingsworth, P.M., Forrest, L.L., Spouge, J.L., Hajibabaei, M., Ratnasingham, S., Fazekas, A.J., 2009. A DNA barcode for land plants. Proc. Natl. Acad. Sci. 106 (31), 12794–12797.

Canter, P.H., Thomas, H., Ernst, E., 2005. Bringing medicinal plants into cultivation: opportunities and challenges for biotechnology. Trends Biotechnol. 23 (4), 180–185.

Carpenter, C., 2005. The environmental control of plant species density on a Himalayan elevation gradient. J. Biogeogr. 32 (6), 999–1018.

Chandra Sekar, K., Manikandan, R., Srivastava, S.K., 2012. In: Invasive alien plants of Uttarakhand Himalaya. Proceedings of the National Academy of Sciences, India Section B, pp. 1–9.

Chaudhuri, A.B., Chaudhuri, A.B., 2003. Megadiversity Conservation: Flora, Fauna and Medicinal Plants of *India's Hot Spots*. Daya Books, New Delhi, India.

Chauhan, R.S., Nautiyal, M.C., 2005. Commercial viability of cultivation of an endangered medicinal herb *Nardostachys jatamansi* at three different agroclimatic zones. Curr. Sci. 89 (9), 1481–1488.

Chauhan, R.S., Tiwari, D., Bisht, A.S., Shukla, A., 2014. Ex situ conservation of medicinal and aromatic plants in Bharsar, Uttarakhand, India. Med. Plants Int. J. Phytomed. Relat. Ind. 6 (4), 282–292.

Chen, S., Yao, H., Han, J., Liu, C., Song, J., Shi, L., Luo, K., 2010. Validation of the ITS2 region as a novel DNA barcode for identifying medicinal plant species. PLoS One 5 (1), e8613.

Chitale, V.S., Behera, M.D., Roy, P.S., 2014. Future of endemic flora of biodiversity hotspots in India. PLoS One 9 (12), e115264.

Choudhary, D., Kala, S.P., Todaria, N.P., Rawat, R.B.S., Kunwar, M.S., Kollmair, M., 2013. Upgrading mountain people in medicinal and aromatic plants value chains: lessons for sustainable management and income generation from Uttarakhand, India. Int. J. Sustain. Dev. World Ecol. 20 (1), 45–53.

Cope, J.S., Corney, D., Clark, J.Y., Remagnino, P., Wilkin, P., 2012. Plant species identification using digital morphometrics: a review. Expert Syst. Appl. 39 (8), 7562–7573.

Corlett, R.T., 1988. The naturalized flora of Singapore. J. Biogeogr. 15, 657–663.

dan Prospek, M., Dar, G., Khuroo, A., 2013. Floristic diversity in the Kashmir Himalaya: progress, problems and prospects. Sains Malaysiana 42 (10), 1377–1386.

Dar, P.A., Reshi, Z.A., Shah, M.A., 2015. Roads act as corridors for the spread of alien plant species in the mountainous regions: a case study of Kashmir Valley, India. Trop. Ecol. 56 (2), 183–190.

Dawa, S., Hussain, A., Tamchos, T., 2013. Medicinal and aromatic plants as a viable livelihood option for the people of Jammu and Kashmir. Indian Hortic. J. 3 (1and2), 40–42.

Devi, B., Bhardwaj, D.R., Panwar, P., Pal, S., Gupta, N.K., Thakur, C.L., 2013. Carbon allocation, sequestration and carbon dioxide mitigation under plantation forests of north western Himalaya, India. Ann. For. Res. 56 (1), 3.

Dhar, U., 2015. Phytogeographic studies on the alpine flora of Kashmir Himalayas. University of Kashmir, http://hdl.handle.net/10603/32937.

Dhiman, M., Soumendra, C., Amar, R., Mokthan, M.W., 2009. Differential approach of germplasm conservation of high value of medicinal plants in North-Eastern Himalayan Region. Int. J. Agric. Environ. Biotechnol. 2 (4), 332–340.

Dhyani, A., Nautiyal, B.P., Nautiyal, M.C., 2010. Importance of Astavarga plants in traditional systems of medicine in Garhwal, Indian Himalaya. Int. J. Biodivers. Sci. Ecosyst. Manage. iFirst, 1–7, https://doi.org/10.1080/21513732.2010.521490.

Dubey, N.K., Kumar, R., Tripathi, P., 2004. Global promotion of herbal medicine: India's opportunity. Curr. Sci. 86 (1), 37–41.

Early, R., Bradley, B.A., Dukes, J.S., Lawler, J.J., Olden, J.D., Blumenthal, D.M., Sorte, C.J., 2016. Global threats from invasive alien species in the twenty-first century and national response capacities. Nat. Commun. 7, 12485.

Eaton, D., 2015. Variability of climate and riverflow in partially glacierised basins in the Karakoram and western Himalaya. Doctoral Dissertation, University of Salford.

Elvin-Lewis, M., 2011. Dereplications can amplify the extent and worth of traditional pharmacopeias. Afr. J. Tradit. Complement. Altern. Med. 8 (5S).

Enan, M.R., Palakkott, A.R., Ksiksi, T.S., 2017. DNA barcoding of selected UAE medicinal plant species: a comparative assessment of herbarium and fresh samples. Physiol. Mol. Biol. Plants 23 (1), 221–227.

Enomoto, T., 1999. Naturalized weeds from foreign countries into Japan. In: Yano, E., Matsuo, K., Shiyomi, M., Andow, D.A. (Eds.), Biological Invasions of Ecosystem Pests and Beneficial Organisms. Yokendo, Tokyo.

Farooquee, N.A., Majila, B.S., Kala, C.P., 2004. Indigenous knowledge systems and sustainable management of natural resources in a high altitude society in Kumaun Himalaya, India. J. Hum. Ecol. 16 (1), 33–42.

Froese, R., Capuli, E., Rañola, M.C., 2000. Challenges to taxonomic information management: how to deal with changes in scientific names. In: Global Environmental Researches on Biological and Ecological Aspects. Center for Global Environmental Research, Tsukuba, Japan, pp. 3–10.

Ghimire, S.K., McKey, D., Aumeeruddy-Thomas, Y., 2004. Heterogeneity in ethnoecological knowledge and management of medicinal plants in the Himalayas of Nepal: implications for conservation. Ecol. Soc. 9 (3), 6.

Ghosh, A.K., 2008. A Comprehensive Handbook on Biodiversity. The Energy and Resources Institute (TERI).

Ghosh, S.P., General, F.D.D., 2013. In: Promotion of medicinal and aromatic plants in the Asia-Pacific region. Expert Consultation on Promotion of Medicinal and Aromatic Plants in the Asia-Pacific Region: Proceedings, p. 33.

Gotelli, N.J., Colwell, R.K., 2001. Quantifying biodiversity: procedures and pitfalls in the measurement and comparison of species richness. Ecol. Lett. 4 (4), 379–391.

Gupta, M.P., 2011. Perspectives of medicinal and aromatic plants and their products in emerging globalization era. Hortic. Bras. 29 (2), 5831–5855.

Gurib-Fakim, A., 2006. Medicinal plants: traditions of yesterday and drugs of tomorrow. Mol. Asp. Med. 27 (1), 1–93.

Hameed, M., Nawaz, T., Ashraf, M., Ahmad, F., Ahmad, K.S., Ahmad, M.S.A., Ahmad, I., 2012. Floral biodiversity and conservation status of the Himalayan foothill region Punjab. Pak. J. Bot. 44, 143–149.

Hamilton, A.C., 2004. Medicinal plants, conservation and livelihoods. Biodivers. Conserv. 13 (8), 1477–1517.

Han, J., Pang, X., Liao, B., Yao, H., Song, J., Chen, S., 2016. An authenticity survey of herbal medicines from markets in China using DNA barcoding. Sci. Rep. 6, 18723.

Hobbs, R.J., Higgs, E., Harris, J.A., 2009. Novel ecosystems: implications for conservation and restoration. Trends Ecol. Evol. 24 (11), 599–605.

Hui, C., Richardson, D.M., 2017. Invasion dynamics. Oxford University Press, UK.

Hundal, S.S., 2004. In: Wildlife conservation strategies and management in India: an overview. Proc. Species at Risk 2004 Pathways to Recovery Conference. vol. 2, pp. 2–6.

Hunt, D.J., Handoo, Z.A., 2009. Taxonomy, identification and principal species. Root-Knot Nematodes 1, 55–88.

Hussain, S., Hore, D.K., 2007. Collection and conservation of major medicinal pants of Darjeeling and Sikkim Himalayas. Indian J. Tradit. Knowl. 6, 352–357.

Jayakumar, S., Kim, S.S., Heo, J., 2011. Floristic inventory and diversity assessment-a critical review. Proc. Int. Acad. Ecol. Environ. Sci. 1 (3–4), 151.

Joshi, P., Rao, N., 2011. Role of indigenous people in conservation of biodiversity of medicinal plants: an Indian case study. Surviv. Sustain. 91–101.

Kaack, K.V., Knudsen, B.F., Rosenberg, L., 2013. In: Guidelines for good agricultural and collection practice for sambucus nigra flowers and berries as active ingredients in traditional herbal medicine and dietary supplements. I International Symposium on Elderberry 1061, June, pp. 225–235.

Kala, C.P., 2000. Status and conservation of rare and endangered medicinal plants in the Indian trans-Himalaya. Biol. Conserv. 93 (3), 371–379.

Kala, C.P., 2015. Medicinal and aromatic plants: boon for enterprise development. J. Appl. Res. Med. Aromatic Plants 2 (4), 134–139.

Kandari, L.S., Phondani, P.C., Payal, K.C., Rao, K.S., Maikhuri, R.K., 2012. Ethnobotanical study towards conservation of medicinal and aromatic plants in upper catchments of Dhauli Ganga in the central Himalaya. J. Mt. Sci. 9 (2), 286–296.

Kanwal, K.S., Joshi, H., 2015. Medicinal plants diversity, indigenous uses and conservation status in Alaknanda Valley of Western Himalaya, Uttarakhand, India. Indian Forester 141 (6), 660–669.

Khan, S.M., Harper, D.M., Page, S., Ahmad, H., 2011. Species and community diversity of vascular flora along environmental gradient in Naran Valley: a multivariate approach through indicator species analysis. Pak. J. Bot. 43 (5), 2337–2346.

Khan, S.M., Page, S.E., Ahmad, H., Harper, D.M., 2013a. Sustainable utilization and conservation of plant biodiversity in montane ecosystems: the western Himalayas as a case study. Ann. Bot. 112 (3), 479–501.

Khan, S.M., Page, S., Ahmad, H., Harper, D., 2013b. Identifying plant species and communities across environmental gradients in the Western Himalayas: method development and conservation use. Eco. Inform. 14, 99–103.

Khuroo, A.A., Reshi, Z., Rashid, I., Dar, G.H., Khan, Z.S., 2008. Operational characterization of alien invasive flora and its management implications. Biodivers. Conserv. 17 (13), 3181–3194.

Khuroo, A.A., Reshi, Z.A., Dar, G.H., 2012a. Plant invasions in Jammu and Kashmir state, India. In: Invasive Alien Plants: An Ecological Appraisal for the Indian Subcontinent. vol, 1. CABI Publishers, UK, pp. 216.

Khuroo, A.A., Reshi, Z.A., Malik, A.H., Weber, E., Rashid, I., Dar, G.H., 2012b. Alien flora of India: taxonomic composition, invasion status and biogeographic affiliations. Biol. Invasions 14 (1), 99.

Kiehn, M., 2011. In: Bramwell, D., Caujapé-Castells, J. (Eds.), Invasive Alien Species and Islands. Cambridge University Press, Cambridge, pp. 365–384.

Kim, H.U., Charusanti, P., Lee, S.Y., Weber, T., 2016. Metabolic engineering with systems biology tools to optimize production of prokaryotic secondary metabolites. Nat. Prod. Rep. 33 (8), 933–941.

Kirkpatrick, M., Barton, N.H., 1997. Evolution of a species' range. Am. Nat. 150 (1), 1–23.

Koh, K.S., Na, J.G., Suh, M.H., Kil, J.H., Ku, Y.B., Yoon, J.H., Oh, H.K., 2000. The Effects of Alien Plants on Ecosystem and Their Management. The Plant Taxonomic Society of Korea, Korea.

Kress, W.J., Wurdack, K.J., Zimmer, E.A., Weigt, L.A., Janzen, D.H., 2005. Use of DNA barcodes to identify flowering plants. Proc. Natl. Acad. Sci. U. S. A. 102 (23), 8369–8374.

Kumar, R., Sharma, D., Tiwari, P., 2017. Floristic diversity assessment of dry temperate and alpine region of Kinnaur district of Himachal Pradesh. Int. J. Farm Sci. 7 (1), 57–60.

Kumar, S., 2016. Understanding altered molecular dynamics in the targeted plant species in Western Himalaya in relation to environmental cues: implications under climate change scenario. In: Abiotic Stress Response in Plants. Wiley-VCH, Germany.

Kumar, V., Babu, S., Revale, A.K., Meena, R.K., Ranjan, M.K., Desai, B.S., 2014. Cultivation of medicinal plants in natural ecosystem in Gujarat (India): constraints and conservation need. J. Plant Dev. Sci. 6 (3), 425–435.

Kuniyal, C.P., Bhatt, V.P., Bhatt, V.P., Butola, J.S., Sundriyal, R.C., 2014. Promoting nursery enterprise in high altitude villages: a participatory approach for conservation and commercialization of Himalayan threatened medicinal plants. J. Med. Plant Res. 8 (48), 1399–1407.

Kunkler, K., 2006. The role of medical simulation: an overview. Int. J. Med. Rob. Comput. Assisted Surg. 2 (3), 203–210.

Laladhas, K.P., Preetha, N., Baijulal, B., Oommen, V., 2015. Conservation of medicinal plant resources through Community Born Biodiversity Management Committee, Kerala, India. In: Biodiversity Conservation-Challenges for the Future. Bentham eBooks, Chennai, India, pp. 27.

Lanker, U., Malik, A.R., Gupta, N.K., Butola, J.S., 2010. Natural regeneration status of the endangered medicinal plant, *Taxus baccata* Hook. F. syn. T. wallichiana, in northwest Himalaya. Int. J. Biodivers. Sci. Ecosyst. Serv. Manage. 6 (1–2), 20–27.

Leppäkoski, E., Gollasch, S., Olenin, S. (Eds.), 2013. Invasive aquatic species of Europe. In: Distribution, Impacts and Management. Springer Science & Business Media, Springer, Netherlands.

Leroux, S.J., Albert, C.H., Lafuite, A.S., Rayfield, B., Wang, S., Gravel, D., 2017. Structural uncertainty in models projecting the consequences of habitat loss and fragmentation on biodiversity. Ecography 40 (1), 36–47.

Lone, P.A., Bhardwaj, A.K., Shah, K.W., Tabasum, S., 2014. Ethnobotanical survey of some threatened medicinal plants of Kashmir Himalaya, India. J. Med. Plant Res. 8 (47), 1362–1373.

Macdonald, D.W., Willis, K.J. (Eds.), 2013. Key Topics in Conservation Biology 2.. John Wiley & Sons, USA.

Maikhuri, R.K., Rawat, L.S., Negi, V.S., Purohit, V.K., Rao, K.S., Saxena, K.G., 2011. Managing natural resources through simple and appropriate technological interventions for sustainable mountain development. Curr. Sci., 992–997.

Maikhuri, R.K., Phondani, P.C., Rawat, L.S., Jha, N.K., Maletha, A., Bahuguna, Y.M., Kandari, L.S., 2016. Conservation and management strategies of medicinal plant resources through action research approaches in Indian Himalaya. Iranian J. Sci. Technol. Trans. A: Sci. 1–7.

Malik, S., Sharma, N., Sharma, U.K., Singh, N.P., Bhushan, S., Sharma, M., Ahuja, P.S., 2010. Qualitative and quantitative analysis of anthraquinone derivatives in rhizomes of tissue culture-raised Rheum emodi Wall. plants. J. Plant Physiol. 167 (9), 749–756.

Mani, M.S. (Ed.), 2012. Ecology and Biogeography in India, vol. 23. Springer Science & Business Media, Springer, Netherlands.

Manish, K., Pandit, M.K., Telwala, Y., Nautiyal, D.C., Koh, L.P., Tiwari, S., 2017. Elevational plant species richness patterns and their drivers across non-endemics, endemics and growth forms in the Eastern Himalaya. J. Plant Res. 130, 829–844.

Marbuah, G., Gren, I.M., McKie, B., 2014. Economics of harmful invasive species: a review. Diversity 6 (3), 500–523.

Maxted, N., Crust, R., 1995. Aids to taxonomic identification. In: Guarino, L., Ramanatha Rao, V., Reid, R. (Eds.), Collecting Plant Genetic Diversity. CAB International, Wallingford, UK.

Mazid, M., Khan, T.A., Mohammad, F., 2012. Medicinal plants of rural India: a review of use by Indian folks. Indo Global J. Pharm. 2 (3), 286–304.

McDougall, K.L., Alexander, J.M., Haider, S., Pauchard, A., Walsh, N.G., Kueffer, C., 2011. Alien flora of mountains: global comparisons for the development of local preventive measures against plant invasions. Divers. Distrib. 17 (1), 103–111.

Measey, G.J., Vimercati, G., Villiers, F.A., Mokhatla, M., Davies, S.J., Thorp, C.J., Kumschick, S., 2016. A global assessment of alien amphibian impacts in a formal framework. Divers. Distrib. 22 (9), 970–981.

Miller, J.H., Schelhas, J., 2008. Adaptive collaborative restoration: a key concept in invasive plant management. In: Invasive Plants and Forest Ecosystems, p. 251.

Miller, J.R., 2010. Survey of Western Tragopan, Koklass Pheasant, and Himalayan Monal populations in the Great Himalayan National Park, Himachal Pradesh, India. Indian Birds 6 (3), 60–65.

Molnar, J.L., Gamboa, R.L., Revenga, C., Spalding, M.D., 2008. Assessing the global threat of invasive species to marine biodiversity. Front. Ecol. Environ. 6 (9), 485–492.

Mukherjee, A., Rakshit, S., Nag, A., Ray, M., Kharbikar, H.L., Shubha, K., Meena, V.S., 2016. Climate change risk perception, adaptation and mitigation strategy: an extension outlook in Mountain Himalaya. In: Conservation Agriculture. Springer, Singapore, pp. 257–292.

Mukherjee, P.K., Wahile, A., 2006. Integrated approaches towards drug development from Ayurveda and other Indian system of medicines. J. Ethnopharmacol. 103 (1), 25–35.

Mukherjee, P.K., Rai, S., Kumar, V., Mukherjee, K., Hylands, P.J., Hider, R.C., 2007. Plants of Indian origin in drug discovery. Expert Opin. Drug Discovery 2 (5), 633–657.

Nandi, S.K., Palni, L.M.S., Pandey, H., Chandra, B., Nadeem, M., 2016. Selection of elites and in vitro propagation of selected high-value Himalayan medicinal herbs for sustainable utilization and conservation. In: Plant Tissue Culture: Propagation, Conservation and Crop Improvement. Springer, Singapore, pp. 15–44.

Nandy, S.N., 2014. Agro-economic indicators—a comparative study of North-Eastern States of India. J. Land Rural Stud. 2 (1), 75–88.

Nandy, S.N., Dhyani, P.P., Samal, P.K., 2006. Resource Information Database of the Indian Himalaya. Environmental Information System on Himalayan Ecology, GP Institute of Himalayan Environment and Development, Uttaranchal, India.

Nautiyal, B.P., Nautiyal, M.C., Khanduri, V.P., Rawat, N., 2009. Floral biology of *Aconitum heterophyllum* Wall.: a critically endangered alpine medicinal plant of Himalaya, India. Turk. J. Bot. 33 (1), 13–20.

Nautiyal, S., Maikhuri, R.K., Semwal, R., Saxena, K., 2002. Agroecosystem function around a Himalayan biosphere reserve. J. Environ. Syst. 29 (1).

Nautiyal, S., Rajan, K.S., Shibasaki, R., 2005. Interaction of biodiversity and economic welfare—a case study from the Himalayas of India. J. Environ. Inform. 6 (2), 111–119.

Negi, G.C.S., Samal, P.K., Kuniyal, J.C., Kothyari, B.P., Sharma, R.K., Dhyani, P.P., 2012. Impact of climate change on the western Himalayan mountain ecosystems: an overview. Trop. Ecol. 53 (3), 345–356.

Negi, V.S., Maikhuri, R.K., Phondani, P.C., Rawat, L.S., 2010. An inventory of indigenous knowledge and cultivation practices of medicinal plants in Govind Pashu Vihar Wildlife Sanctuary, Central Himalaya, India. Int. J. Biodivers. Sci. Ecosyst. Serv. Manage. 6 (3-4), 96–105.

Negi, V.S., Maikhuri, R.K., Rawat, L.S., Parshwan, D., 2013. Protected cultivation as an option of livelihood in mountain region of central Himalaya, India. Int. J. Sustain. Dev. World Ecol. 20 (5), 416–425.

Norton, B.G., 2014. The Preservation of Species. Princeton University Press, USA.

Pandit, M.K., Manish, K., Koh, L.P., 2014. Dancing on the roof of the world: ecological transformation of the Himalayan landscape. Bioscience 64 (11), 980–992.

Parihaar, R.S., Bargali, K., Bargali, S.S., 2015. Status of an indigenous agroforestry system: a case study in Kumaun Himalaya, India. Indian J. Agric. Sci. 85 (3), 442–447.

Patel, D.K., 2014. Some medicinal and aromatic shrubs introduced in herbal garden for their propagation and ex-situ conservation. J. Med. Plants 2 (6), 9–12.

Pyšek, P., Richardson, D.M., 2010. Invasive species, environmental change and management, and health. Annu. Rev. Environ. Resour. 35, 25–55.

Qiu, J., 2008. Scientific publishing: identity crisis. Nat. News 451 (7180), 766–767.

Ramanan, C., Ranjith, P.V., Mehta, M.J.. Future prospects of herbal medicines in India, http://www.siescoms.edu/images/pdf/FUTURE%20PROSPECTS%20OF%20HERBAL%20MEDICINES%20IN%20INDIA.pdf.

Rana, M.S., Samant, S.S., 2009. Prioritization of habitats and communities for conservation in the Indian Himalayan Region: a state-of-the-art approach from Manali Wildlife Sanctuary. Curr. Sci., 326–335.

Rana, R.S., Bhagat, R.M., Kalia, V., Lal, H., Sen, V., 2013. Indigenous perceptions of climate change vis-a-vis mountain agricultural activities in Himachal Pradesh, India. J. Tradit. Knowl. 12 (4), 596–604.

Rawat, S., Andola, H., Giri, L., Dhyani, P., Jugran, A., Bhatt, I.D., Rawal, R.S., 2014. Assessment of nutritional and antioxidant potential of selected vitality strengthening Himalayan medicinal plants. Int. J. Food Prop. 17 (3), 703–712.

Rawat, S., Jugran, A.K., Bahukhandi, A., Bahuguna, A., Bhatt, I.D., Rawal, R.S., Dhar, U., 2016. Anti-oxidant and anti-microbial properties of some ethno-therapeutically important medicinal plants of Indian Himalayan Region. 3 Biotech 6 (2), 154.

Reddy, C.S., 2008. Catalogue of invasive alien flora of India. Life Sci. J. 5 (2), 84–89.

Richardson, D.M., 2004. Plant invasion ecology—dispatches from the front line. Divers. Distrib. 10 (5–6), 315–319.

Rokaya, M.B., Münzbergová, Z., Timsina, B., Bhattarai, K.R., 2012. *Rheum australe* D. Don: a review of its botany, ethnobotany, phytochemistry and pharmacology. J. Ethnopharmacol. 141 (3), 761–774.

Saha, D., Sundriyal, R.C., 2012. Utilization of non-timber forest products in humid tropics: implications for management and livelihood. Forest Policy Econ. 14 (1), 28–40.

Sajem, A.L., Rout, J., Nath, M., 2008. Traditional tribal knowledge and status of some rare and endemic medicinal plants of North Cachar Hills district of Assam, Northeast India. Ethnobot. Leaflets 2008 (1), 31.

Samant Pant, S., Singh, M., Lal, M., Singh, A., Sharma, A., Bhandari, S., 2007. Medicinal plants in Himachal Pradesh, north western Himalaya, India. Int. J. Biodivers. Sci. Manage. 3 (4), 234–251.

Samant, S.S., Dhar, U., 1997. Diversity, endemism and economic potential of wild edible plants of Indian Himalaya. Int. J. Sustain. Dev. World Ecol. 4 (3), 179–191.

Samant, S.S., Dhar, U., Palni, L.M.S., 1998. Medicinal Plants of Indian Himalaya. Himavikas Publ.-13, GBPIHED Kosi, Almora, India.

Schori, M., Showalter, A.M., 2011. DNA barcoding as a means for identifying medicinal plants of Pakistan. Pak. J. Bot. 43, 1–4.

Sekar, K.C., 2012. Invasive alien plants of Indian Himalayan region—diversity and implication. Am. J. Plant Sci. 3 (02), 177.

Semwal, D.P., Saradhi, P.P., Nautiyal, B.P., Bhatt, A.B., 2007. Current status, distribution and conservation of rare and endangered medicinal plants of Kedarnath Wildlife Sanctuary, Central Himalayas, India. Curr. Sci., 1733–1738.

Shah, M.A., Reshi, Z.A., 2014. Characterization of alien aquatic flora of Kashmir Himalaya: implications for invasion management. Trop. Ecol. 55 (2), 143–157.

Shah, M.A., Reshi, Z.A., Lavoie, C., 2011. Predicting plant invasiveness from native range size: clues from the Kashmir Himalaya. J. Plant Ecol. 5 (2), 167–173.

Shaheen, H., Khan, S.M., Harper, D.M., Ullah, Z., Allem Qureshi, R., 2011. Species diversity, community structure, and distribution patterns in western Himalayan alpine pastures of Kashmir, Pakistan. Mt. Res. Dev. 31 (2), 153–159.

Sharma, N., Raina, A.K., 2013. Composition, structure and diversity of tree species along an elevational gradient in Jammu province of north-western Himalayas, Jammu and Kashmir, India. J. Biodivers. Environ. Sci. 3 (10), 12–23.

Sharma, P., Samant, S.S., 2014. Diversity, distribution and indigenous uses of medicinal plants in Parbati Valley of Kullu district in Himachal Pradesh, Northwestern Himalaya. Asian J. Adv. Basic Sci. 2, 77–98.

Sharma, P., Samant, S.S., Lal, M., Sharma, A., 2014. Diversity, indigenous uses, threat categorization and conservation prioritization of medicinal plants: a case study from Himachal Pradesh, India. J. Biodivers. Endanger. Species 2, 134. https://doi.org/10.4172/2332-2543.1000134.

Sheikh, M.A., 2014. Phytochemical investigation and uses of some of the medicinal plants of Kashmir Himalaya-a review. Int. J. Biol. Pharm Allied Sci 3, 462–489.

Singh, G., 2016. Plant Systematics, 3/ed.: An Integrated Approach. CRC Press, Boca Raton.

Singh, G., Rawat, G.S., 2012. Depletion of oak (*Quercus* spp.) forests in the western Himalaya: grazing, fuelwood and fodder collection. In: Global Perspectives on Sustainable Forest Management. InTech, UK.

Singh, K.M., 2009. Scope of medicinal and aromatic plants farming in Eastern India. https://doi.org/10.2139/ssrn.2019789.

Singh, K.M., Jha, A.K., 2008. Medicinal and Aromatic Plants Cultivation in Bihar, India: Economic Potential and Condition for Adoption. Environ. Ecol. 32 (4B), 1642–1646.

Singh, M.P., Singh, B.S., Dey, S., 2002. Plant Biodiversity and Taxonomy. Daya Books, New Delhi.

Singh, R.K., Singh, A., Pandey, C.B., 2014. Agro-biodiversity in rice-wheat-based agroecosystems of eastern Uttar Pradesh, India: implications for conservation and sustainable management. Int. J. Sustain. Dev. World Ecol. 21 (1), 46–59.

Singh, V., Singh, B., Kaul, V.K., 2003. Domestication of wild marigold (*Tagetes minuta* L.) as a potential economic crop in western Himalaya and north Indian plains. Econ. Bot. 57 (4), 535–544.

Siwach, M., Siwach, P., Solanki, P., Gill, A.R., 2013. Biodiversity conservation of Himalayan medicinal plants in India: a retrospective analysis for a better vision. Int. J. Biodivers. Conserv. 5 (9), 529–540.

Srivastava, T.N., Kapahi, B.K., Kirn, H.S., Sarin, Y.K., 2001. Threatened plants of medicinal and aromatic value of North Western Himalaya. J. Non-Timber Forest Prod. 7 (3/4), 165–179.

Steele, P.R., Pires, J.C., 2011. Biodiversity assessment: state-of-the-art techniques in phylogenomics and species identification. Am. J. Bot. 98 (3), 415–425.

Stuessy, T.F., 2009. Plant Taxonomy: The Systematic Evaluation of Comparative Data. Columbia University Press, New York.

Thakur, M.K., Chauhan, R., Pant, K.S., 2013. Bioresources for productivity enhancement in kutki, *Picrorrhiza kurrooa* Royle ex Benth. Intl. J. Bioresour. Stress Manage 4 (4), 482–486.

Theis, N., Lerdau, M., 2003. The evolution of function in plant secondary metabolites. Int. J. Plant Sci. 164 (S3), S93–S102.

Thomas, J., Joy, P.P., Mathew, S., Skaria, B.P., 2000. Plant sources of aroma chemicals and medicines in India. Chem. Ind. Digest (Special Millennium Issue), 104–108.

Thomas, K.J., Nicholl, J.P., Coleman, P., 2001. Use and expenditure on complementary medicine in England: a population based survey. Complement. Ther. Med. 9 (1), 2–11.

Thompson, I.D., Okabe, K., Parrotta, J.A., Brockerhoff, E., Jactel, H., Forrester, D.I., Taki, H., 2014. Biodiversity and ecosystem services: lessons from nature to improve management of planted forests for REDD-plus. Biodivers. Conserv. 23 (10), 2613.

Tiwari, V.K., Kale, R.R., Mishra, B.B., Singh, A., 2008. A facile one-pot MW approach for 3-heteroaryl-2-thioxo-2, 3-dihydroquinazolin-4 (1H)-one. Arkivoc 14, 27–36.

United Nations Office on Drugs & Crime, 2004. World Drug Report 2004. vol. 1. United Nations Publications, USA.

Uniyal, S.K., Awasthi, A., Rawat, G.S., 2002. Current status and distribution of commercially exploited medicinal and aromatic plants in upper Gori valley, Kumaon Himalaya, Uttaranchal. Curr. Sci., 1246–1252.

Uprety, Y., Poudel, R.C., Asselin, H., Boon, E.K., Shrestha, K.K., 2011. Stakeholder perspectives on use, trade, and conservation of medicinal plants in the Rasuwa District of Central Nepal. J. Mt. Sci. 8 (1), 75–86.

Verma, A., Jana, G.K., Sen, S., Chakraborty, R., Sachan, S., Mishra, A., 2010. Pharmacological evaluation of *Saracaindica* Leaves for central nervous system depressant activity in mice. J Pharm. Sci. Res. 2 (6), 338–343.

Vidyarthi, S., Samant, S.S., Sharma, P., 2013. Traditional and indigenous uses of medicinal plants by local residents in Himachal Pradesh, North Western Himalaya, India. Int. J. Biodivers. Sci. Ecosyst. Serv. Manage. 9 (3), 185–200.

Vijayan, N., Swapna, T.S., Kiran, G.S., Abdulhameed, S., 2017. Medicinal plants of India: implications in modern medicine. In: Bioresources and Bioprocess in Biotechnology. Springer, Singapore, pp. 421–443.

Wani, N.R., Mushtaq, S.M., 2012. Status of exotic flora in Kashmir Valley. J. For. Environ. Sci. 28 (1), 8–11.

Weber, E., 2017. Invasive Plant Species of the World: A Reference Guide to Environmental Weeds. CABI.

Weber, E., Sun, S.G., Li, B., 2008. Invasive alien plants in China: diversity and ecological insights. Biol. Invasions 10 (8), 1411–1429.

Wu, S.H., Hsieh, C.F., Chaw, S.M., Rejmánek, M., 2004. Plant invasions in Taiwan: insights from the flora of casual and naturalized alien species. Divers. Distrib. 10 (5-6), 349–362.

Yadav, N.P., Dixit, V.K., 2008. Recent approaches in herbal drug standardization. Int. J. Integr. Biol. 2 (3), 195–203.

Zerbe, S., Choi, I.K., Kowarik, I., 2004. Characteristics and habitats of non-native plant species in the city of Chonju, southern Korea. Ecol. Res. 19, 91–98.

Zhang, X., Yang, D., Zhou, G., Liu, C., Zhang, J., 1996. Model expectation of impacts of global climate change on biomes of the Tibetan Plateau. In: Climate Change and Plants in East Asia. Springer, Japan, pp. 25–38.

CHAPTER

3

Conservation and Cultivation of Herbs: Challenge and Policy Making

OUTLINE

Himalayan Phytochemicals
https://doi.org/10.1016/B978-0-08-102227-6.00003-6

3.1 INTRODUCTION

Indian Himalaya is a hot spot of diversity harboring immense wealth of forest products that has potential to establish them as prime herbal hub for the entire world (Samant et al., 1998; Agrawal and Chhatre, 2006). But with increasing urbanization and development of modern medicine, there is rapid dwindling of indigenous knowledge and depletion of natural resources (Kala, 2007; Dhyani et al., 2010). Medicinal and aromatic plants besides being as alternate medicine provide livelihood to underprivileged communities (Kala, 2015; Banjaw et al., 2016). Enormous conventional knowledge pertaining to utilization of diverse medicinal and aromatic plants is still intact with tribal people of Indian Himalaya due to inaccessibility to modern technology and relatively lower development rate (Phondani et al., 2013; Lone et al., 2015). The increasing proclivity of modern civilization toward herbal medicine is due to extensive toxicity effects of allopathic medicines (Li et al., 2015). Due to escalating demands of herbal-based medicines, there has been a parallel increase in overexploitation and indiscriminate harvesting of medicinal and aromatic herbs (Baishya et al., 2013; Chhetri, 2014). Majority of medicinal and aromatic plants exhibit slow growth rate and low population densities and are confined to restricted geographic boundaries making them more liable to extinction (Sharma and Thokchom, 2014; Singh et al., 2017). There is dearth of information pertaining to documentation, collection, and conservation of MAPs in Indian Himalaya. On the contrary, therapeutic potential of MAPs thriving in Indian Himalaya has not been harnessed completely due to unavailability of proper knowledge pertaining to practice and cultivation of MAPs (Kuniyal et al., 2015 Phondani et al., 2016).Though, this knowledge has been passed verbally from one generation to the next generation, and as time advanced, indigenous knowledge became obsolete due to current socioeconomic changes (Zargar et al., 2014; Maikhuri et al., 2016). Conventional knowledge pertaining to rare and endemic MAPs is rapidly decreasing because of instant decline in conventional practice of folk medicine (Sharma and Thakur, 2015; Mahato et al., 2016). The conservation and sustainable utilization of MAPs have been illustrated expansively (Bhat et al., 2013; Kuniyal et al., 2014). Several recommendation have been put forward relating to conservation including development of inventories for species documentation and status assessment and requirement for coordinated conservation practices based on both ex situ and in situ approaches (Kuniyal et al., 2015).

If we establish link between conventional knowledge regarding the sustainable use of MAPs and use of medicinal plant diversity, we can take conventional medicine toward advancement of modern medicine. This chapter aims to discover the use of diversity in medicinal plant sector, preparation and preservation of conventional knowledge pertaining to

MAP use, and policies that execute its proper utilization and commercialization. This chapter illustrates global trends, advancements, and scenarios for the approaches and methodologies relating to conservation and sustainable utilization of MAPs to provide authentic reference for conservation and sustainable utilization of MAPs. We will highlight conservation approaches both in situ and ex situ conservation and cultivation practices as well as resource management.

3.2 TAXONOMY AS IMPERATIVE TOOL

Taxonomy is an imperative discipline that functions as primary contrivance for deciphering biodiversity of MAPs in order to execute their conservation and management more effectively (Costello et al., 2015; Ely et al., 2017). Moreover, taxonomy is significant for effective control and management of invasive alien species (IAS) by government authorities. In order to conserve and manage invasion by exotic species, there is immediate requirement for identifying them with the help of taxonomy (Shah and Reshi, 2014; Adhikari et al., 2015). Misidentification among plant species can lead to rapid dwindling of native species (Cronk and Fuller, 2014). For rapid control of IAS, effective eradication strategies should be developed to detect and prevent invasion that declines nativity of MAPs. Taxonomists can provide their valuable suggestions to frame economically and ecologically feasible plan for control and management of IAS. Taxonomists can integrate their knowledge of identification and document and monitor invasion in order to control current and available lists of already identified IAS (Tali et al., 2015; Rana and Rastogi, 2017). It is essential for taxonomists to provide baseline information pertaining to native species distribution and identification both at national and international level so that IAS can be distinguished from native species (Gairola et al., 2013). Taxonomy aids in sustainability of varied ecosystem around globe displaying assorted play of ecological plasticity and compliance. This could improve validity of ecological studies and enhance precision of simulation modeling. Taxonomy impedes in elucidating the origin, evolution, and diversification by combining varied discipline like plant morphology, anatomy, and embryology (Endress et al., 2000; Taia, 2005; Ao and Tobe, 2015). Numerous species are being utilized for anthropogenic purposes and economic development, and there still remains more to be unraveled. Detailed biodiversity map with illustrative biogeographic distribution will be a valuable tool for description and identification of bioresources (Badola and Aitken, 2003; Semwal et al., 2007; Khuroo et al., 2010). Proper documentation of world biota is yet to be carried out; there is only about 10% of biota being documented, so there is an immediate conserve; not only resources but also information pertaining to their identification and cultivation needs to be assessed (Khuroo et al., 2011; Siwach et al., 2013). Every plant species having distinctive features is assigned particular name so as to facilitate proper identification and nomenclature (Bhattarai and Vetaas, 2003; Genovesi et al., 2013). Inspite of advanced technological interventions, there has been no advancement in taxonomic study though there is still enormous biota yet to be identified. Hence, timely documentation of taxonomic evaluation on regular basis should be carried out in communion with proper funding and support integrated with recently developed technologies advanced to hasten the taxonomic process (Riedel et al., 2013; Dell'Anno et al., 2015). Hence, the most articulate and regular description can be provided by teaming up expert taxonomists with new innovative technological interventions. We need to integrate taxonomy with automation, syntax pattern recognition, bioinformatics, and 3-D imaging so as to capture and analyze species description (Granjou et al., 2014; Bik, 2017). Numerous other impediments in development of taxonomy include specimen collection; mounting and processing require automation as well if possible. Taxonomic evaluation will ensure proper documentation of native species and improve resource management, biosecurity, predictive modeling, and predictive classification (Guisan et al., 2013; Singh et al., 2015). However, we should not keep our documentation descriptive but rather make evolutionary pertinent (Kumschick et al., 2014). Taxonomist must offer solid hypothesis regarding natural entities and their linkage with related taxa and should integrated comparative and molecular biology so as to improve the validity of taxonomy and biological collection (reference). Automation of taxonomy will be an imperative contrivance for future development of taxonomy so as to overcome taxonomic hindrances (Wilson et al., 2014; McGeoch et al., 2016). With the help of digital technology, informatics tool, we can effectively assemble and store information and improve networking between field and farmer. Due to advanced technological interventions, there has been immense decline in image cost and data storage so as to record and store maximum and high-resolution image of specimens from varied angles (Walker, 2015). Utilization of robotic automated pest ID (RAPID) will impede in sorting field-generated specimens, image analysis, and relational databases (Downie et al., 2015; Ampatzidis et al., 2017). There should be development of cybertaxonomic applications, artificial intelligence, and structured language of descriptive morphology, and other pattern recognition contrivances will facilitate characterization and homologue determination (Rehman and Saba, 2014; Srivastava and Kumar, 2017). This can facilitate identification and extraction of morphological features from source in order to develop character matrix for cataloging specimens into categorized and distinct species (Geiser et al., 2013; Brady et al., 2017). The applications of electronic documentation of biota and open access for species description are noticeably obvious. All the taxonomic data will be accessible online, and specimen character matrices will be utilized to robotically generate diagnosis, keys, monographs, and phylogenetic trees (Remagnino et al., 2016). Hence, taxonomists could manage extensive data with enormous species in comparatively short span of time, and web-based revisions can be frequently done with digitization of novel or new specimens combining data on their biogeographic distribution, ecology, biology, and tropical associations (Ellis et al., 2010; Granjou et al., 2014). Such schemes will hasten the taxonomic research and delivery, and eventually, an enormous biodiversity information bank will be accessible for policy management and research development.

3.3 DOCUMENTATION OF FLORA

For accurate and validated estimation of plant biodiversity in Indian Himalaya, there must be illustrative documentation and exhaustive surveys. Due to increasing technological interventions and urbanization, there is a marked effect on biodiversity that needs to be monitored (Rotherham, 2015). There are numerous research reports documenting data pertaining to estimation, enumeration, and biogeographic biodiversity of Indian Himalaya. Preexisting reported and prevalidated analysis methods were followed for sample collection and data analysis (Costello and Wieczorek, 2014). Though, detailed data pertaining to plant diversity in particular region need to be recorded from the baseline data of that particular area (Zhan et al., 2014; Theobald et al., 2015). Hence, for documentation of flora, there are following steps.

3.3.1 Duration of Survey

For validated documentation of flora, survey is a prerequisite step. Survey is then followed by exploration, collection, and preparation of specimens for developing inventory containing floristic diversity of area (Saha et al., 2014; Victor, 2016). Phytosociological studies are to be illustrated in order to evaluate biogeographic distribution, diversity, status, and abundance of MAPs in a particular area (Kala and Mathur, 2002; Guisan et al., 2006; Jayakumar et al., 2011). During surveys, we have to take phenology of MAPs into consideration and record reproductive cycles based on field observation (Butola and Malik, 2012; Máthé, 2015). Once we get all this information, we need to cross-check our field evaluation with conventional information available with people of that particular region (Kunwar et al., 2006; Bisht and Badoni, 2009). Traditional knowledge pertaining to plant species together with its status recorded in IUCN needs to be emphasized (Kala, 2002; Khan et al., 2013).

3.3.2 Collection Methods

Collection is a significant aspect for documenting floral diversity. During collection of specimen, we need to carefully observe and note down even minor morphological features, its habit, and total geographic distribution in order to develop descriptors for identifying particular MAPs (Kaul, 1997; Luintel, 2002). Details like date of collection, local name of species, family, locality of the collection, altitude, habit and habitat, scientific name, vegetation, local use, and collector's name need to be recorded and preserved with vouchers in herbaria (Maiti and Maiti, 2017). For MAPs having special characteristics like production of essential oil/oil glands, color production/dye needs to highlighted (Rao et al., 2004; Bogers et al., 2006). The collection of plant specimen is significant for developing herbarium. In the case of flowering MAPs, flowers are characteristic features of specimen while collecting herbs. While in the case of shrubs, flowering and fruiting twigs were collected, and for grass and creepers, underground parts are to be collected (Behera et al., 2008; Maity et al., 2015). In the case of collection of minute herbs, species need to be collected and preserved, and methods for developing herbarium are different from higher plants (Mann, 1997; Cunningham, 2001). The herbaria of minute herbs cannot be prepared using generalized methods. Small herbs were dried and preserved in packets mounted on herbarium sheet, for example, *Picrorhiza kurroa* and *Hypericum perforatum*. However, larger shrubs having irregular leaf shape are difficult to preserve in herbarium like artichoke. So, in order to preserve their description, we should take high-resolution images of scale trunk, inflorescence, and leaves for records (Jain and Rao. 1977; Gwalwanshi et al., 2014).

3.3.3 Data Analysis

Numerous methodologies have been proposed to estimate species richness, their composition, abundance, and diversity. The most generalized variables used for estimating species diversity include basal area, relative dominance, and relative frequency (Whittaker, 1972; Magurran, 2013); sum of relative dominance and relative frequency determine important value indexes (IVI) for diverse species (Lande, 1996). GIS and GPS applications were also utilized for developing land-use and land-cover change (Treitz et al., 1992).

3.4 DIVERSITY AND SIGNIFICANCE OF MEDICINAL PLANTS

There are about 1717 medicinal herbs among the total of 17,000 higher plant species distributed across Indian Himalaya (Kala et al., 2006; Chettri et al., 2014). These MAPs contribute highest fraction of medicinal plants from any country across the globe. The oldest medicinal system of Indian Ayurveda relies primary on natural medicinal and aromatic plants documenting about 2000 species followed by Siddha and Unani system (Mukherjee, 2001; Mukherjee et al., 2007; Kumar et al., 2011).The oldest Hindi scripture, Charak Samhita, documents about 340 herbal formulations exhibiting folklore uses (Joshi, 2000). At present, there are about 32% plant-based drugs and numerous other drugs containing medicinal constituents isolated from plants (Petrovska, 2012). Indian Himalaya harbors enormous wealth of MAPs particularly at elevation between 18,003 and 500 m (Kharkwal et al., 2005; Sharma et al., 2009). So far, majority of MAPs are procured from central Himalaya (Uttaranchal, Sikkim, and North Bengal), while Transhimalaya embraces meager but endemic medicinal species approximately about 337 MAPs that is appreciably lower than rest part of Indian Himalaya due to extensive microclimatic variations (Gairola et al., 2008; Sharma et al., 2010).

Indian Himalaya, besides its floristic diversity, shelters about 46% of endemic plants out of which 62 and 208 in adjacent areas of Himalaya reported to exhibit medicinal properties (reference). Majority of MAPs in Himalaya are consumed as spices, and for *Myrica esculenta* and *Terminalia bellirica* known for medicinal properties are consumed as fruits. Various MAPs from Himalaya are used in perfumery due to their essential oil; about 81 species from Himalaya are essential-oil-bearing plants (Chauhan, 1999, 2010; Joshi et al., 2016). There are about 671 edible MAPs used in traditional system of medicines (Pant and Samant, 2006; Mehra et al., 2014). There are numerous other MAPs exhibiting spiritual healing properties and are related with religious activities and in some places of India are revered as gods and goddesses represented in Table 3.1. In Northern India, *T. chebula* is an important medicinal plant, and its fruit revered as scared fruit in Buddhism. The religious faith related to plants encourages local people to preserve herbs (Tewari, 2001; Uniyal et al., 2011; Mazid et al., 2012).

TABLE 3.1 Description of Himalayan Herbs Exhibiting Spiritual Healing Power

Himalayan MAPs exhibiting spiritual healing properties		
Origanum vulgar	*Saussurea obvallata*	*Ocimum sanctum*
Cedrus deodara	*Cynodon dactylon*	*Aegle marmelos*
Juniperus communis	*Musa paradisiaca*	*Nardostachys grandiflora*
Zanthoxylum armatum	*Ficus benghalensis*	*Ficus religiosa*

3.5 CHALLENGES IN MEDICINAL PLANTS SECTOR

Overincreasing human population is one of the significant factors that cause enormous pressure on forests products that are used as prime source of food and medicines (Kaul, 1997; Nautiyal et al., 2001; Rao et al., 2004). Due to overexploitation of forest resources, there has been an increasing dwindling of bioresources particularly MAPs that have become integral part of not only medicine but also food and resource of livelihood (Olsen and Larsen, 2003; Larsen and Olsen, 2007). Hence, there is an immediate need to preserve MAPs, as there is rapid addition of new species into doctrine of "*Materia medica* (Nadkarni, 1927, 1996). Expansion in medicinal plant sector is not in parallel with correct identification of MAPs (Srivastava et al., 1996). Local markets are incapable of providing exact figures of profits, supply, and demand; hence, there is an immediate need to identify prime characteristics and challenges in medicinal plant sector.

3.5.1 Rising Demand

Increased demand of MAPs as estimated by WHO that more than US $14 billion per year is required for the development of effective medicinal and health system (Subrat et al., 2002; Kala et al., 2006). According to WHO in 2050, the increased demand of raw material will rise from 15% to 25% annually to supply more than US $5 trillion demand of MAPs (Kumar et al., 2011). Indian system of medicine, Ayurveda, has increased trade of MAPs to be approximately US $1 billion per year as consumption and export of ayurvedic products have increased three timers from year 2000 to 2005 (Kapoor, 2000; Khare, 2008). But due to the increasing demand of ayurvedic products, there has been an escalating demand of medicinal herbs due to overharvesting of many plants from wild habitat leading to rapid decline in native MAPs (Kala, 2009). An expansive amount of Himalayan yew (*Taxus baccata*) has been collected from wild habitats as it yields anticancer drug taxol utilized in varied carcinogenic tissues like ovarian cancer (Dar and Dar, 2006; Malik et al., 2011). The other herbs that are extensively exploited for procurement of phytoconstituents are enlisted in rare and endangered species as represented in Table 3.2. Majority of these MAPs are used against varied diseases resulting in their overharvesting exerting increased pressure on their productivity (Ghimire, 2008; Jain et al., 2012). Overexploitation of MAPs has resulted in the loss of diversity but hampered livelihood of Himalayan tribal communities living on forests as only source of economy (Rao et al., 2004; Khandekar and Srivastava, 2014). Current herbal-based drug industry is dependent on more than 400 MAPs procured from Indian Himalaya. Majority of these MAPs are harvested for underground parts like rhizome, inflorescence, and seeds, which are prime modes of reproduction

TABLE 3.2 Medicinal and Aromatic Plants Having High Market Demand

MAPs with highest market demands		
Picrorhiza kurroa (kutki)	*Saussurea lappa* (kuth)	*Swertia chirayita*
Podophyllum	*Streblus asper*	*Valeriana jatamansi*
Viola odorata	*Aconitum heterophyllum*	*Aloe barbadensis*
Ocimum sanctum	*Asparagus racemosus*	*Acorus calamus*
Rauwolfia serpentina	*Curcuma angustifolia*	

(Prakash and Nirmalaa, 2013). Indiscriminate harvesting of reproductive organs is indispensable factors that need to be addressed for sustainable development of MAPs thriving in India (Rai et al., 2000; Nasi et al., 2008). Survival of MAPs is further threatened by indiscriminate harvesting of aerial parts and leaves hampering photosynthetic efficiency and their effective propagation (Olsen, 2001; Lohani et al., 2013; Lone et al., 2014). Marked microclimatic variations in Indian Himalaya exert more pressure on survival of MAPs leading to decline of total cover of MAPs in Indian Himalaya (Bhattacharyya et al., 2014; Jan et al., 2016a; Yadav et al., 2016).

3.5.2 Increasing Rarity

The overexploitation of MAPs and the lack of proper regulations and customary laws exert an immense pressure on natural resources of Himalaya rendering them in rare list (Khan et al., 2014; Laladhas et al., 2015). The regulatory laws and legalized principles underlying conservation of MAPs are effortlessly diluted by existing socioeconomic pressures (Kala, 2005). There are numerous reasons for rarity of MAPs, for example, habitat specificity, limited range of distribution, introduction of exotic species, habitat loss, climatic changes, overgrazing, overurbanization, and genetic drift (Kala and Sajwan, 2007; Kala and Silori, 2013; Kuniyal et al., 2014). Moreover, biotic stress like pathogens, herbivore attack, and seed predators further led to substantial decline in abundance of rare MAPs (Dhar et al., 2000). Local herbs used for varied disease are dwindling at faster rate like different species of *Swertia* like *S. chirayita*, *S. angustifolia*, and *S. cordata* used for curing malarial fever, and different species of *Berberis* like *B. aristata*, *B. asiatica*, *B. lycium*, *B. chitria*, and *B. jaeschkeana* that are significant source of berberine are used in the treatment of ophthalmic disorders (Shameem et al., 2010; Pradhan and Badola, 2012; Tahir et al., 2016).

There is disparity in accumulation of active constituents in different species of same genera; hence, there is parallelism in exploitation of particular species with respect to level of metabolite (Jan et al., 2016b). Hence, these species having higher levels of metabolites are at increased risk of extinction (Jan et al., 2013). There are about 4000–10,000 MAPs at risk of maximum threat exerting highly depressing effect on economy and health system (Karki et al., 2003; Sher et al., 2012). There are very few research reports on rare and endangered MAPs of Indian Himalaya, and none of these research reports present all-inclusive information even for single species of MAPs thriving in Indian Himalaya (Semwal et al., 2007). World Conservation Unit has estimated about 71 rare and endangered MAPs from Indian Himalaya in 2003 (Kala et al., 2004). Indian Himalaya embraces wealth of rarest MAPs including *Aconitum* having five species and *Rheum* with four species. Majority of rare MAPs species are commercially utilized about 71; among them, 92% are in active trade, and 74 and 34% are exported at national and international level (Ray et al., 2011; Bhat et al., 2013; Singh et al., 2016). There is dearth of data pertaining to population dynamics and herbarium collection of rare MAPs from Indian Himalaya (Singh et al., 2014a,b; Maikhuri et al., 2016). Assessment of rare species in higher altitudes of Indian Himalaya is even more intricate due to remoteness of topography, harsh climatic conditions, and short life cycle of plants (Sati, 2014; Singh et al., 2016). Majority of MAPs are procured from easily accessible areas of these mountains invaded by local traders and marginal farmers for collection of MAPs (Kuniyal et al., 2015; Singh, 2017). Hence, the population density of rare MAPs is not exact but differs within areas that are still unexplored (Bhushan et al., 2016; Sundriyal and Sharma, 2016).

3.5.3 Cultivation of Medicinal Plants

Cultivation of MAPs is not still able to cover even meager fraction of their overall population. There are about 10% data pertaining to micropropagation of medicinal herbs, and only for 1% of MAPs, there is development of agrotechnology (Mishra et al., 2013; Chhetri, 2014). Hence, the development of agrotechnology must be one of prime areas for research. Keeping in view of the rising market demands of MAPs, we need to conserve and mass propagate them integrating conventional farm practices and modern technological innovations (Singh, 2006; Phondani et al., 2011; Dawa et al., 2013). This could fetch extensive supply of MAPs to markets and preserve genetic diversity of MAPs (Bawa et al., 2007; Sen et al., 2011). Development of standardized farm practices can reproduce consistent level of standardized products (Dubey et al., 2004; Shaheen et al., 2012; Butola and Vashistha, 2013). Cultivation of MAPs would facilitate proper species identification, improve quality control, and enhance potential for genetic improvement (Luintel, 2002; Kala, 2009; Choudhary and Sekhon, 2011). For mass propagation and production of quality products from MAPs, the first step is to select high-quality plant material. The plant material should be selected on the basis of its high productivity potential, disease-free, metabolically rich, and ecologically feasible (Debnath et al., 2006). For better production of MAPs, one needs to find out whether monoculture or polyculture is a preferable model. Since MAPs prefer wide forest cover, agroforestry would be preferable option for its mass cultivation of MAPs (Rao et al., 2004; Kala et al., 2006). Agroforestry provides suitable approach for cultivation and conservation of MAPs as it facilitates establishment of shade-tolerant MAPs in lower strata of mutistrata system. Intercropping of cultivating short-cycled MAPs with forest tree species, cultivating MAPs as shade providers and boundary markers and interplanting of MAPs with food crops (Thakur and Vaishnu, 2007; Acharya et al., 2012). Farming practices of MAPs are quite intricate; local and marginal farmers from Northern India replaced MAPs like *Saussurea costus* with common crops like *Pisum sativum*, *Solanum tuberosum*, and *Humulus lupulus* (Kala, 2015). In some states of Indian Himalaya, MAPs have lower costs than seasonal vegetables, which is a major reason for their inadequate cultivation (Dhami and Mishra, 2015). Efforts are to be made by diverse organizations to cultivate MAPs both rare and endangered species. Different organizations in Northern India have developed agrotechnology for about 20 species of MAPs (Siwach et al., 2013; Juneja et al., 2017). The expenditure for cultivation of MAPs varies with respect to diverse species of MAPs (Yadav, 2008; Sen et al., 2011; Upadhyay and Saikia, 2012). *Rheum emodi* is the most beneficial cash crop among rare and endangered medicinal herbs of Northern India (Kumaria et al., 2013; Lone et al., 2014). Currently, the expenditure incurred in the cultivation of MAPs is executed only on small scale and limited to few hectares of Northern India due to restriction imposed by government agencies for mass cultivation of MAPs (Sreeramu, 2004; Krishnan et al., 2011). Moreover, people are oblivious pertaining to permission issuing authorities, as farmers fail to get permission for cultivation of MAPs hence liable to sell products illegally leading to their indiscriminate exploitation (Canter et al., 2005; Lubbe and Verpoorte, 2011).

3.5.4 Bio-Prospecting and Bio-Piracy

Forest and forest products are emerging as most dynamic, economy-generating sector that demands pluralistic politics globally. Hence, it is quite cumbersome to curtail demands on production of forest products like MAPs for both policy makers and for economists. Conservation of MAPs is of prime significance as proposed by Convention on Biological Diversity at Rio that provides organized framework that could facilitate proper utilization of biodiversity (Raustiala and Victor, 1996; Nasi et al., 2008). Bioprospecting and biopiracy are prime activities right now functioning in illegalized markets and practiced by local growers for maximum benefits shared between developed and developing countries (Torri, 2011; Gupta et al., 2014). Majority of issues associated with legal status of conventional knowledge and compensation of local herbal practitioners for information are highly intricate. There are opinions for the current state of compensation or benefit sharing under the intellectual property rights, which is recognized as novel legal form of biopiracy by one group, while other groups argue that the intellectual property right is a legal contrivance to safeguard the rights of knowledge holders (Sahoo et al., 2011; Singh et al., 2014a,b). Diverse methods for awarding patents on the medicinal plants in India, the United States, Europe, Canada, and other countries have expanded the commotion further (Subbian, 2007). In some countries, the plants and plant-based products like seeds, flowers, gums, and resins are legible for filing patent (Maheswari, 2011).While in the United States, any new development of variety by conventional or modern biotechnological intervention from any living organism is eligible for filing patent (Dewan, 2011). The Indian Protection of Plant Varieties and Farmers Rights Act of 2001 identified the role of farmers in the development of breeding programs (Swaminathan, 2002; Brahmi et al., 2004). Moreover, this act possesses provisions for benefit sharing whereby local communities are recognized as a provider of plants (Dhar and Anuradha, 2004; Morgera and Tsioumani, 2010). However, there is an extensive gap between developed and developing countries, for instance, India on patenting the products. There are about 301,177 or 9.6% registered for patent in developing countries among 3,125,603 patents filed in 91 countries. Almost 97.7% of total patents filed are from nonresidents who have exclusively control export market in developing countries that belong to nonresidents (Kala, 2007; Kala et al., 2008). Now, scientists and herbal sector stakeholders are demanding specialized rules about documentation of native MAPs in order to share benefits between their native habitat, inventor, and commercial sponsor (Sagar, 2005; Eritja, 2017). However, developed nations are reluctant to allow such conditions. These issues have perturbed several scientists who believe that natural products remain the most valuable resources of new drugs (Rout et al., 2009; Subramoniam, 2014). To alleviate such issues and prepare to discover novel sources of drugs, the legislature of 188 nations at Kuala Lumpur Conference in 2004 approved to develop such outline that would be suitable to all participants, and thus, the anticipated plan was further up considered and developed into concrete plan in 2006 in Brazil (Osman et al., 2008; Chaturvedi, 2011).

3.5.5 Strengthening Legalized Market System

In addition to government authorities, there are numerous stakeholders ranging from local herb growers, marginal famers, urban traders, herbal exporters to state, and national international market traders in medicinal plant sector (Singh, 2006; Mazid et al., 2012). Marketing system in medicinal plant sector is extensively free and in discriminatory (Kala et al., 2006; Aneesh et al., 2009). Generally, MAPs are collected by marginal farmers, and local laborers paid wages to meet their requirements of livelihood by selling MAPs (Phondani et al., 2011, 2014). Local herb growers are oblivious of actual market prices of numerous MAPs; generally, in medicinal plant sector, there is top-down strategy, and numerous stakeholders are oblivious of increasing market demands of products and their market accessibility (Sheth, 2011). Cultivations of kuth (*S. costus*) and pumb chalan (*R. emodi*) in some villages of Uttaranchal fetch no benefits to farmers due to scarcity of information related to its marketing (Semwal et al., 2010; Bisht et al., 2011). On the contrary, major fraction of MAPs are traded through illegal channels, and other limitations hindering development and expansion of herbal sector include slow productivity rate, prolonged gestation period, inadequacy of appropriate cultivation technology, native harvesting modes, incompetent processing technologies, underproduction of active constituents and varied demand supply ratio, inadequate quality control measures, and limited cooperation between stakeholders (Kala et al., 2006; Ved and Goraya, 2007; Sahoo et al., 2011). Most of the time, market demands more wild medicinal plants than cultivated ones as the former yields more active constituents. Levels of active constituents in MAPs vary with respect to season and phenology of species (Prasad et al., 2007; Gairola et al., 2010). The data pertaining to standardized harvesting for each MAP are hindering development of herbal sector leading to meager economical benefits (Negi et al., 2010; Pan et al., 2014). The economy and management of wild medicinal plant-based herbal sector is poorly structured due their partial commercialization.

3.6 OPPORTUNITIES IN DEVELOPING THE MEDICINAL PLANTS SECTOR

Indian Himalaya embraces huge floristic diversity especially in Northern India diversity of MAPs is varied along different altitudinal gradients. MAPs thriving at higher altitudes exhibit higher active constituents contributing to their application in drug discovery and their role as functional food following proper value addition (Nikam et al., 2012; Kala, 2015). Indian Himalaya is already recognized as low-cost manufacturer of high-quality generic drugs in global market (Agarwal et al., 2007; Harilal, 2009).

3.6.1 Existing Policies

Indian Himalaya has an immense scope in developing herbal enterprise by integrating native knowledge of herbs with modern technological interventions (Phondani et al., 2013). Medicinal plant sector is already recognized as an imperative part of Indian System of Medicine and Homeopathy as established by National Five-Year Plans of India (Sahoo and Manchikanti, 2013). Integrative approach among conventional and modern medicine system was initiated in 1983 by National Health Policy in order to carry forward complete utilization of health man power in Ayurveda, Siddha, Unani, Homeopathy, and Naturopathy (Jansen, 2016; Kala, 2017). Commercialization and advancement of herbal sector were further improved by establishing policy document by Planning Commission and the National Medicinal Plants Board (NMPB) (Samal, 2016; Sen and Chakraborty, 2017). Moreover, NMPB and Planning Commission enlisted some MAPs for research and development on priority basis to meet market demands as represented in Table 3.2. Sustainable utilization of MAPs was further ensured by farming numerous rules like Biodiversity Act 2002 and alleviating any probability of biopiracy (Pisupati and Bavikatte, 2014; Devi and Padmavati, 2016). In Northern Himalaya, Forest Department recognized two main regions in each forest division explicitly, the conservation area and developmental area selected on the basis of species richness and abundance and discernible for in situ conservation and inclusive fortification in the associated Forest Division (Dhiman et al., 2016). In developmental areas, MAPs are not only conserved but also introduced and cultivated on large scale (Purohit et al., 2016; Juneja et al., 2017) and rest of forest cover utilized for harvesting of MAPs. Harvesting is carried forward by joint effort of forest officials, local farmers and herbal experts together decide the extent of annual harvesting of desired MAPs (Nandi et al., 2016; Rokaya et al., 2017). The diverse state and national level policies and their consequent implementation will contribute to outstanding prospect in the expansion of herbal sector that could lead to the development of new Herbal industries together with potential for conservation of rare and endangered MAPs.

3.6.2 Institutional Support

Several government and nongovernmental organizations are engaged in the development and expansion of herbal enterprise from floristic diversity of Indian Himalaya as enlisted in Table 3.3. Numerous opportunities for funding have been generated to facilitate capacity building of medicinal plant sector at state and national level (Kala and Sajwan, 2007; Maikhuri et al., 2016). One of the significant mandates of NMPB entails funding for promotion and commercialization of MAPs. Promotional schemes include identification and documentation of MAPs, conservation both in situ and ex situ conservation of MAPs, production of metabolically rich and disease-free plant material, distribution of knowledge via education and communication, establishment of research, development, and man power. In commercialization scheme, main activities

TABLE 3.3 Description of Government and Nongovernmental Organizations Engaged in Expansion of Herbal Enterprise

Government and nongovernmental organizations engaged in expansion of herbal enterprise		
AICRP MAPNRCMAP, Anand	CIMAP, Lucknow	CDRI, Lucknow
NBRI, Lucknow	State government ISM Directorates	Ministry of Environment and Forestry
State Forest Departments	State Government Department of S&T	NABARD
World Bank	Exim Bank of India	M. S. Swaminathan Research Foundation
Research Foundation	Indian Institute of Integrative Medicine, Srinagar	Indian Institute of Integrative Medicine, Jammu
Pragya	CSIR Institute of Himalayan Bioresource Technology	Defence Institute of High Altitude Research (DIHAR)

include mass production plant material to maintain supply and demand ratio, plantation of rare and endangered MAPs, postharvest technology development like value addition and processing, and finally inventive marketing of MAPs (Singh, 2006; Kala, 2009). Besides this, NMPB cooperates with diverse departments and organizations so as to enhance and support herbal sector. NMPB has established about 35 state medicinal plant boards for expansion and development of herbal sector (Kala, 2009; Bhat et al., 2013). NMPB has brought about 35,000 ha of land under cultivation in order to cultivate commercially significant MAPs on mass level. NMPB disseminates huge funds for cultivating MAPs on contract basis providing about 30% funding to any group or organization having at least 3 years of expertise in medicinal plant sector (Kumar et al., 2011; Kala, 2015). NMPB has made significant contribution by funding about 4254 projects over past 5 years out of which 732 projects are sanctioned to northern Indian Himalaya (Kala et al., 2006; Kala, 2009). These projects further advance and expand MAPs to overpass gaps and to meet challenges in advancing this medicinal plant sector. Different funding organizations, G. B. Pant Institute of Himalayan Environment and Development, provided grants on projects via Ministry of Environment and Forests for conservation of MAPs. The council of scientific and industrial research is expanding conventional knowledge digital library that comprises about 35,000 herbal medicinal formulations utilized in ayurvedic system of medicine (Ramawat and Goyal, 2008; Singh et al., 2013). The National Bank for Agriculture and Rural Development

(NABARD) provides financial support to institutes for facilitating large-scale production of about 50 MAPs. Moreover, NABARD facilitates capacity building of potential entrepreneurs via Rural Entrepreneurs Development Programs by providing 100% refinancing (Sharma et al., 2012; Sharma, 2013). NABARD has strong connections with NMPB and diverse state medicinal plant boards. Efforts are made by diverse organizations at diverse levels to ascertain and promote the medicinal plant sector (Kumar et al., 2011; Mazid et al., 2012).

3.6.3 Recommendations for Developing the Medicinal Plants Sector

Promotion and commercialization of plant-based medicine from Indian origin need to regulate through proper and highly organized framework of policies that expand both research and development along with its large-scale export at international levels (Smith et al., 2010; Subramoniam, 2014). There is an urgent need to frame coordinated attempts for polishing research, cultivation, processing, production, and release of market products procured from MAPs. Selection of appropriate or suitable plant species for mass cultivation is a prime step toward advancement of medicinal plant sector. In order to bring MAPs into cultivation, economic suitability is the main focus, and Planning Commission has prioritized some rare and endangered wild MAPs for large-scale cultivation in order to meet their demand and supply (Simmonds, 2009; Rasheed et al., 2013; Biradar, 2015). Because cultivation of abundant species is not economical as these can be collected easily, hence, rare species need to be conserved and cultivated at mass level. Cultivation of selected species should be carried out using suitable and reliable cultivation technology. Northern Himalaya harbors immense floristic diversity due to varied microclimatic conditions that are to be cultivated as per niche ecology. For establishment of expansive medicinal sector, we need the following:

(1) Documentation of indigenous knowledge pertaining to utilization of MAPs.
(2) Certification of quality raw material.
(3) Development of suitable agrotechnology of commercially important MAPs.
(4) Identifying and safeguarding policies and customary laws of tribal people.
(5) Framing comprehensible policy for granting permission for cultivation within specific time.
(6) Establishing training programs on harvesting and processing technologies.
(7) Identifying abiotic and biotic factors affecting MAPs.

(8) Farming and marketing of plant-based products using community setup.
(9) Evaluation of status of MAPs and analysis of market policy using community-based approach.
(10) Conservation of habitats embracing threatened and rare MAPs.
(11) Sharing of benefits that arise from tribal knowledge.

These efforts may lower reliance on wild or native resources of MAPs and generate substantial income opportunities for rural and impoverished communities (Kala, 2015; Ahmad et al., 2016). Expansion and development of MAPs can be brought through joint efforts of plant growers and research institutes that develop agrotechnologies for their mass cultivation (Bisht et al., 2015; Deb et al., 2016). Proper dissemination of farming information, governmental support, market accessibility, profit levels, accessibility to agrotechnologies, and availability of skilled labour are determinant aspects for booming cultivation of MAPs (Senapati, 2014; Kumar, 2016). Currently, there are numerous MAPs with pharmaceutical potential, but due to the lack of proper documentation, their utilization is not achieved. So there is an urgent requirement for disseminating information related to MAPs through proper media and network. The conventional consumption of underutilized profile of MAPs should be documented and disseminated for their pharmaceutical potential by formulating suitable phytomedicine and also to lower demands on overexploited species (Mishra et al., 2015; Bhatt et al., 2017). Herbal formulations need to be standardized before their market release for curing diverse diseases. Proper and standardized herbal formulations will be placed in global market. Certification of raw plant material should be done for quality control using sophisticated extraction and isolation techniques like high-performance liquid chromatography (HPLC), volumetric analysis, gas chromatography, column chromatography, and gravimetric determinations (Andola et al., 2010; Rawal et al., 2010; Meena, 2015). Prior to standardization, herbs should be microscopically examined for incidence of pathogenic microbes like bacteria, virus, fungi, and nematodes causing diseases such as mildew, rust, necrosis, spots, blight, rot, wilt, galls, curl, and canker affecting farm produce (Pandey and Tripathi, 2014; Ameh et al., 2014; Govindaraghavan and Sucher, 2015). Indiscriminate collection of rare and endangered MAPs from natural habitats for diverse experimental purposes poses an immense threat on their nativity of these herbs. Researchers should be well acquainted with germination potential, survival rates of seedlings and rhizomes of MAPs harvested from their native habitats for experimental evaluations so as to ensure replantation of similar species. Communication between farmers and researchers should be transparent and lucid so that conservation of MAPs even in remote areas is carried out with ease. Because conservation of

MAPs is now carried out at species level, for example, Valley of Flowers in Northwestern Himalaya and the Tipi Orchid Sanctuary in Northeast Himalaya are protected areas within Himalayan community boundaries (De and Medhi, 2014; De et al., 2016). There has not been established even a single protected area for MAPs that solely focus on conservation of medicinal herbs (Kumar et al., 2014; Chen et al., 2016). In order to conserve MAPs, there is a need to identify and mark areas for their conservation. There are numerous social issues associated with medicinal plant sector. The conventional information related to harvesting, processing, and utilization of MAPs needs to be assembled to advance herbal sector (Máthé, 2015; Jangdey et al., 2016). Advancement of farming practices, improvement in conventional utilization of MAPs, expansion of conservation areas, ascertaining Social Capital Trust for herbal healers, ascertaining linkages among varied stakeholders are among few of the societal issues that need to be flattered and tackled accurately (Siddiqui and Shukla, 2015; Srivastava et al., 2016). The utilization of folklore and development of formulation from ethnic communities can be brought into validation using modern contrivances.

Conservation and utilization of MAPs are maintained by three-tier structures wherein equity and ethical issues related to biological resources are preserved by National Biodiversity Authority (NBA). NBA is trying to establish local panchayats and municipal corporations to manage MAPs at regional levels. There is an immense debate on IPR and biodiversity that are meant for protecting patent and plant breeder right. Significance of geographic indicators and utility model patents should be unraveled in order to protect indigenous knowledge and protect rich heritage of Indian Himalaya and monitor evolution in biogeographic distribution (reference). There must be proper sharing of benefits between tribal communities and research foundations. Tribal community of Thiruvananthapuram, Kerala, along with Tropical Botanical Garden and Research Institute has developed drug from plant named arogyapacha (*Trichopus zeylanicus* Gaertn), which has proved to be stamina-inducing herb. This drug was then commercially produced by Coimbatore-based *Arya Vaidya Pharmacy* for rupees 10 lakh as license fee, and royalty obtained from drug sale was utilized in health and education of community (Madhavan, 2014; Abrol et al., 2016). Tribal people were given reward of 50,000 for sharing conventional knowledge with TBGRI scientists for drug development. In the current administrative of WTO, there is a discriminatory threat of biopiracy connecting intellectual robbery, resource, and economic loss. Proper framing of legislative policies pertaining to Article 27.2 of TRIPS contributes to protection of indigenous knowledge since Articles 7 and 8 TRIPS can be efficiently utilized for stopping biopiracy.

3.7 BIODIVERSITY STRATEGY

In the recent years, India has done immense work toward preservation of biodiversity. Main steps are taken by Indian government to conserve Indian biodiversity establishment of following organizations:

1. Convention on Biological Diversity (CBD):
   It legally binding international treaty approved on June 1992. This is the primary inclusive global agreement dealing with all characteristics of biodiversity. This convention works for three main aims of conventions: first, conservation of biodiversity; second, sustainable utilization of its components; and third, fair and equitable sharing of benefits developed from utilization of bioresources. India signed partnership with CBD on 5 June 1992, approved on 18 February 1994. The two main protocols are to be adopted on Cartagena Protocol on Biosafety in 2000 and Nagoya Protocol on Access and Benefit Sharing (ABS) in 2010.
2. Nagoya Protocol on Access and Benefit Sharing (ABS).
   NP is a legally obligatory, balancing accord to CBD approved at CoP-10 held in Nagoya in October 2010 after 6 years of extreme and inconsiderate discussions, under the aegis of CBD. India along with megadiverse countries is imperative in determining the protocol. NP plans to advance ABS framework provided by the CBD. NP was implemented on 12 October 2014, 90 days following 50th authorization. India assisted implementation of NP by accelerating indispensable authorization during the presidency of CoP-11. Identifying stellar role played by India in its entry into force, India was invited to chair CoP/MoP-1. First CoP/MoP was held in Pyeongchang in October 2014 simultaneously with CoP-12 to CBD. CoP/MoP-2 of NP was held in Cancun, Mexico, in December 2016 in concert with CoP-13 to CBD. So far, 74 parties have authorized the NP. India has elected MOEFCC as National Focal Point and NBA as Competent National Authority for NP. India published the first IRCC on ABS Clearing House on 1 October 2015. So far, 25 IRCCs out of 28 are published; each one is from Ecuador and South Africa.
3. National Biodiversity Action Plan (NBAP) and National Biodiversity Targets (NBTs).
   NBA deals with requests for access to bioresources and associated TK by foreign individuals/institutions for research and commercial utilization and for relocating of results of Indian research by Indians to foreigners. Although conceding approvals, NBA enforces settings that secure equitable sharing of benefits from utilization of biological resources and allied data. Benefits include monetary gains, grant of joint ownership of IPRs, transfer of technology, association of Indian

scientists in R&D, and setting up of venture capital funds. NBA approval is also required prior to looking for IPR based on biological material and TK obtained from India. NBA has power to combat grant of IPRs in other countries on bioresources/TK obtained from India. NBA advises central government on biodiversity associated issues.

4. National Reports to CBD.
   Parties associate with convention to present their reports and initiatives taken for execution of convention and their competency in meeting convention's aims:
   - National reporting is a regular obligation.
   - Five cycles of national reporting have been terminated.
   - India yielded NR 5 in March 2014. NR 6 would be due in March 2018.
5. Intergovernmental Platform on Biodiversity and Ecosystem Services (IPBES).

It is an autonomous intergovernmental body recognized to reinforce the science-policy interface for biodiversity and ecosystem services for the conservation and sustainable utilization of biodiversity, continuing human well-being, and sustainable development. It was recognized in Panama City, on 21 April 2012 by 94 governments. IPBES is positioned under the patronage of four United Nations entities—UNEP, UNESCO, FAO, and UNDP—and administered by UNEP.

All the member countries of the United Nations can connect the platform, and its members are dedicated to construct IPBES as the important intergovernmental body for evaluating the state of the planet's biodiversity, its ecosystems, and the necessary facilities they contribute to society. The IPBES provides a mechanism renowned by both the scientific and policy communities to create, reassess, consider, and significantly estimate pertinent data and facts produced all-inclusive by governments, academia, scientific organizations, and nongovernmental organizations in addition to native people and local communities. This entails a convincing cluster of experts in demeanoring evaluation of such data and awareness in an obvious way. IPBES is exclusive in that it will aspire to reinforce competence for the effectual utilization of science in management at all levels.

3.7.1 Biodiversity Act 2002

An act to offer for preservation of biological diversity, sustainable utilization of its machinery, and fair and reasonable sharing of the benefits developed from the utilization of biological resources, information, and for matters connected therewith or incidental thereto. India is a party to the United Nations Convention on Biological Diversity signed at Rio de Janeiro on the 5 June 1992, and the said convention came into force on 29 December 1993; and whereas the said convention reaffirms the sovereign rights of the states over their biological resources; and whereas the said convention has the main objective of conservation of biological diversity, sustainable use of its components, and fair and equitable sharing of the benefits arising out of utilization of genetic resources; and whereas it is considered necessary to provide for conservation, sustainable utilization and equitable sharing of the benefits arising out of utilization of genetic resources and also to give effect to the said Convention. BE it enacted by Parliament in the Fifty-third Year of the Republic of India as follows:

Present status of implementation of BD Act

- All states (29) have established SBBs; Rules informed in 22 states.
- Over 41,000 BMCs started in 26 states (none in Bihar, Haryana, and J&K).
- Over 2880 PBRs prepared in 17 states.
- 313 agreements signed so far, maximum in 2015–16 (92).
- Benefit sharing received by NBA; Rs. 34.26 cores.
- Notification on normally traded commodities (NTCs) expanding these from 190 to 385 species issued in April 2016.
- Guidelines on ABS issued in November 2014, which illustrate scheme of processing of applications together with template and terms of benefit sharing.

Exemptions under the Act

- To local people and community for free access to use bioresources within India
- To growers and cultivators and vaids and hakims (practitioners of traditional medicinal systems) to use bioresources
- Through notification by central government of normally traded commodities
- To collaborative research through government-sponsored institutes subject to overall guidelines and approval of central government
- To access value-added products
- For research by Indians in India

3.8 CONCLUSION

From the past few years, the conventional knowledge pertaining to conservation and management of MAPs has been ignored due to globalization and evolved socioeconomic scenario. Advancement of medicinal plant sector is determined by shared interests and complete understanding between stakeholders, accommodating governmental policies, market accessibility, and agrotechnological conveniences. Inclusive expansion of medicinal plant sector will not only conserve and preserve biological resources but also raise living standards, improve commercial status, and contribute to societal up gradation in Indian Himalaya. Numerous legislative policies have been framed in order to conserve MAPs thriving across varied states, union territories, and biogeographic zones; for example, Indian Forest Act 1927, Wildlife (Protection) Act 1972, Forest (Conservation) Act 1980, and Biological Diversity Act 2002 are prime examples of conserving MAPs. In addition to framing of legislative policies, there are different set of challenges in conserving MAPs, hence, establishment of about 35 state medicinal plant boards set up by NMPB, India, which develops monographs and e-portals for disseminating information pertaining to raising quality planting material, development of resistant varieties, and development of herbal gardens all over India including schools. Keeping in view of the significance of MAPs from Indian Himalaya in international market, there is a huge prospect for international collaboration and trade in Himalayan MAPs.

References

Abrol, D., Sundararaman, T., Madhavan, H., Joseph, K.J., 2016. Building inclusive health innovation systems: lessons from India. Cadernos Saude Publ. 32 (Suppl 2), e00045215.

Acharya, B., Anupam, S., Mishra, R.K., Patel, S.P., Vashistha, R.K., Ajay, S., Parul, S., 2012. Astavarga plants-threatened medicinal herbs of the North-West Himalaya. Int. J. Med. Arom. Plants 2 (4), 661–676.

Adhikari, D., Tiwary, R., Barik, S.K., 2015. Modelling hotspots for invasive alien plants in India. PloS One 10 (7), e0134665.

Agarwal, S.P., Gupta, A., Dayal, R., 2007. Technology transfer perspectives in globalising India (drugs and pharmaceuticals and biotechnology). J. Technol. Transfer 32 (4), 397–423.

Agrawal, A., Chhatre, A., 2006. Explaining success on the commons: community forest governance in the Indian Himalaya. World Dev. 34 (1), 149–166.

Ahmad, S., Gangoo, S.A., SULTAN, S.M., Dar, M.D., 2016. Ethnoveterinary practices and use of herbal medicine by pastoralists of Himalaya: a review. Ann. Biol. 32 (2), 260–268.

Ameh, S. J., Toge, B. K., Ebeshi, B. U., Ambi, A. A., Garba, M., Alfa, J., & Gamaniel, K. S., 2014. Herbal Drug Development from Traditional Formulations: Refocusing Pharmaceutics and Posology for Accelerated Validation.

Ampatzidis, Y., De Bellis, L., Luvisi, A., 2017. iPathology: robotic applications and management of plants and plant diseases. Sustainability 9 (6), 1010.

Andola, H.C., Rawal, R.S., Rawat, M.S.M., Bhatt, I.D., Purohit, V.K., 2010. Analysis of berberine content using HPTLC fingerprinting of root and bark of three Himalayan Berberis species. Asian J. Biotechnol. 2 (4), 239–245.

Aneesh, T., Hisham, M., Sekhar, M., Madhu, M., Deepa, T., 2009. International market scenario of traditional Indian herbal drugs-India declining. Int. J. Green Pharm. 3 (3), 184.

Ao, C., Tobe, H., 2015. Floral morphology and embryology of *Helwingia* (Helwingiaceae, Aquifoliales): systematic and evolutionary implications. J. Plant Res. 128 (1), 161–175.

Badola, H.K., Aitken, S., 2003. The Himalayas of India: a treasury of medicinal plants under siege. Biodiversity 4 (3), 3–13.

Baishya, R.A., Sarma, J., Begum, A., 2013. Forest-based medicinal plants rendering their services to the rural community of Assam, India. Int. J. Appl. Biol. Pharm. Technol. 4 (4), 10–20.

Banjaw, D.T., Dikir, W., Gebre, A., Geja, W., Tsegaye, D., 2016. Aromatic and medicinal plants in wondogenet agricultural research center botanical Garden, South Ethiopia. Med. Aromat. Plants (Los Angel) 5 (278), 2167-0412.

Bawa, K.S., Joseph, G., Setty, S., 2007. Poverty, biodiversity and institutions in forest-agriculture ecotones in the Western Ghats and Eastern Himalaya ranges of India. Agric. Ecosyst. Environ. 121 (3), 287–295.

Behera, K.K., Sahoo, S., Patra, S., 2008. Floristic and medicinal uses of some plants of Chandaka denudated forest patches of Bhubaneswar, Orissa, India. Ethnobot. Leaflets 2008 (1), 138.

Bhat, J.A., Kumar, M., Bussmann, R.W., 2013. Ecological status and traditional knowledge of medicinal plants in Kedarnath Wildlife Sanctuary of Garhwal Himalaya, India. J. Ethnobiol. Ethnomed. 9 (1), 1.

Bhatt, I.D., Rawat, S., Badhani, A., Rawal, R.S., 2017. Nutraceutical potential of selected wild edible fruits of the Indian Himalayan region. Food Chem. 215, 84–91.

Bhattacharyya, S., Adhikari, B.S., Rawat, G.S., 2014. Influence of microclimate on the activity of Royle's pika in the western Himalaya, India. Zool. Stud. 53 (1), 73.

Bhattarai, K.R., Vetaas, O.R., 2003. Variation in plant species richness of different life forms along a subtropical elevation gradient in the Himalayas, east Nepal. Global Ecol. Biogeogr. 12 (4), 327–340.

Bhushan, I., Kumar, A., Patel, J.S., Yadav, R.P., Singh, S., Meena, R., Meena, V.S., 2016. The Indian Himalayan ecosystem as source for survival. In: Conservation Agriculture. Springer, Singapore. pp. 367–380.

Bik, H.M., 2017. Let's rise up to unite taxonomy and technology. PLoS Biol. 15 (8), e2002231.

Biradar, D.P., 2015. Medicinal plants and phytomedicines. Ann. Phytomed. 4 (1), 1–5.

Bisht, A.K., Bhatt, A., Dhar, U., Bhatt, A., Dhar, U., 2015. Note on somatic embryogenesis and synthetic seed production in *Angelica glauca*: a valuable medicinal plant of Himalaya. J. Med. Plants Res. 9 (12), 419–425.

Bisht, C., Badoni, A., 2009. Medicinal strength of some alpine and sub-alpine zones of western Himalaya, India. New York Sci. J. 2, 41–46.

Bisht, V.K., Kathait, A.S., Negi, J.S., Bhandari, A.K., Rana, C.S., 2011. Development of Market Supply chain for Picrorhiza kurrooa and Saussurea costus in the district Chamoli of the Uttarakhand state, India: a case study. Rep. Opin. 3, 14–17.

Bogers, R.J., Craker, L.E., Lange, D., 2006. Medicinal and Aromatic Plants: Agricultural, Commercial, Ecological, Legal, Pharmacological and Social Aspects. Springer, Dordrecht, The Netherlands, 75–92.

Brady, C., Arnold, D., McDonald, J., Denman, S., 2017. Taxonomy and identification of bacteria associated with Acute Oak Decline. World J. Microbiol. Biotechnol. 33 (7), 143.

Brahmi, P., Saxena, S., Dhillon, B.S., 2004. The protection of plant varieties and farmers' rights act of India. Curr. Sci. 86 (3), 392–398.

Butola, J.S., Malik, A.R., 2012. Phenology and Survival of some Himalayan medicinal plants domesticated at different altitudes. Int. J. Med. Arom. Plants 2 (4), 683–687.

Butola, J.S., Vashistha, R.K., 2013. An overview on conservation and utilization of *Angelica glauca Edgew.* in three Himalayan states of India. Emerg. Med. Plant 5 (3), 171–178.

Canter, P.H., Thomas, H., Ernst, E., 2005. Bringing medicinal plants into cultivation: opportunities and challenges for biotechnology. Trends Biotechnol. 23 (4), 180–185.

Chaturvedi, S., 2011. South-South Cooperation in health and pharmaceuticals: emerging trends in India-Brazil Collaborations. In: Research and Information System for Developing Countries. RIS Publications, New Delhi.

Chauhan, N.S., 1999. Medicinal and Aromatic Plants of Himachal Pradesh. Indus Publishing, New Delhi.

Chauhan, R.S., 2010. Socioeconomic improvement through medicinal and aromatic plants (MAPs) cultivation in Uttarakhand, India. J. Sustain. Agric. 34 (6), 647–658.

Chen, S.L., Yu, H., Luo, H.M., Wu, Q., Li, C.F., Steinmetz, A., 2016. Conservation and sustainable use of medicinal plants: problems, progress, and prospects. Chinese Med. 11 (1), 37.

Chettri, N., Rasul, G., Sharma, E., 2014. Managing ecosystem services for enhancing climate change adaptation in the Hindu Kush Himalayas. Impact of Global Changes on Mountains: Responses and Adaptation. 208.

Chhetri, D.R., 2014. Medicinal Plants of the Himalaya: Production Technology and Utilization. Agrobios.

Choudhary, N., Sekhon, B.S., 2011. An overview of advances in the standardization of herbal drugs. J. Pharm. Educ. Res. 2 (2), 55.

Costello, M.J., Wieczorek, J., 2014. Best practice for biodiversity data management and publication. Biol. Conserv. 173, 68–73.

Costello, M.J., Vanhoorne, B., Appeltans, W., 2015. Conservation of biodiversity through taxonomy, data publication, and collaborative infrastructures. Conserv. Biol. 29 (4), 1094–1099.

Cronk, Q.C., Fuller, J.L., 2014. Plant Invaders: The Threat to Natural Ecosystems. Routledge, Earthscan Publications Limited, UK.

Cunningham, A.B., 2001. Applied Ethnobotany: People, Wild Plant Use and Conservation. Earthscan Publication Limited, USA.

Dar, A.R., Dar, G.H., 2006. The wealth of Kashmir Himalaya—gymnosperms. Asian J. Plant Sci. 5 (2), 251–259.

Dawa, S., Hussain, A., Tamchos, T., 2013. Medicinal and aromatic plants as a viable livelihood option for the people of Jammu and Kashmir. Indian Horticult. J. *3* (1 and 2), 40–42.

De, L.C., Medhi, R.P., 2014. Diversity and conservation of rare and endemic orchids of North East India-A Review. Indian J. Hill Farm. 27 (1), 138–153.

De, L.C., Rao, A.N., Singh, D.R., 2016. Endangered orchids and their conservation in North East India. In: Bioprospecting of Indigenous Bioresources of North-East India. Springer, Singapore, pp. 61–75.

Deb, D., Ghosh, A., Singh, J.P., Chaurasia, R.S., 2016. A study on general allometric relationships developed for biomass estimation in regional scale taking the example of *Tectona grandis* grown in Bundelkhand region of India. Curr. Sci. 110 (3), 414.

Debnath, M., Malik, C.P., Bisen, P.S., 2006. Micropropagation: a tool for the production of high quality plant-based medicines. Curr. Pharm. Biotechnol. 7 (1), 33–49.

Dell'Anno, A., Carugati, L., Corinaldesi, C., Riccioni, G., Danovaro, R., 2015. Unveiling the biodiversity of deep-sea nematodes through metabarcoding: are we ready to bypass the classical taxonomy? PLoS One 10 (12), e0144928.

Devi, S., Padmavati, M., 2016. Biodiversity monitoring: a pre-condition to access and benefit sharing under the Indian Biological Diversity Act, 2002. J. Intellect. Property Rights 21, 288–294.

Dewan, M., 2011. IPR protection in agriculture: an overview. J. Intellect. Property Rights 16, 131–138.

Dhami, N., Mishra, A.D., 2015. Phytochemical variation: how to resolve the quality controversies of herbal medicinal products? J. Herbal Med. 5 (2), 118–127.

Dhar, B., Anuradha, R.V., 2004. Access, benefit-sharing and intellectual property rights. J. World Intell. Property 7 (5), 597–639.

Dhar, U., Rawal, R.S., Upreti, J., 2000. Setting priorities for conservation of medicinal plants—a case study in the Indian Himalaya. Biol. Conserv. 95 (1), 57–65.

Dhiman, N., Patial, V., Bhattacharya, A., 2016. Vitro approaches for conservation and sustainable utilization of *Podophyllum hexandrum* and *Picrorhiza kurroa*: endangered medicinal herbs of Western Himalaya. In: Plant Tissue Culture: Propagation, Conservation and Crop Improvement. Springer, Singapore. pp. 45–69.

Dhyani, A., Nautiyal, B.P., Nautiyal, M.C., 2010. Importance of Astavarga plants in traditional systems of medicine in Garhwal, Indian Himalaya. Int. J. Biodivers. Sci. Ecosyst. Serv. Manage. 6 (1–2), 13–19.

Downie, H.F., Adu, M.O., Schmidt, S., Otten, W., Dupuy, L.X., White, P.J., Valentine, T.A., 2015. Challenges and opportunities for quantifying roots and rhizosphere interactions through imaging and image analysis. Plant Cell Environ. 38 (7), 1213–1232.

Dubey, N.K., Kumar, R., Tripathi, P., 2004. Global promotion of herbal medicine: India's opportunity. Curr. Sci. 86 (1), 37–41.

Ellis, R., Waterton, C., Wynne, B., 2010. Taxonomy, biodiversity and their publics in twenty-first-century DNA barcoding. Public Understand. Sci. 19 (4), 497–512.

Ely, C.V., de Loreto Bordignon, S.A., Trevisan, R., Boldrini, I.I., 2017. Implications of poor taxonomy in conservation. J. Nat. Conserv. *36*, 10–13.

Endress, P.K., Baas, P., Gregory, M., 2000. Systematic plant morphology and anatomy: 50 years of progress. Taxon 401–434.

Eritja, M.C., 2017. Bio-prospecting in the arctic: an overview of the interaction between the rights of indigenous peoples and access and benefit sharing. BC Envtl. Aff. L. Rev. 44, 223–573.

Gairola, S., Rawal, R.S., Todaria, N.P., 2008. Forest vegetation patterns along an altitudinal gradient in sub-alpine zone of west Himalaya, India. Afr. J. Plant Sci. 2 (6), 042–048.

Gairola, S., Shariff, N.M., Bhatt, A., 2010. Influence of climate change on production of secondary chemicals in high altitude medicinal plants: issues needs immediate attention. J. Med. Plants Res. 4 (18), 1825–1829.

Gairola, S., Procheş, Ş., Rocchini, D., 2013. High-resolution satellite remote sensing: a new frontier for biodiversity exploration in Indian Himalayan forests. Int. J. Remote Sensing 34 (6), 2006–2022.

Geiser, D., Kang, S., Ward, T., O'Donnell, K., Robert, V., Crous, P., 2013. In: The impact of culture collections on molecular identification, taxonomy, and solving real problems. Meeting Abstract.

Genovesi, P., Butchart, S.H., McGeoch, M.A., Roy, D.B., 2013. Monitoring trends in biological invasion, its impact and policy responses. In: Collen, B., Pettorelli, N., Baillie, J.E.M., Durant, S.M. (Eds.), Biodiversity Monitoring and Conservation: Bridging the Gap Between Global Commitment and Local Action. Wiley-Blackwell, Oxford, UK, pp. 138–158.

Ghimire, S.K., 2008. Sustainable harvesting and management of medicinal plants in the Nepal Himalaya: current issues, knowledge gaps and research priorities. Med. Plants Nepal: Anthol. Contemp. Res., 25–44.

Govindaraghavan, S., Sucher, N.J., 2015. Quality assessment of medicinal herbs and their extracts: criteria and prerequisites for consistent safety and efficacy of herbal medicines. Epilepsy Behav. 52, 363–371.

Granjou, C., Mauz, I., Barbier, M., Breucker, P., 2014. Making taxonomy environmentally relevant. Insights from an all taxa biodiversity inventory. Environ. Sci. Policy *38*, 254–262.

Guisan, A., Tingley, R., Baumgartner, J.B., Naujokaitis-Lewis, I., Sutcliffe, P.R., Tulloch, A.I., Martin, T.G., 2013. Predicting species distributions for conservation decisions. Ecol. Lett. 16 (12), 1424–1435.

Guisan, A., Broennimann, O., Engler, R., Vust, M., Yoccoz, N.G., Lehmann, A., Zimmermann, N.E., 2006. Using niche-based models to improve the sampling of rare species. Conserv. Biol. 20 (2), 501–511.

Gupta, P.D., Daswani, P.G., Birdi, T.J., 2014. Approaches in fostering quality parameters for medicinal botanicals in the Indian context. Indian J. Pharmacol. 46 (4), 363.

Gwalwanshi, D.R., Salunkhe, O., Shukla, A., Bishwas, A.J., Vyas, D., 2014. Indigenous knowledge and documentation of ethno-medicinal plants of Panna District, Central India: a case study. J. Ethnobiol. Tradit. Med. Photon 122, 868–876.

Harilal, M.S., 2009. "Commercialising Traditional Medicine": Ayurvedic Manufacturing in Kerala. Econ. Polit. Week. 44–51.

Jain, M., Johnson, T.S., Krishnan, P., 2012. Biotechnological approaches to conserve the wealth of nature: endangered and rare medicinal plant species, a review. J. Nat. Remed. 12 (2), 93–102.

Jain, S.K., Rao, R.R., 1977. A Handbook of Field and Herbarium Methods. Today and Tomorrow's Printers and Publishers xvi, New Delhi. 157 pp., Illus. General (KR, 197700062).

Jan, S., Kamili, A.N., Hamid, R., Parray, J.A., 2013. Variation in adaptation mechanisms of medicinal herbs to the extreme winter conditions across the North Western Himalaya. Israel J. Plant Sci. 61 (1–4), 1–11.

Jan, S., Kamili, A.N., Parray, J.A., Bedi, Y.S., Ahmad, P., 2016a. Microclimatic variation in UV perception and related disparity in tropane and quinolizidine alkaloid composition of *Atropa acuminata*, *Lupinus polyphyllus* and *Hyoscyamus niger*. J. Photochem. Photobiol. B: Biol. 161, 230–235.

Jan, S., Kamili, A.N., Parray, J.A., Bedi, Y.S., 2016b. Differential response of terpenes and anthraquinones derivatives in *Rumex dentatus* and *Lavandula officinalis* to harsh winters across north-western Himalaya. Nat. Prod. Res. 30 (5), 608–612.

Jangdey, M.S., Gupta, A., Kaur, C.D., Saraf, S., 2016. Assessment of utilization, value addition and characterization of Tamarind: a natural gum of Chhattisgarh. Int. J. Pharm. Res. Allied Sci. 5 (2).

Jansen, E., 2016. Naturopathy in South India: Clinics Between Professionalization and Empowerment. Brill, South India.

Jayakumar, S., Kim, S.S., Heo, J., 2011. Floristic inventory and diversity assessment-a critical review. Proc. Int. Acad. Ecol. Environ. Sci. 1 (3–4), 151.

Joshi, R.K., Satyal, P., Setzer, W.N., 2016. Himalayan aromatic medicinal plants: a review of their ethnopharmacology, volatile phytochemistry, and biological activities. Medicines 3 (1), 6.

Joshi, S.G., 2000. Medicinal Plants. Oxford and IBH publishing, New Delhi, 216.

Juneja, K., Saini, S.S., Gaid, M., Beerhues, L., Sircar, D., 2017. Production of life-saving drugs from himalayan herbs. Sustain. Util. Nat. Resources, 333.

Kala, C.P., 2002. Indigenous knowledge of Bhotiya tribal community on wool dyeing and its present status in the Garhwal Himalaya, India. Curr. Sci. 83 (7), 814–817.

Kala, C.P., 2007. Local preferences of ethnobotanical species in the Indian Himalaya: implications for environmental conservation. Curr. Sci. 1828–1834.

Kala, C.P., 2009. Medicinal plants conservation and enterprise development. Med. Plants 1 (2), 79–95.

Kala, C.P., 2015. Medicinal and aromatic plants: boon for enterprise development. J. Appl. Res. Med. Arom. Plants 2 (4), 134–139.

Kala, C.P., 2017. Traditional health care systems and herbal medicines. Eur. J. Environ. Publ. Health 1 (1), 03.

Kala, C.P., 2005. Ethnomedicinal botany of the Apatani in the Eastern Himalayan region of India. J. Ethnobiol. Ethnomed. 1 (1), 11.

Kala, C.P., Mathur, V.B., 2002. Patterns of plant species distribution in the trans-Himalayan region of Ladakh, India. J. Veg. Sci. 13 (6), 751–754.

Kala, C.P., Sajwan, B.S., 2007. Revitalizing Indian systems of herbal medicine by the National Medicinal Plants Board through institutional networking and capacity building. Curr. Sci. 797–806.

Kala, C.P., Silori, C.S. (Eds.), 2013. Biodiversity, Communities and Climate Change. The Energy and Resources Institute (TERI), New Delhi.

Kala, C.P., Farooquee, N.A., Dhar, U., 2004. Prioritization of medicinal plants on the basis of available knowledge, existing practices and use value status in Uttaranchal, India. Biodivers. Conserv. 13 (2), 453–469.

Kala, C.P., Dhyani, P.P., Sajwan, B.S., 2006. Developing the medicinal plants sector in northern India: challenges and opportunities. J. Ethnobiol. Ethnomed. 2 (1), 32.

Kala, C.P., Dollo, M., Farooquee, N.A., Choudhury, D.C., 2008. Land-use management and wet-rice cultivation (Jebi Aji) by the Apatani People in Arunachal Pradesh, India: traditional knowledge and practices. Outlook Agric. 37 (2), 125–129.

Kapoor, L.D., 2000. Handbook of Ayurvedic Medicinal Plants: Herbal Reference Library. Vol. 2. CRC Press, Boca Raton, Florida.

Karki, M., Tiwari, B., Badoni, A., Bhattarai, N., 2003. In: Creating livelihoods enhancing medicinal and aromatic plants based biodiversity-rich production systems: preliminary lessons from South Asia. Oral paper presented at The 3rd World Congress on Medicinal and Aromatic Plants for Human Welfare (WOCMAP III). Chiang Mai, Thailand (Vol. 10).

Kaul, M.K., 1997. Medicinal Plants of Kashmir and Ladakh: Temperate and Cold Arid Himalaya. Indus Publishing, New Delhi.

Khan, S.M., Page, S.E., Ahmad, H., Harper, D.M., 2013. Sustainable utilization and conservation of plant biodiversity in montane ecosystems: the western Himalayas as a case study. Ann. Bot. 112 (3), 479–501.

Khan, S.M., Page, S., Ahmad, H., Harper, D., 2014. Ethno-ecological importance of plant biodiversity in mountain ecosystems with special emphasis on indicator species of a Himalayan Valley in the northern Pakistan. Ecol. Indicat. 37, 175–185.

Khandekar, V., Srivastava, A., 2014. Ecosystem biodiversity of India. In: Gorillo, O. (Ed.), Biodiversity—The Dynamic Balance of the Planet. InTech, Rijeka, Croatia.

Khare, C.P., 2008. Indian Medicinal Plants: An Illustrated Dictionary. Springer Science & Business Media, New Delhi, India.

Kharkwal, G., Mehrotra, P., Rawat, Y.S., Pangtey, Y.P.S., 2005. Phytodiversity and growth form in relation to altitudinal gradient in the Central Himalayan (Kumaun) region of India. Curr. Sci. 873–878.

Khuroo, A.A., Weber, E., Malik, A.H., Dar, G.H., Reshi, Z.A., 2010. Taxonomic and biogeographic patterns in the native and alien woody flora of Kashmir Himalaya, India. Nordic J. Bot. 28 (6), 685–696.

Khuroo, A.A., Reshi, Z.A., Rashid, I., Dar, G.H., 2011. Towards an integrated research framework and policy agenda on biological invasions in the developing world: a case-study of India. Environ. Res. 111 (7), 999–1006.

Krishnan, P.N., Decruse, S.W., Radha, R.K., 2011. Conservation of medicinal plants of Western Ghats, India and its sustainable utilization through in vitro technology. In Vitro Cell. Dev. Biol. Plant 47 (1), 110–122.

Kumar, S., 2016. The Changing Future of Rural Marketing in India. Imp. J. Interdiscipl. Res. 2 (4).

Kumar, S., Kumar, R., Khan, A., 2011. Medicinal plant resources: manifestation and prospects of life-sustaining healthcare system. Cont. J. Biol. Sci 4 (1), 19–29.

Kumar, V., Babu, S., Revale, A.K., Meena, R.K., Ranjan, M.K., Desai, B.S., 2014. Cultivation of medicinal plants in natural ecosystem in Gujarat (India): constraints and conservation need. J. Plant Dev. Sci. 6 (3), 425–435.

Kumaria, S., Kayang, H., Tandon, P., 2013. Conservation of Medicinal and Aromatic Plants of Northeast India. Capital Publishing Company, New Delhi, India.

Kumschick, S., Gaertner, M., Vilà, M., Essl, F., Jeschke, J.M., Pyšek, P., Evans, T., 2014. Ecological impacts of alien species: quantification, scope, caveats, and recommendations. BioScience 65 (1), 55–63.

Kuniyal, C.P., Bhatt, V.P., Bhatt, V.P., Butola, J.S., Sundriyal, R.C., 2014. Promoting nursery enterprise in high altitude villages: a participatory approach for conservation and commercialization of Himalayan threatened medicinal plants. J. Med. Plants Res. 8 (48), 1399–1407.

Kuniyal, C.P., Bisht, V.K., Negi, J.S., Bhatt, V.P., Bisht, D.S., Butola, J.S., Singh, S.K., 2015. Progress and prospect in the integrated development of medicinal and aromatic plants (MAPs) sector in Uttarakhand, Western Himalaya. Environ. Dev. Sustain. 17 (5), 1141–1162.

Kunwar, R.M., Nepal, B.K., Kshhetri, H.B., Rai, S.K., Bussmann, R.W., 2006. Ethnomedicine in Himalaya: a case study from Dolpa, Humla, Jumla and Mustang districts of Nepal. J. Ethnobiol. Ethnomed. 2 (1), 27.

Laladhas, K.P., Preetha, N., Baijulal, B., Oommen, V., 2015. Conservation of medicinal plant resources through Community Born Biodiversity Management Committee, Kerala, India. In: Biodiversity Conservation-Challenges for the Future. p. 27. New Delhi, India.

Lande, R., 1996. Statistics and partitioning of species diversity, and similarity among multiple communities. 76, 5–13. Oikos.

Larsen, H.O., Olsen, C.S., 2007. Unsustainable collection and unfair trade? Uncovering and assessing assumptions regarding Central Himalayan medicinal plant conservation. Biodivers. Conserv. 16 (6), 1679–1697.

Li, X., Chen, Y., Lai, Y., Yang, Q., Hu, H., Wang, Y., 2015. Sustainable utilization of traditional chinese medicine resources: systematic evaluation on different production modes. Evid.-Based Complement. Alternat. Med. 2015.

Lohani, N., Tewari, L.M., Kumar, R., Joshi, G.C., Kumar, S., Upreti, B.M., 2013. Population studies, habitat assessment and threat categorization of *Polygonatum verticillatum* (L.) Allioni in Kumaun Himalaya. J. Ecol. Nat. Environ. 5 (5), 74–82.

Lone, P.A., Bhardwaj, A.K., Shah, K.W., Tabasum, S., 2014. Ethnobotanical survey of some threatened medicinal plants of Kashmir Himalaya, India. J. Med. Plants Res. 8 (47), 1362–1373.

Lone, P.A., Bhardwaj, A.K., Bahar, F.A., 2015. Study of indigenous/traditional medicinal plant knowledge—an endeavour towards new drug discovery. Afr. J. Tradit. Complement. Alternat. Med. 12 (2), 73–95.

Lubbe, A., Verpoorte, R., 2011. Cultivation of medicinal and aromatic plants for specialty industrial materials. Ind. Crops Prod. 34 (1), 785–801.

Luintel, H., 2002. Issues and options of sustainable management of Himalayan medicinal herbs. J. For. Livelihood Vol. 2, 1.

Madhavan, H., 2014. Innovation system and increasing reformulation practices in the Ayurvedic pharmaceutical sector of South India. Asian Med. 9 (1–2), 236–271.

Magurran, A.E., 2013. Ecological Diversity and its Measurement. Springer Science & Business Media, Netherlands.

Mahato, S., Dasgupta, S., Todaria, N.P., Singh, V.P., 2016. Agroforestry mapping and characterization in four districts of Garhwal Himalaya. Energy Ecol. Environ. 1 (2), 86–97.

Maheswari, J., Patenting Indian medicinal plants and products. Indian J. Sci. Technol. 4 (3), 2011. 298–301.

Maikhuri, R.K., Phondani, P.C., Rawat, L.S., Jha, N.K., Maletha, A., Bahuguna, Y.M., Kandari, L.S., Conservation and management strategies of medicinal plant resources through action research approaches in Indian Himalaya. Iran. J. Sci. Technol. Trans. A: Sci. 2016. 1–7.

Maiti, P.K., Maiti, P., 2017. Biodiversity: Perception, Peril and Preservation. PHI Learning Pvt. Ltd., New Delhi.

Maity, D., Dey, S.K., Chatterjee, S., Maiti, G.G., 2015. Ethnobotany and environmental management by the tribal communities of Patratu, Hazaribagh district, Jharkhand. Explor. Anim. Med. Res. 5 (1), 44–61.

Malik, S., Cusidó, R.M., Mirjalili, M.H., Moyano, E., Palazón, J., Bonfill, M., 2011. Production of the anticancer drug taxol in Taxus baccata suspension cultures: a review. Process Biochem. 46 (1), 23–34.

Mann, D. (1997). The economics of botanical collections. The value and valuation of natural science collections, Manchester, 1995, pp. 68–82.

Máthé, Á., 2015. Botanical aspects of medicinal and aromatic plants. In: Medicinal and Aromatic Plants of the World. Springer, Netherlands, pp. 13–33.

Mazid, M., Khan, T.A., Mohammad, F., 2012. Medicinal plants of rural India: a review of use by Indian folks. Indo Global J. Pharm. Sci. 2 (3), 286–304.

McGeoch, M.A., Genovesi, P., Bellingham, P.J., Costello, M.J., McGrannachan, C., Sheppard, A., 2016. Prioritizing species, pathways, and sites to achieve conservation targets for biological invasion. Biol. Invasions 18 (2), 299–314.

Meena, R.L., 2015. Standardization of agrotechnique of Aloe vera in Mid hills of Western Himalaya. Glob. J. Res. Med. Plants Indigen. Med. 4 (5), 88.

Mehra, A., Bajpai, O., Joshi, H., 2014. Diversity, utilization and sacred values of ethnomedicinal plants of Kumaun Himalaya. Trop. Plant Res. 1, 80–86.

Mishra, D., Singh, R.K., Srivastava, R.K., 2013. Agri-business and entrepreneurship development through medicinal and aromatic plants: an indian state of affairs. Int. J. Food Agric. Vet. Sci. 3 (1), 238–246.

Mishra, S.R., Neupane, D., Kallestrup, P., 2015. Integrating complementary and alternative medicine into conventional health care system in Developing Countries: an example of Amchi. J. Evid.-Based Complement. Altern. Med. 20 (1), 76–79.

Morgera, E., Tsioumani, E., 2010. The evolution of benefit sharing: linking biodiversity and community livelihoods. Rev. Eur. Comparat. Int. Environ. Law 19 (2), 150–173.

Mukherjee, P.K., 2001. Evaluation of Indian traditional medicine. Drug Inform. J. 35 (2), 623–632.

Mukherjee, P.K., Rai, S., Kumar, V., Mukherjee, K., Hylands, P.J., Hider, R.C., 2007. Plants of Indian origin in drug discovery. Exp. Opin. Drug Discov. 2 (5), 633–657.

Nadkarni, K.M., 1927. The Indian *Materia medica*. The Indian Materia Medica, Bombay, India.

Nadkarni, K.M. (Ed.), 1996. [Indian *Materia medica*]; Dr. KM Nadkarni's Indian *Materia medica*: With Ayurvedic, Unani-Tibbi, Siddha, Allopathic, Homeopathic, Naturopathic & Home Remedies, Appendices & Indexes. 1. Vol. 1. Popular Prakashan, Mumbai, India.

Nandi, S.K., Palni, L.M.S., Pandey, H., Chandra, B., Nadeem, M., 2016. Selection of elites and in vitro propagation of selected high-value Himalayan medicinal herbs for sustainable utilization and conservation. In: Plant Tissue Culture: Propagation, Conservation and Crop Improvement. Springer, Singapore, pp. 15–44.

Nasi, R., Brown, D., Wilkie, D., Bennett, E., Tutin, C., Van Tol, G., & Christophersen, T. (2008). Conservation and Use of Wildlife-Based Resources: The Bushmeat Crisis. Secretariat of the Convention on Biological Diversity, Montreal. Center for International Forestry Research (CIFOR), Bogor. Technical Series, 50.

Nautiyal, S., Maikhuri, R.K., Rao, K.S., Saxena, K.G., 2001. Medicinal plant resources in Nanda Devi Biosphere Reserve in the Central Himalayas. J. Herbs Spices Med. Plants 8 (4), 47–64.

Negi, V.S., Maikhuri, R.K., Phondani, P.C., Rawat, L.S., 2010. An inventory of indigenous knowledge and cultivation practices of medicinal plants in Govind Pashu Vihar Wildlife Sanctuary, Central Himalaya, India. Int. J. Biodivers. Sci. Ecosyst. Serv. Manage. 6 (3–4), 96–105.

Nikam, P.H., Kareparamban, J., Jadhav, A., Kadam, V., 2012. Future trends in standardization of herbal drugs. J. Appl. Pharm. Sci. 02 (06), 38–44.

Olsen, C.S., 2001. Trade in the Himalayan medicinal plant product kutki. Med. Plant Conserv. 7, 11–13.

Olsen, C.S., Larsen, H.O., 2003. Alpine medicinal plant trade and Himalayan mountain livelihood strategies. Geograph. J. 169 (3), 243–254.

Osman, N. A. A., Ibrahim, F., Abas, W. A. B. W., Rahman, H. S. A., & Ting, H. N. (Eds.). (2008). 4th Kuala Lumpur International Conference on Biomedical Engineering 2008: BIOMED 2008, 25–28 June 2008, Kuala Lumpur, Malaysia (Vol. 21). Springer Science & Business Media, Springer, Berlin, Heidelberg.

Pan, S.Y., Litscher, G., Gao, S.H., Zhou, S.F., Yu, Z.L., Chen, H.Q., Ko, K.M., 2014. Historical perspective of traditional indigenous medical practices: the current renaissance and conservation of herbal resources. Evid. Based Complement. Alternat. Med. 2014.

Pandey, A., Tripathi, S., 2014. Concept of standardization, extraction and pre phytochemical screening strategies for herbal drug. J. Pharmacogn. Phytochem. 2 (5).

Pant, S., Samant, 2006. Diversity, distribution, uses and conservation status of plant species of the Mornaula Reserve Forests, West Himalaya, India. Int. J. Biodivers. Sci. Manage. 2 (2), 97–104.

Petrovska, B.B., 2012. Historical review of medicinal plants' usage. Pharmacog. Rev. 6 (11), 1.

Phondani, P.C., Negi, V.S., Bhatt, I.D., Maikhuri, R.K., Kothyari, B.P., 2011. Promotion of medicinal and aromatic plants cultivation for improving livelihood security: a case study from West Himalaya, India. Int. J. Med. Arom. Plants 1 (3), 245–252.

Phondani, P.C., Maikhuri, R.K., Bisht, N.S., 2013. Endorsement of ethnomedicinal knowledge towards conservation in the context of changing socio-economic and cultural values of traditional communities around Binsar Wildlife Sanctuary in Uttarakhand, India. J. Agric. Environ. Ethics 26 (3), 573–600.

Phondani, P.C., Bhatt, I.D., Negi, V.S., Kothyari, B.P., Bhatt, A., Maikhuri, R.K., 2016. Promoting medicinal plants cultivation as a tool for biodiversity conservation and livelihood enhancement in Indian Himalaya. J. Asia-Pacific Biodivers. 9 (1), 39–46.

Phondani, P.C., Maikhuri, R.K., Saxena, K.G., 2014. The efficacy of herbal system of medicine in the context of allopathic system in Indian Central Himalaya. J. Herbal Med. 4 (3), 147–158.

Pisupati, B., Bavikatte, K.S., 2014. Access and benefit sharing as an innovative financing mechanism. Dev. Rev. 16 (2), 53–70.

Pradhan, B.K., Badola, H.K., 2012. Effects of microhabitat, light and temperature on seed germination of a critically endangered Himalayan medicinal herb, Swertia chirayita: Conservation implications. Plant Biosyst.—Int. J. Deal. Aspects Plant Biol. 146 (2), 345–351.

Prakash, K., Nirmalaa, A., 2013. A review on conservation of endangered medicinal plants in India. Int. J. Ethnomed. Pharmacol. Res. 1 (1), 21–33.

Prasad, V.K., Badarinath, K.V.S., Eaturu, A., 2007. Spatial patterns of vegetation phenology metrics and related climatic controls of eight contrasting forest types in India—analysis from remote sensing datasets. Theor. Appl. Climatol. 89 (1-2), 95.

Purohit, V.K., Nautiyal, M.C., Nautiyal, A.R., Gairola, K.C., 2016. Successful cultivation of highly valued and endangered medicinal herb Kutki (Picrorrhiza kurrooa): a case study from Ghesh village of Uttarakhand, Himalaya. Med. Plants Int. J. Phytomed. Rel. Ind. 8 (3), 207–212.

Rai, L.K., Prasad, P., Sharma, E., 2000. Conservation threats to some important medicinal plants of the Sikkim Himalaya. Biol. Conserv. 93 (1), 27–33.

Ramawat, K.G., Goyal, S., 2008. The Indian herbal drugs scenario in global perspectives. Bioactive Mol. Med. Plants 325–347.

Rana, S., Rastogi, J., 2017. Occurrence and floral details of four new invasive alien species in Uttarakhand, India. Arch. Agric. Environ. Sci. 2 (2), 113–118.

Rao, M.R., Palada, M.C., Becker, B.N., 2004. Medicinal and aromatic plants in agroforestry systems. In: New Vistas in Agroforestry. Springer, Netherlands, pp. 107–122.

Rasheed, N.M.A., Nagaiah, K., Waheed, M.A., 2013. Recent analytical techniques in quality control of indigenous system of medicine. Ann. Phytomed. 2 (1), 44–58.

Raustiala, K., Victor, D.G., 1996. Biodiversity since Rio: the future of the convention on biological diversity. Environ.: Sci. Policy Sustain. Dev. 38 (4), 16–45.

Rawal, R.S., Bhatt, I.D., Garhwal, S., 2010. Chromatographic and spectral fingerprinting standardization of traditional medicines: an overview as modern tools. Res. J. Phytochem. 4 (4), 234–241.

Ray, R., Gururaja, K.V., Ramchandra, T.V., 2011. Predictive distribution modeling for rare Himalayan medicinal plant *Berberis aristata* DC. J. Environ. Biol. 32 (6), 725.

Rehman, A., Saba, T., 2014. Evaluation of artificial intelligent techniques to secure information in enterprises. Artif. Intell. Rev. 42 (4), 1029–1044.

Remagnino, P., Mayo, S., Wilkin, P., Cope, J., Kirkup, D., 2016. Computational Botany: Methods for Automated Species Identification. Springer-Verlag, Berlin Heidelberg.

Riedel, A., Sagata, K., Suhardjono, Y.R., Tänzler, R., Balke, M., 2013. Integrative taxonomy on the fast track-towards more sustainability in biodiversity research. Front. Zool. 10 (1), 15.

Rokaya, M.B., Münzbergová, Z., Dostálek, T., 2017. Sustainable harvesting strategy of medicinal plant species in Nepal—results of a six-year study. Folia Geobot., 1–14.

Rotherham, I.D., Bio-cultural heritage and biodiversity: emerging paradigms in conservation and planning. Biodivers. Conserv. 24 (13), 2015. 3405–3429.

Rout, S.P., Choudary, K.A., Kar, D.M., Das, L., Jain, A., 2009. Plants in traditional medicinal system-future source of new drugs. Int. J. Pharm. Pharm. Sci. 1 (1), 1–23.

Sagar, R., 2005. intellectual property, benefit-sharing and traditional knowledge. J. World Intell. Property. 8 (3), 383–400.

Saha, M.R., De Sarker, D., Kar, P., Gupta, P.S., Sen, A., 2014. Indigenous knowledge of plants in local healthcare management practices by tribal people of Malda district, India. J. Intercult. Ethnopharmacol. 3 (4), 179.

Sahoo, N., Manchikanti, P., 2013. Herbal drug regulation and commercialization: an Indian industry perspective. J. Alternat. Complement. Med. 19 (12), 957–963.

Sahoo, N., Manchikanti, P., Dey, S.H., 2011. Herbal drug patenting in India: IP potential. J. Ethnopharmacol. 137 (1), 289–297.

Samal, J., 2016. Medicinal plants and related developments in India: a peep into 5-year plans of India. Indian J. Health Sci. Biomed. Res. (KLEU) 9 (1), 14.

Samant, S.S., Dhar, U., Palni, L.M.S., 1998. Medicinal Plants of Indian Himalaya. Gyanodaya Prakashan, New Delhi, India.

Sati, V.P., 2014. Towards Sustainable Livelihoods and Ecosystems in Mountain Regions. Springer International Publishing, Switzerland.

Semwal, D.P., Saradhi, P.P., Nautiyal, B.P., Bhatt, A.B., 2007. Current status, distribution and conservation of rare and endangered medicinal plants of Kedarnath Wildlife Sanctuary, Central Himalayas, India. Curr. Sci. 1733–1738.

Semwal, D.P., Saradhi, P.P., Kala, C.P., Sajwan, B.S., 2010. Medicinal plants used by local Vaidyas in Ukhimath block, Uttarakhand. Indian J. Tradit. Knowl. 9 (3), 480–485.

Sen, S., Chakraborty, R., 2017. Revival, modernization and integration of Indian traditional herbal medicine in clinical practice: importance, challenges and future. J. Tradit. Complement. Med. 7 (2), 234–244.

Sen, S., Chakraborty, R., De, B., 2011. Challenges and opportunities in the advancement of herbal medicine: India's position and role in a global context. J. Herbal Med. 1 (3), 67–75.

Senapati, P., 2014. A study of agricultural inputs marketing in India. Indian Streams Res. J. 4 (5), 1–6.

Shah, M.A., Reshi, Z.A., 2014. Characterization of alien aquatic flora of Kashmir Himalaya: implications for invasion management. Trop. Ecol. 55 (2), 143–157.

Shaheen, H., Shinwari, Z.K., Qureshi, R.A., Ullah, Z., 2012. Indigenous plant resources and their utilization practices in village populations of kashmir himalayas. Pak. J. Bot. 44 (2), 739–745.

Shameem, S.A., Soni, P., Bhat, G.A., 2010. Comparative study of herb layer diversity in lower Dachigam National Park, Kashmir Himalaya, India. Int. J. Biodivers. Conserv. 2 (10), 308–315.

Sharma, A., Dua, S., Hatwal, V., 2012. Micro enterprise development and rural women entrepreneurship: way for economic empowerment. Arth Prabhand: J. Econ. Manage. 1 (6), 114–127.

Sharma, A.K., Thakur, N.S., 2015. Resource potential and development of small hydro power projects in Jammu and Kashmir in the western Himalayan region: India. Renew. Sustain. Energy Rev. 52(C), 1354–1368.

Sharma, C.M., Suyal, S., Gairola, S., Ghildiyal, S.K., 2009. Species richness and diversity along an altitudinal gradient in moist temperate forest of Garhwal Himalaya. J. Am. Sci. 5 (5), 119–128.

Sharma, C.M., Suyal, S., Ghildiyal, S.K., Gairola, S., 2010. Role of physiographic factors in distribution of Abies pindrow (Silver Fir) along an altitudinal gradient in Himalayan temperate forests. Environmentalist 30 (1), 76–84.

Sharma, P., 2013. Rural entrepreneurship, science and technology and innovations in farm based entrepreneurship venture. A case study of a farmer inventing accustomed machine for tillage farming. Int. J. Innovat. Res. Dev. 2 (4), 6–14.

Sharma, S., Thokchom, R., 2014. A review on endangered medicinal plants of India and their conservation. J. Crop Weed 10 (2), 205–218.

Sher, H., Ali, H., Rehman, S., 2012. Identification and conservation of important plant areas (IPAS) for the distribution of medicinal, aromatic and economic plants in the Hindukush-Himalaya mountain range. Pak. J. Bot. 44 (SI), 187–194.
Sheth, J.N., 2011. Impact of emerging markets on marketing: rethinking existing perspectives and practices. J. Market. 75 (4), 166–182.
Siddiqui, A., Shukla, S., 2015. Conservation of plant genetic resources and their utilisation in global perspective. LS—Int. J. Life Sci. 46.
Simmonds, M.S., 2009. Opportunities and challenges for ethnobotany at the start of the twenty-first century. In: Plant-Derived Natural Products. Springer, US. pp. 127–140.
Singh, A., Singh, M.K., Singh, R., 2013. Traditional medicinal flora of the district Buxar (Bihar, India). J. Pharmacogn. Phytochem. 2 (2).
Singh, A.K., 2017. Medicinal and aromatic plants. In: Wild Relatives of Cultivated Plants in India. Springer, Singapore, pp. 165–176.
Singh, H., 2006. Prospects and challenges for harnessing opportunities in medicinal plants sector in India. Law Env. Dev. J. 2, 196.
Singh, H., Husain, T., Agnihotri, P., Pande, P.C., Khatoon, S., 2014a. An ethnobotanical study of medicinal plants used in sacred groves of Kumaon Himalaya, Uttarakhand, India. J. Ethnopharmacol. 154 (1), 98–108.
Singh, K.N., Lal, B., Todaria, N.P., 2016. Ecological status and conservation of rare plants in high altitude landscapes of Indian western Himalaya. Glob. J. Res. Med. Plants Indigen. Med. 5 (1), 1.
Singh, M., Ali, A.A., Qureshi, M.I., 2017. Unravelling the impact of essential mineral nutrients on active constituents of selected medicinal and aromatic plants. In: Essential Plant Nutrients. Springer, Cham. pp. 183–209.
Singh, R.D., Mody, S.K., Patel, H.B., Devi, S., Modi, C.M., Kamani, D.R., 2014b. Pharmaceutical Biopiracy and Protection of Traditional Knowledge. Int. J. Res. Dev. Pharm. Life Sci. 3 (2), 866–871.
Singh, S.K., Ash, G.J., Hodda, M., 2015. Keeping "one step ahead" of invasive species: using an integrated framework to screen and target species for detailed biosecurity risk assessment. Biol. Invasions 17 (4), 1069–1086.
Siwach, M., Siwach, P., Solanki, P., Gill, A.R., 2013. Biodiversity conservation of Himalayan medicinal plants in India: a retrospective analysis for a better vision. Int. J. Biodivers. Conserv. 5 (9), 529–540.
Smith, A., Voß, J.P., Grin, J., 2010. Innovation studies and sustainability transitions: the allure of the multi-level perspective and its challenges. Res. Policy 39 (4), 435–448.
Sreeramu, B.S., 2004. Cultivation of Medicinal and Aromatic Crops. Universities Press (India) Private Limited, Hyderabad, India.
Srivastava, A.K., Kumar, S., 2017. An effective computational technique for taxonomic position of security vulnerability in software development. J. Comput. Sci. in press. https://sci-hub.bz/10.1016/j.jocs.2017.08.003.
Srivastava, J., Lambert, J., Vietmeyer, N., 1996. Medicinal Plants: An Expanding Role in Development. Vol. 320. World Bank Publications, Washington, USA.
Srivastava, P., Singh, R., Tripathi, S., Raghubanshi, A.S., 2016. An urgent need for sustainable thinking in agriculture–An Indian scenario. Ecological Indicators 67, 611–622.
Subbian, A., 2007. Intellectual Property Rights: Heritage Science and Society Under International Treaties. Deep and Deep Publications, New Delhi, India.
Subramoniam, A., 2014. Present scenario, challenges and future perspectives in plant based medicine development. Ann. Phytomed. 3 (1), 31–36.
Subrat, N., Iyer, M., Prasad, R., 2002. The Ayurvedic Medicine Industry: Current Status and Sustainability. Ecotech Services (India) Pvt. Ltd., New Delhi, India.
Sundriyal, M., Sharma, B., 2016. Status of biodiversity in Central Himalaya. Appl. Ecol. Environ. Sci. 4 (2), 37–43.

Swaminathan, M.S., 2002. The protection of plant varieties and farmers' rights act: from legislation to implementation. Curr. Sci. 82 (7), 778–780.

Tahir, N., Bibi, Y., Iqbal, M., Hussain, M., Laraib, S., Safdar, I., Bibi, G., 2016. Overview of *Dioscorea deltoidea* Wall. Ex Griseb: an endangered medicinal plant from Himalaya Region. J. Biodivers. Environ. Sci. 9, 13–24.

Taia, W.K., 2005. Modern trends in plant taxonomy. Asian J. Plant Sci. 4 (2), 2005.

Tali, B.A., Ganie, A.H., Nawchoo, I.A., Wani, A.A., Reshi, Z.A., 2015. Assessment of threat status of selected endemic medicinal plants using IUCN regional guidelines: a case study from Kashmir Himalaya. J. Nat. Conserv. 23, 80–89.

Tewari, L., 2001. Traditional Himalayan Medicine System and its *Materia medica*. HIST, India.

Thakur, P.S., Vaishnu, D., 2007. Cultivation of medicinal and aromatic herbs in agroforestry for diversification under submontane conditions of western Himalayas. Indian J. Agrofor. 9 (2), 67–76.

Theobald, E.J., Ettinger, A.K., Burgess, H.K., DeBey, L.B., Schmidt, N.R., Froehlich, H.E., Parrish, J.K., 2015. Global change and local solutions: tapping the unrealized potential of citizen science for biodiversity research. Biol. Conserv. 181, 236–244.

Torri, M.C., 2011. Bioprospecting and commercialisation of biological resources by Indigenous Communities in India: moving towards a new paradigm? Sci. Technol. Soc. 16 (2), 123–146.

Treitz, P.M., Howarth, P.J., Gong, P., 1992. Application of satellite and GIS technologies for land-cover and land-use mapping at the rural-urban fringe: a case study. Photogram. Eng. Remote Sensing 58 (4), 439–448.

Uniyal, S.K., Sharma, V., Jamwal, P., 2011. Folk medicinal practices in Kangra district of Himachal Pradesh, western Himalaya. Hum. Ecol. 39 (4), 479–488.

Upadhyay, S., Saikia, L.R., 2012. Influence of eco-friendly manures on leaf biomass production of a traditional medicinal plant *Adhatoda vasica* Nees under upper Brahmaputra valley agroclimatic zone. J. Agric. Sci. Technol. *A*, *2* (1A), 115.

Ved, D.K., Goraya, G.S., 2007. Demand and Supply of Medicinal Plants in India. NMPB, New Delhi & FRLHT, Bangalore, India, p. 18.

Victor, J.E., 2016. Strategies for the development of plant systematics in a floristically diverse region. Doctoral Dissertation.

Walker, R., 2015. From Big Data to Big Profits: Success With Data and Analytics. Oxford University Press, USA.

Whittaker, R.H., 1972. Evolution and measurement of species diversity. Taxon, 213–251.

Wilson, J.R., Caplat, P., Dickie, I.A., Hui, C., Maxwell, B.D., Nunez, M.A., Spear, D., 2014. A standardized set of metrics to assess and monitor tree invasions. Biol. Invasions 16 (3), 535–551.

Yadav, B.K.V., 2008. Problems and Prospects of Medicinal and Aromatic Plants' Cultivation and Marketing (A Case Study from Sarlahi District). Institute of Forestry, Pokhara, Nepal.

Yadav, V.K., Sharma, S.K., Rao, V.K., Yadav, R., Radhakrishna, A., 2016. Assessment of morphological and biochemical diversity in Seabuckthorn (*Hippophae salicifolia* D. Don.) populations of Indian Central Himalaya. Proc. Natl. Acad. Sci. India Sect. B: Biol. Sci. 86 (2), 351–357.

Zargar, S.M., Sharma, A., Sadhu, A., Agrawal, G.K., Rakwal, R., 2014. Exploring genetic diversity in common bean from unexploited regions of Jammu & Kashmir-India. Mol. Plant Breed. 5.

Zhan, A., He, S., Brown, E.A., Chain, F.J., Therriault, T.W., Abbott, C.L., MacIsaac, H.J., 2014. Reproducibility of pyrosequencing data for biodiversity assessment in complex communities. Methods Ecol. Evol. 5 (9), 881–890.

Further Reading

Bhatt, I.D., Rawat, S., Badhani, A., Rawal, R.S., 2017. Nutraceutical potential of selected wild edible fruits of the Indian Himalayan region. Food Chem. 215, 84–91.

Rao, K.S., Saxena, K.G., Tiwari, B.K., 2015. Biodiversity, Climate Change and Socio-economic Development in the Indian Himalaya. Shiva Offset Press, Dehradun, India.

Tamang, J.P., 2009. Himalayan Fermented Foods: Microbiology, Nutrition, and Ethnic Values. CRC Press.

Verpoorte, R., Memelink, J., 2002. Engineering secondary metabolite production in plants. Curr. Opin. Biotechnol. 13 (2), 181–187.

Way, M.J., 2003. Collecting seed from non-domesticated plants for long-term conservation. Seed Conserv.: Turn. Sci. Pract. 165–201.

CHAPTER

4

Chemistry of Himalayan Phytochemicals

OUTLINE

4.1 INTRODUCTION

In addition to the bioactive compounds, plant foods are important source of nutrients like vitamins and minerals. The phytochemicals produced from the medicinal plants are synthesized in very meager amounts. Some of the phytochemicals found to be beneficial to the human health belong to the groups like carotenoids, phenolic compounds (flavonoids, phytoestrogens, and phenolic acids), phytosterols and phytostanols, tocotrienols, organosulfur compounds (allium compounds and glucosinolates), and nondigestible carbohydrates (dietary fiber and prebiotics). Recently, a lot of research all

Himalayan Phytochemicals
https://doi.org/10.1016/B978-0-08-102227-6.00004-8

over the world has been carried out on the medicinal properties of plants. In most of the developing countries, herbal medicines are used widely because of their potent therapeutic efficacy, antioxidant activities, no side effect, and cost-effectiveness and as a raw material for drugs, thus serving as effective and reasonable health care for people.

The pharmacologically effective metabolites have been isolated from the various medicinal plants. These plant metabolites are organic compounds and are classified as primary and secondary metabolites. Primary metabolites include glucose, starch, polysaccharide, protein, lipids, and nucleic acid that are advantageous for the growth and development of our body. While secondary metabolites synthesized by plants are alkaloids, flavonoids, saponins, terpenoids, steroids, glycosides, tannins, volatile oils, etc., these secondary metabolites are very efficient therapeutically and are used for curing many diseases. Among the various phytochemicals, alkaloids have antispasmodic, antimalarial, analgesic, diuretic activities while terpenoids have antiviral, anthelmintic, antibacterial, anticancer, antimalarial, anti-inflammatory properties. Phenols and flavonoids were known to have an antioxidant, antiallergic, antibacterial properties, etc. Glycosides are reported for their antifungal and antibacterial properties, and saponins are reported to have anti-inflammatory, antiviral plant defense activities (Maurya et al., 2008; Chopra and Doiphode, 2002). Nowadays, a lot of chemical compounds used in modern medicine are derived from various plant sources like digoxin, aspirin, ephedrine, atropine, and colchicines (Ram et al., 2015; Moteriya et al., 2015). Medicinal plants have been used as natural healing system without proper guidance and scientific knowledge about thousands of years ago. People have been using these herbal medicines for some common treatment like colds, allergy, upset stomachs, and toothaches, and the trend is constantly increasing. Thus, there is the shift in the universal trend from synthetic to natural medicine, which can be said as return to nature for the prevention of various ailments. According to WHO, about 4 billion people (80% of the world's population) use some herbal medicines for aspect of primary health care (Fabricant and Farnsworth, 2001), and about 252 drugs that constitute 11% are derived from plant (Taylor, 2000). In spite of nutrients and various phytochemicals provided by the vegetable foods, they also serve as a reservoir of several pharmacological constituents that has antiviral, antioxidant, anticancer, antifungal, antimicrobial, and antiparasitic properties. The phenolics, flavonoids, anthocyanins, and vitamins synthesized in the plants have free radical scavenging molecules by virtue of which they show strong antioxidant activity (Chopra and Doiphode, 2002). The various herbal formulations that are either directly or indirectly derived from plants are used nowadays as a remedy for various human diseases like hepatorenal diseases, cardiovascular disease, cancers, diabetes, and neurodegenerative disorders (Modak et al., 2007;

Shakya and Shukla, 2011). The development of natural product chemistry with the isolation, the purification, and the chemical characterization of bioactive molecules from plants have opened new window for the synthesis of plant-based drugs. These drugs are effective and inexpensive with lesser side effects than the allopathic medicine. Thus, herbal plants and their formulations play an important role in the development of new drugs (WHO, 2008; Heinrich, 2000; Shakya et al., 2012). Despite the accomplishment of medication improvement from the medicinal plants in the past 2–3 decades, many challenges are further faced in the future endeavors. The major drawback to the herbal industry is the quality of the herbal products (Yadav et al., 2014; Patwardhan et al., 2004). Herbal plants can be simply adulterated during growth, processing, and collection. Adulteration and heavy metal contamination are the two major problems reported in herbal medicines. For that reason, it is necessary to advance the quality and quantity of bioactive compounds for developing new herbal drug and to keep pace with other drug discovery efforts (Dash and Sharma, 2001; Clark, 1996).

4.1.1 Major Classes of Phytochemicals

On the basis of the role in plant metabolism, the phytochemicals are classified into two major classes as primary and secondary constituents. The common sugars, amino acids, proteins, purines, and pyrimidines of nucleic acids, chlorophylls, etc. comprise primary constituents, while secondary constituents include remaining plant chemicals like alkaloids, terpenes, flavonoids, lignans, plant steroids, curcumins, saponins, phenolics, flavonoids, and glucosides (Hahn, 1998). The various secondary phytoconstitutents are discussed as under.

Phenolics

Plant phenolics are the largest category of phytochemicals and the most widely distributed in the plant kingdom. Within the plants, these metabolites are produced from either the shikimate pathway, which directly provides phenylpropanoids, or the polyketide acetate/malonate pathway, which produces simple phenols, or both. Thus, by both pathways, monomeric and polymeric phenols and polyphenols can be produced. In order to cope with the continuously changing environmental challenges over evolutionary time, plants have been gifted with the ability of producing thousands of different types of phenolic compounds. These phenolic compounds play an important role in the defense mechanism by scavenging the free radicals produced during the various biotic and abiotic stresses like pathogen infection, herbivores, nutrient deficiency, high light, and low temperatures (Lattanzio, 2013). Flavonoids, phenolic acids, and polyphenols are the three most vital groups of dietary phenolics. Phenolics are the class of chemical compounds where hydroxyl groups are directly attached to an aromatic hydrocarbon group. Among the group, phenol (C_6H_5OH) is the simplest class, and flavonoid is the largest and most widely studied group of natural compounds (Walton et al., 2003; Dai and Mumper, 2010). Tannins are also an important class of phenolic polymers with the high molecular weight and are further classified into two classes: hydrolyzable and condensed tannins. Phenolic acids, the phenols with the carboxylic group, form a varied group and consist of two widely distributed hydroxybenzoic and hydroxycinnamic acids.

Phenolic Acids

This type of phenolics comprises phenols in which one of the functional group is carboxylic acid. The naturally occurring phenolic acids contain two distinctive carbon skeletons: hydroxycinnamic and hydroxybenzoic acids. Hydroxycinnamic acid is produced as simple esters with glucose or hydroxycarboxylic acids. The structure of some of the naturally occurring phenolic acids is given in Fig. 4.1. The major function of phenolic acids is the prevention against the oxidative damage that is the main problem associated with the various degenerative diseases, like cancer, inflammation, and cardiovascular diseases. It has been found that the tumorous or leukemia cells are more prone to oxidative stress as compared with normal cells because of the higher levels of reactive oxygen species (ROS) (Mandal et al., 2010). Mostly, the phenolic acids are obtained directly from the consumed food or can be supplied indirectly by the metabolic activity of the stomach, intestine, and liver (Battisti et al., 2008).

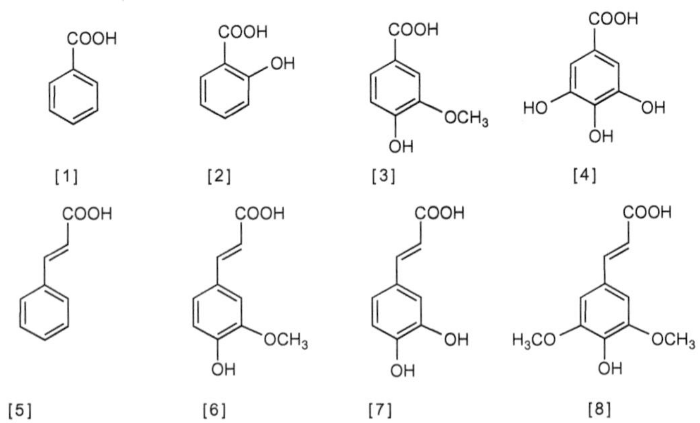

Hydroxybenzoic acid are Benzoic acid [1]. Salicylic acid [2]. Vailinilic acid [3]. Gallic acid [4] and Hydroxycinnamic aid are Cinnamic acid [5]. Ferulic acid [6]. Sinapic acid [7] and Caffeic acid [8].

FIG. 4.1 Structure of naturally occurring phenolic acids (Saxena et al., 2013).

Flavonoids

Flavonoids are polyphenolic compounds with ubiquitous occurrence and form the large group of natural products. To date, more than 8000 different flavonoids have been documented and are mostly present within the cells or surface of different organs of plant tissue. The chemical structure of flavonoids depends upon the basic skeleton C6-C3-C6. On the basis of position of benzopyrone moiety to the aromatic rings, flavonoids are divided into three classes like the flavonoids (2-phenylbenzopyrans), the isoflavonoids (3-benzopyrans), and the neoflavonoids (4-benzopyrans) (Fig. 4.2). On the basis of hydroxylation pattern and the saturation pattern of heteroatomic ring C, flavonoids are divided into various subgroups. Flavonoids are modified by hydroxylation, methoxylation, or O-glycosylation of hydroxyl groups over and above C-glycosylation directly to carbon atom of the flavonoid skeleton. Besides, alkyl groups and some other additional rings may also be covalently attached to the central skeleton of the flavonoid moiety. Further, based on the substitution pattern and the degree of saturation and oxidation present in the heterocyclic ring, flavonoids are divided into 13 subgroups and isoflavonoids into 11 subgroups. The neoflavonoids on the other hand are structurally and biogenetically closely related to the flavonoids, and the isoflavonoids comprise 4-arylcoumarins (4-aryl-2H-1-benzopyran-2-ones), 3,4-dihydro-4-arylcoumarins, and neoflavenes (Andersen and Markham, 2005; Veitch, 2010; Harborne and Williams, 1975; Hahlbrock, 1981; Harborne, 1994;

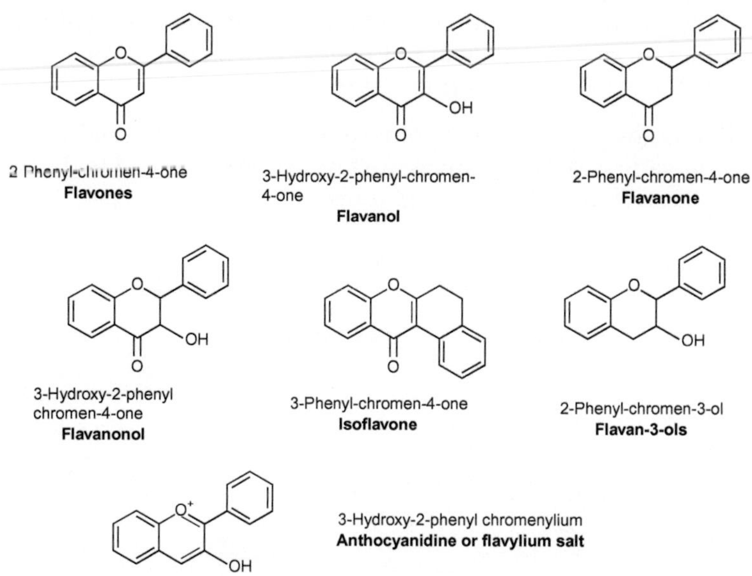

FIG. 4.2 Chemical structure of various flavonoids (Saxena et al., 2013).

Harborne and Williams, 2000; Grotewold, 2006; Veitch, 2007). Two major classes of complex plant phenolics include proanthocyanidins and bi- and triflavonoids. These compounds arise as a result of oxidative coupling of a variety of flavonoid structures and thus mostly contain carbonyl group at C-4 or its equivalent in every constituent unit. On the basis of the chemical structure and the position of hydroxyl groups of flavonoids, these compounds have the strong capacity to act as powerful antioxidants. Thus, the free radical scavenging activity of flavonoids protects the human body from free radicals and ROS. In short, flavonoids possess multiple biological properties like anti-inflammatory, enzyme inhibition, antimicrobial, estrogenic, antiallergic, antioxidant, and cytotoxic antitumor activities (Tapas et al., 2008).

Tannin

The plant tannins are heterogeneous group of high-molecular-weight phenolic compounds that have the ability to form reversible and irreversible complexes with proteins (mainly), polysaccharides (cellulose, hemicellulose, pectin, etc.), alkaloids, nucleic acids, minerals, etc. (Schofield et al., 2001; Van Soest, 1994; Mueller-Harvey and McAllan, 1992). On the basis of metabolite profile in higher plants, tannins consist of two major groups: the hydrolyzable tannins and condensed tannins. Recently, phlorotannins have been found to be the third class of tannins that have been isolated from various algae. Hydrolyzable tannins are those types of tannins that can be split into sugars (usually D-glucose) or related polyols and a phenolic acid by acids, bases, or hydrolytic enzymes (tannase). Structurally, tannins are divided into four major groups like gallotannins, ellagitannins, complex tannins, and condensed tannins (Mangan, 1988; McLeod, 1974; Mole and Waterman, 1986) (Fig. 4.3).

(1). Gallotannins are all those tannins in which a single galloyl unit or their metadepsidic derivatives are glycosidically linked to diverse polyol, catechin, or triterpenoid units.

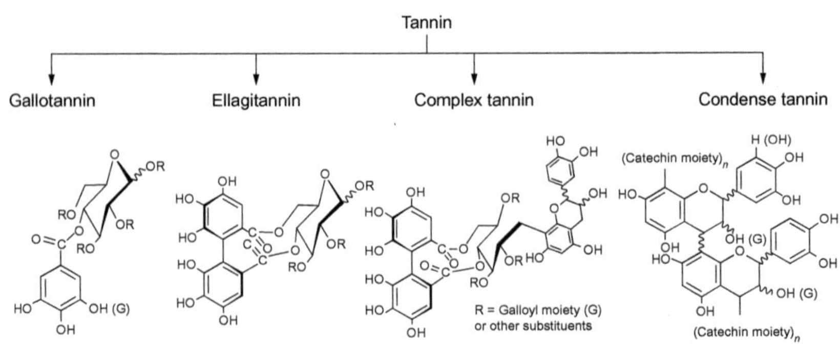

FIG. 4.3 Classification of tannins (Saxena et al., 2013).

(2). Ellagitannins are those tannins in which at least two galloyl units are C-C coupled to each other, but do not contain a glycosidically linked catechin unit.
(3). Complex tannins are tannins in which a catechin unit is bound glycosidically to a gallotannin or an ellagitannin unit.
(4). Condensed tannins include all oligomeric and polymeric proanthocyanidins that are formed by linkage of C-4 of one catechin with C-8 or C-6 of the next monomeric catechin.

Tannins are mostly found in fruits such as grapes, blueberry, tea, chocolate, legume forages, and legume trees like *Acacia* spp. and *Sesbania* spp. and in grasses, that is, sorghum, corn, etc. (Giner-Chavez, 1996). Tannins are very important, are found to be an important use in curing various types of diseases like diarrhea and stomach and duodenal tumors (De Bruyne et al., 1999), and are also used as anti-inflammatory, antiseptic, antioxidant, and hemostatic pharmaceuticals (Dolara et al., 2005). Tannins are also used in the dyestuff industry as caustics for cationic dyes (tannin dyes) and also in the production of inks (iron gallate ink). Besides, in food industry, tannins are used to clear up wine, beer, and fruit juices. Recently, the tannins have attracted the attention of pharmaceutical industries because of the potential use of tannins for the cure of deadly diseases like AIDS and various cancers (Blytt et al., 1988).

Alkaloids

Alkaloids are natural products derived from plants, bacteria, and even fungi; are basic in nature; and are composed of heterocyclic nitrogen atoms (Mueller-Harvey and McAllan, 1992). Alkaloids were perhaps the first natural products to be isolated from medicinal plants and were used as local anesthetic and stimulant, for example, cocaine (Rao et al., 1978). The peculiar nature of alkaloids is their bitter taste, and it has been found that quinine has the bitterest taste and has a significant bitter taste at the concentration of even 10^{-5} molarity (Mishra, 1989). Due to the enormous diversity of alkaloids and the complexity in molecular structure, the rational classification of alkaloids is very difficult. Thus, they have been classified into the various families on the basis of heterocyclic ring system present within the compound (Krishnan et al., 1983). Further, naming to the individual member is given based on the characteristic physiological activity and the plant from which they are derived. The classification of alkaloids on the basis of their heterocyclic ring system is enlisted as and is shown in Fig. 4.4.

Pyrrolidine alkaloids: They consist of pyrrolidine or tetrahydropyrrole ring system, for example, hygrine isolated from *Erythroxylum coca* leaves.

Pyridine alkaloids: They consist of piperidine or hexahydropyridine ring system, for example, coniine, piperine, and isopelletierine.

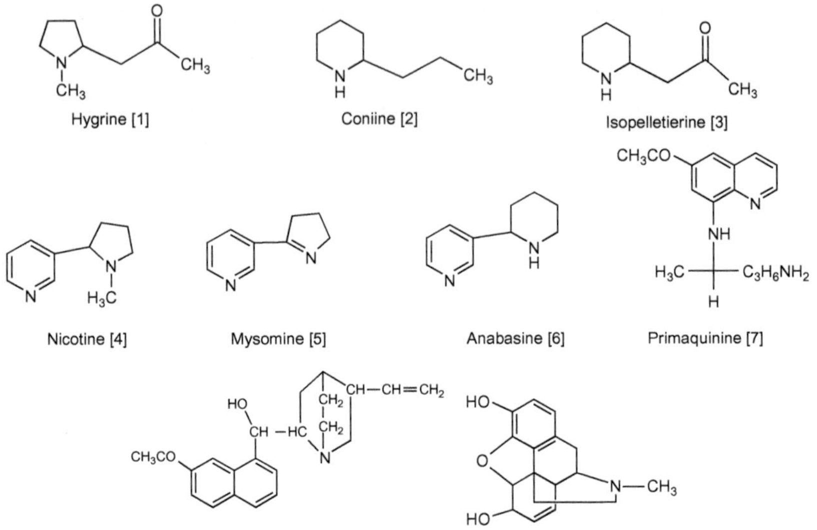

FIG. 4.4 Structure of some of the naturally found alkaloids (Saxena et al., 2013).

Pyrrolidine-pyridine alkaloids: The heterocyclic ring system present in these alkaloids is pyrrolidine-pyridine, for example, myosmine and nicotine alkaloid found in tobacco (*Nicotiana tabacum*) plant.

Pyridine-piperidine alkaloids: In this family of alkaloids, a pyridine ring system is joined to a piperidine ring system. The simplest member is anabasine alkaloid isolated from poisonous Asiatic plant *Anabasis aphylla*.

Quinoline alkaloids: These alkaloids have the basic heterocyclic ring system quinoline, for example, quinine that has been used since prehistoric times for treatment of malaria occurs in the bark of cinchona tree.

Isoquinoline alkaloids: They contain heterocyclic ring system isoquinoline, for example, opium alkaloids like narcotine, papaverine, morphine, codeine, and heroine.

Alkaloids play an important role for the protection of plants because they confirm their existence against microbes (antibacterial and antifungal activities), insects, and herbivores (feeding deterrents) and also against other plants by means of allelopathically active chemicals (Molyneux et al., 1996). Pharmacologically alkaloids are effective against various diseases like hypertension (many indole alkaloids), malaria (quinine), and cancer (dimeric indoles, vincristine, and vinblastine).

Terpenoids

Terpenoids (isoprenoids) are the largest and most diverse class of natural products among the innumerable compounds produced by plants. These are derived from the five-carbon isoprene units and consist of multicyclic structures that differ from one another on the basis of carbon skeleton and different functional groups. The terpenoids are volatile substances that generally give plants and flowers their flavor and fragrance. They are widely found even in the leaves and fruits of higher plants, conifers, citrus, and eucalyptus. They are the important constituent of essential oils and consist of isoprene unit with molecular formula (C_5H_8) as basic building block (Langenheim, 1994). The structure of some individual terpenes of some classes is given in Fig. 4.5. Depending upon the number of isoprene units, the terpenoids are classified as follows:

Hemiterpenoids: The terpenoids with a single isoprene unit are classified as hemiterpenoids. The only hemiterpene is the isoprene itself, but oxygen-containing derivatives of isoprene such as isovaleric acid and prenol are also included in this class.

Monoterpenoids: Monoterpenoids consist of two isoprene units, and their biochemical modifications such as oxidation or rearrangement generate

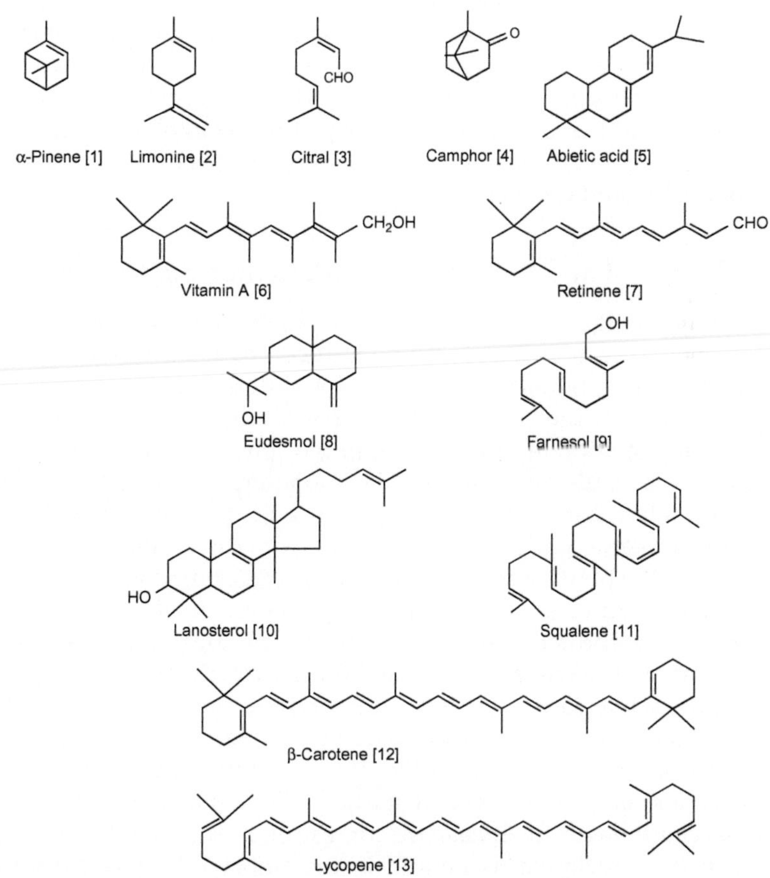

FIG. 4.5 Structure of important terpenes of each class (Saxena et al., 2013).

the related monoterpenoids. Monoterpenes further can be grouped into different types, for example, linear or acyclic and cyclic that contain rings, for example, geranyl pyrophosphate, eucalyptol, limonene, citral, camphor, and pinene.

Sesquiterpenes: Sesquiterpenes are composed of three isoprene units, for example, artemisinin, bisabolol and fernesol, oil of flowers, or cyclic compounds, such as eudesmol, found in eucalyptus oil.

Diterpenes: These are composed of four isoprene units and are derived from geranylgeranyl pyrophosphate. Some peculiar diterpenes are cembrene, taxadiene, cafestol, etc. Some of the biologically important compounds that are based on the diterpenes are retinol, retinal, and phytol.

Triterpenes: They are composed of six isoprene units, for example, lanosterol and squalene found in wheat germ and olives.

Tetraterpenoids: These terpenes are composed of eight isoprene units that may be acyclic like lycopene, monocyclic like gammacarotene, and bicyclic like alpha- and beta-carotenes.

Terpenoids are structurally the most diverse group and play an important role in plant defense as phytoalexins (McCaskill and Croteau, 1998). Terpenoids also have vital medicinal properties like anticarcinogenic (e.g., perilla alcohol and taxol), antimalarial (e.g., artemisinin), antiulcer, hepaticidal, antimicrobial, or diuretic (e.g., glycyrrhizin) activity (Langenheim, 1994; McCaskill and Croteau, 1998).

4.1.2 Temporal and Spatial Accumulation of Phytochemicals

Endogenous signals (e.g., hormones) can regulate the accumulation of phytochemicals, and it can also be induced by exogenous factors also known as environmental factors. Among different plant species, there occurs variation of the compounds of specific type, their levels of accumulation, and spatiotemporal distribution of phytochemicals as well. For instance, the volatile compounds like monoterpenes or sesquiterpenes are accumulated during the development and growth phase. They are released from aerial parts of plants and play a crucial role in the interaction between the plants and the surrounding environment (e.g., pathogen defense) (Aharoni et al., 2003). Glucosinolates in *Arabidopsis* usually accumulates in reproductive tissues (seeds and siliques) and young leaves. Roots, stems, and leaves are accounted for intermediate amounts of these compounds, while lower levels are incorporated with senescing rosette leaves (Brown et al., 2003). Flavonoids being one large class of phenolic compounds are located in most plant tissues, while single flavonoid subclasses have most specialized distribution. Three major subclasses of flavonoids accumulated in *Arabidopsis* are as follows: (1) Anthocyanins are located in the vegetative tissue and in the embryo, (2) flavonols are located in vegetative and reproductive tissues, and (3) proanthocyanidins (polymers of flavan-3-ol, PAs) are located only in the endothelium of developing seed coats. The precursors of the brick-red phlobaphene polymers, namely, flavan-4-ol 3-deoxyflavonoids apiferol and luteoferol, are located in floral tissues (e.g., pericarp and cob glumes) of maize and other monocots (Lepiniec et al., 2006). *Arabidopsis* that belongs to Brassicacea family have the presence of glucosinolates exclusively, while neither the phlobaphenes nor the isoflavones are found in *Arabidopsis*, when highlighting the taxon-specific distribution of some phytochemicals. A colorless compound, namely, isoflavones, accumulates mainly in the embryo and in the seed coat of soybean (Halkier and Gershenzon, 2006).

Plants usually synthesize and store active phytochemicals in specific organs. For instance, glandular trichomes found on the surface of leaves and stems in various plants provide vast location of phytochemical synthesis, storage, and secretion (Amme et al., 2005). Genes of many metabolic pathways and proteins associated with defense responses (e.g., superoxide dismutase in tobacco leaves) (Amme et al., 2005) or sulfur metabolism in *Arabidopsis* (Wienkoop et al., 2004) are mainly expressed within these differentiated organs. The biosynthesis of many benzylisoquinoline-derived alkaloids in opium poppy takes place in specific cells located in sieve elements, and the final products accumulate in specialized cell type, the laticifers, that accompany the vascular tissues throughout the plant (Bird et al., 2003; Weid et al., 2004). After synthesis, metabolites can be immediately used by the plant. For instance, after photosynthetic tissues receive high-energy quanta, photoinhibition results, and chloroplasts generate ROS damaging the cell at multiple levels. Anthocyanins accumulate mainly in higher plant in response to light (Chalker-Scott, 1999; Winkel-Shirley, 2002). Anthocyanin from the leaves protects the photosynthetic apparatus by reducing photoinhibition and eradicating ROS (Gould, 2004). This function can also be operated by other phytochemicals as anthocyanins are usually not present. Metabolites can also be stored inside the cell, mainly in specialized compartments until being "activated." For instance, *Arabidopsis* seed coat proanthocyanidins accumulate in the storage vacuole after their synthesis in the cytoplasm (Debeaujon et al., 2003) that are released by vacuole to perform their protective functions after the seed coat collapses (involving programmed cell death) or wounding (Pourcel et al., 2007).

Products of glucosinolate metabolism (such as isothiocyanates, oxazolidine-2-thiones, nitrile, epithionitriles, and thiocyanates), which are the reasons for many of the biological activity shown by members of this family of phytochemicals, are released after myrosinase-catalyzed hydrolysis of the respective glucosinolate (Halkier and Gershenzon, 2006). In intact plants, myrosinase and glucosinolates accumulate in distinct subcellular compartments or in distinct cells, and the glucosinolate and corresponding metabolizing enzymes come into contact after the cell damage (Grubb and Abel, 2006).

4.1.3 Environmental Regulation of Phytochemicals

It has been observed that same plant species growing under different environmental conditions usually show significant differences in the accumulation and production of the primary and secondary metabolites (Pavarini et al., 2012; Akula and Ravishankar, 2011; Gutbrodt et al., 2012; Edreva et al., 2008; Oh et al., 2009; Theis and Lerdau, 2003; Bennett and Wallsgrove, 1994; Wink, 1988). Under the influence of various environmental factors, the particular class of secondary metabolites acts as a chemical interface between the plant and its environment. Hence, the biosynthesis of secondary metabolites is mainly mediated by chemical interaction between plants and their environment that act as a plastic adaptive response to their environment. Because of this chemical interaction, there is an obvious variation in the production of plant metabolites (Szakiel et al., 2011; Miranda et al., 2015; Treutter, 2005; Winkel-Shirley, 2002). It has been found that there is enormous variation in metabolite profiling of the plants of the same species belonging to the different geographic locations (Stashenko et al., 2010; Lukas et al., 2009; Vilela et al., 2013). However, major variations in the metabolite levels have been observed in the individual plant species because of the various factors like long-term acclimation or local adaptation, environmental changes in the biotic and abiotic factors, seasonal differences related to phenology, geographic differences involving different populations (genetic differences within a plant species), or different environmental conditions of the growth location of the species individuals, especially when they have genetic homogeneity (i.e., cultivars and/or clones) (Telascrea et al., 2007; Rahimmalek et al., 2009). Altogether, it has been found that metabolomics holds the key role in studying the effect of different environmental factors on the metabolism of the plant. Thus, metabolomic profile data have been employed to compare different species of the same family, individuals of a single species growing under different environmental conditions, or variation in the metabolite profiling of individuals within the same population at different seasons (Jones et al., 2013; Kim et al., 2010; Bundy et al., 2009; Arbona et al., 2013).

The environmental factors quantitatively affect the metabolic processes of plant by affecting the development, growth rates, and separation of assimilates into vital metabolites. Besides, these factors trigger the qualitative changes in secondary metabolite production of the plants (Laughlin, 1993; Lommen et al., 2008; Perez-Estrada et al., 2000). Being sessile, plants cannot escape from the extreme environmental factors like light, temperature, and drought, and in order to adapt to the conditions to which they are exposed, plants have developed highly complex mechanisms to integrate physiology and metabolism. Thus, secondary metabolites are one of the important constituents by which plant adapts with the environment.

During the course of growth and development, plants are exposed to various biotic and abiotic factors to which they respond with an activation of their defense system.

Light/Solar Radiation

For the proper growth and development of plants, solar radiation is one of the most important environmental factors. The plants can sense extreme of light intensities and also elevated doses of UV radiations and have evolved biochemical protective mechanisms to cope with these extreme factors. Among the various radiations perceived by plants, UV-B (280–315 nm) has been sensed as an environmental stressor that promotes UV acclimation and survival in sunlight (Rozema et al., 1997). UV-B radiation damages the plants by altering the transcription and translation of genes and photosynthesis (Jansen et al., 1998). UV-B radiation has a major impact on the secondary metabolites of the plant like phenolics, terpenoids, and alkaloids (Kazan and Manners, 2011; Rozema et al., 1997). Phenylpropanoid derivatives are elevated during higher radiations of UV-B and protect the plant by selectively absorbing UV-B spectral region without having any effect on the penetration of photosynthetic radiation. Thus, the phenolics like hydroxycinnamic acids, flavonoids, and complex polymeric lignin- or tannin-like compounds play an important defensive role by absorbing UV-B radiations (Burchard et al., 2000; Kliebenstein, 2004). Besides giving protection against ultraviolet radiation, these compounds also prevent plants against pathogenic and herbivorous attack (Harborne and Williams, 2000). In view of this, solar radiation seems to have an especially profound contribution to the content of flavonoids and phenolic acids in plants.

Soil Nutrients

The soil nutrients have a great impact on the plant growth and physiology and also play an important role in stress tolerance. It has been found that alteration in resource availability to the plant results in the alteration in the availability of secondary metabolites (Coley et al., 1985). By decreasing the phosphate levels of soil, there was an increase in the levels of proanthocyanidins (Kouki and Manetas, 2002). Similarly, low iron levels stimulated the biosynthesis of phenolic compounds within the plant (Dixon and Paiva, 1995). Salidroside, a phenolic glycoside obtained from the roots of *Rhodiola sachalinensis* (Chinese medicinal herb), was highly elevated with the increase in the organic matter and nitrogen and decrease in the soil pH (Ming et al., 1988; Yan et al., 2004). Thus, the shift in the soil nutrition promptly leads to the variation in the production of both primary and secondary metabolites within the plants.

Moisture Stress

During the drought stress, there is the severe reduction in the photosynthetic rate as a result of which the growth rate is reduced. It has been found that with the reduction, the bioavailability of water and increase in temperature result in the high production of phenolic compounds in plants (Alonso-Amelot et al., 2007; Glynn et al., 2004). Thus, there is a negative correlation between water shortage and synthesis of secondary metabolites by the plants. In the medicinal plants, the levels of phenolic and saponin levels showed great variation with seasonal change (Ncube et al., 2011). In the winter season, there were higher levels of phenolic content in all the species of plants.

Temperature

Temperature stress in plants is usually known to induce the ROS-scavenging enzymes like superoxide dismutase, catalase, peroxidase, and several antioxidants. The stress due to the variation in temperature leads to a number of physiological, biochemical, and molecular changes in plant metabolism such as protein denaturation or perturbation of membrane integrity. These changes can lead to the variation in secondary metabolites in the plant tissues that are often used as an indicator of stress injury in the plant (Zobayed et al., 2005). It has been found that high temperature (35°C) resulted in the elevation of total peroxidase activity along with the increase in hypericin, pseudohypericin, and hyperforin concentrations in the shoot tissues of St. John's wort (Zobayed et al., 2005). Various studies have found an exponential increase in a variety of volatile organic compounds, with a linear increase in temperature in a range of plant species (Parker, 1977; Sharkey and Loreto, 1993; Sharkey and Yeh, 2001). On the other hand, cold stress has been shown to stimulate an increased production of phenolic compounds with subsequent incorporation into the cell wall (Christie et al., 1994). The exposure of maize seedlings and oilseed rape plants to low temperature has resulted into the increased levels of PAL, which in turn elevated the corresponding phenolic content (Christie et al., 1994; Solecka and Kacperska, 1995). In particular, levels of anthocyanins increase following cold stress (Pennycooke et al., 2005).

4.1.4 Extraction and Isolation Techniques

Separation and characterization are vital for the desired chemical components of a medicinal plant for which extraction marks the initial step in the analysis. During the preparation of extract from the plant samples, proper measures should be taken in order to maintain the active constituents, which otherwise might be distorted or lost. The crucial conditions needed to maintain the active components were prewashing, drying of plant materials or freeze drying, proper grinding to obtain a uniform sample, often improving the kinetics of analytic extraction, and increasing the contact of sample surface with the solvent system. In case the plant was elected on the basis of traditional uses (Fabricant and Farnsworth, 2001), then it is required to prepare the extract as per the traditional healer that in turn mimics the traditional "herbal" drug. The nature of bioactive compounds that are under consideration decide the selection of solvent system as various solvent systems are available to extract the bioactive compound from natural products. Polar solvents such as methanol, ethanol, or ethyl acetate are used for extraction of hydrophilic compounds, whereas dichloromethane or a mixture of dichloromethane/methanol in ratio of 1:1 is used for extraction of more lipophilic compounds. In some extractions, chlorophyll is removed using hexane (Cos et al., 2006).

The target compounds can vary from nonpolar to polar and maybe thermally labile; thus, suitability of the methods of extraction must be taken into consideration. For extraction of plant samples, wide methods, such as sonification, heating under reflux, and Soxhlet extraction, are utilized (Pharmacopeia, 2002; Zhonghua, 2000; Pharmacopoeia, 2001). Maceration or percolations of fresh green plants or dried powdered plant material in water and/or organic solvent systems are some other techniques used for preparation of extracts from plants. A brief summary of the experimental conditions for the various methods of extraction is shown in Table 4.1.

Solid-phase microextraction, supercritical-fluid extraction, pressurized-liquid extraction, microwave-assisted extraction, solid-phase extraction, and surfactant-mediated techniques are some advanced extraction

TABLE 4.1 Extraction of Plant Material by Various Types of Experimental Methods Used

	Soxhlet extraction	Sonification	Maceration
Common solvents used	Methanol, ethanol, or a mixture of alcohol and water	Methanol, ethanol, or a mixture of alcohol and water	Methanol, ethanol, or a mixture of alcohol and water
Temperature (°C)	Depending on solvent used	Can be heated	Room temperature
Pressure applied	Not applicable	Not applicable	Not applicable
Time required	3–18 h	1 h	3–4 days
Volume of solvent required (mL)	150–200	50–100	Depending on the sample size
References	Zygmunt and Namieśnik (2003) and Huie (2002)	Zygmunt and Namieśnik (2003) and Huie (2002)	Phrompittayarat et al. (2007), Cunha et al. (2004), and Woisky and Salatino (1998)

techniques that include many merits, which mainly include reduction in organic solvent consumption and in sample degradation, elimination of additional sample cleanup and concentration steps before chromatographic analysis, improvement in extraction efficiency, selectivity, and/kinetics of extraction. The easy availability and ready-to-use features of these techniques also lead their usage to the extraction of plants materials (Huie, 2002).

Identification and Characterization

Separation of bioactive compounds remains a crucial step for identification and characterization as the plant extracts mainly occur as a mixture of various types of bioactive compounds or phytochemicals with different polarities. A wide number of separation techniques such as thin-layer chromatography (TLC), column chromatography, flash chromatography, Sephadex chromatography, and high-performance liquid chromatography (HPLC) must be utilized for pure extraction of these bioactive compounds. These pure compounds are lined for the determination of structure and biological activity. Immunoassays, which use monoclonal antibodies (MAbs), phytochemical screening assay, and Fourier-transform infrared spectroscopy (FTIR), are some nonchromatographic techniques that can be utilized to acquire and initiate the identification of the bioactive compounds.

Chromatographic Techniques

Thin-Layer Chromatography and Bio-Autographic Methods

In order to investigate the number of components in a mixture, TLC provides an easy and fast procedure to a researcher. TLC also supports the identity of a compound in a mixture when the R_f of a compound is compared with the R_f of a known compound. Phytochemical screening reagents are sprayed causing change in color according to existing phytochemical in the extract, or visualizing under UV comes under additional tests. Purity and identity of isolated compounds are also confirmed through it.

Bioautography is a resourceful technique to determine bioactive compound with antimicrobial activity from plant extract. TLC bioautographic technique combines chromatographic separation and in situ activity determination initiating the localization and target-directed isolation of active constituents in a mixture. Traditionally, bioautographic technique has utilized the growth inhibition of microorganisms to detect antimicrobial components of extracts chromatographed on a TLC layer. This technique has been deemed as the most efficient assay for the detection of antimicrobial compounds (Shahverdi et al., 2007). Bioautography localizes antimicrobial activity on a chromatogram using three approaches: (i) direct bioautography, in which the microorganism grows directly on the TLC plate; (ii) contact bioautography, in which the antimicrobial compounds are transferred from the TLC plate to an inoculated agar plate through direct contact; and (iii) agar overlay bioautography, in which a seeded agar medium is applied directly onto the TLC plate (Hamburger and Cordell, 1987; Rahalison et al., 1991). The inhibition zones created on TLC plates by one of the above bioautographic techniques will be used to visualize the position of the bioactive compound with antimicrobial activity in the TLC fingerprint with reference to R_f values (Homans and Fuchs, 1970). Preparative TLC plates with a thickness of 1 mm were prepared using the same stationary and mobile phases as above, with the goal of isolating the bioactive components that exhibited the antimicrobial activity against the test strain. These areas were scraped from the plates and the substance eluted from the silica with ethanol or methanol. Eluted samples were then purified utilizing the above preparative chromatography method. Finally, the components were identified by HPLC, LCMS, and GCMS. It may have high sensitivity; its applicability is limited to microorganisms that easily grow on TLC plates. Other problems encountered are the need for complete removal of residual low-volatile solvents, such as *n*-BuOH, trifluoroacetic acid, and ammonia, and the transfer of the active compounds from the stationary phase into the agar layer by diffusion (Cos et al., 2006). As bioautography allows localizing antimicrobial activities of an extract on the chromatogram, it supports a fast search for new antimicrobial agents through bioassay-guided isolation (Cos et al., 2006).

The bioautography agar overlay method is meritorious in that, initially, it utilizes very little quantity of sample when compared with the normal disc diffusion method, and therefore, it can be utilized for bioassay-guided isolation of compounds. Secondly, as the crude extract is resolved into its different components, this technique simplifies the process of identification and isolation of the bioactive compounds (Rahalison et al., 1991).

High-Performance Liquid Chromatography

HPLC is a flexible, powerful, and commonly used technique for the isolation of natural products (Cannell, 1998). Presently, this technique is attaining fame among various analytic techniques as the initial choice for fingerprinting study for the quality control of herbal plants. Natural products are periodically isolated, followed by the evaluation of a relatively crude extract in a biological assay in order to fully characterize the active component. The biologically active component is usually present in minor quantity in the extract, and the resolving power of HPLC is best suited to the fast processing of such multicomponent samples on both analytic and preparative scale. Various bench top HPLC instruments are now modular in design and comprise a solvent delivery pump, a sample introduction device such as autosampler or manual injection valve, an analytic column, a guard column, a detector, and a recorder or a printer.

Chemical separations can be achieved using HPLC by using the fact that certain compounds have different migration rates given a particular column and mobile phase. The choice of stationary phase and mobile phase is usually determined by the extent or degree of separation. Basically, the identification and separation of phytochemicals can be achieved using isocratic system (using single unchanging mobile phase system). Gradient elution where the proportion of organic solvent to water is changed with time may be desirable in the case of studying more than one sample component and differ from each other significantly in retention under the conditions employed.

HPLC is defined as a technique where the compound of interest is purified via separation or extraction of the target compound from other (possibly structurally related) compounds or contaminants. Each compound must have a characteristic peak under certain chromatographic conditions. To get an optimum separation, a chromatographer chooses the conditions such as the proper mobile phase, flow rate, suitable detectors and columns depending on what needs to be separated and how closely related the samples are.

Identification of compounds by HPLC is a vital part of any HPLC assay. In order to identify any compound by HPLC, a detector selection is the initial priority. A separation assay develops after the selection of detector and the implementation of the optimal detection settings. The parameters of this assay should display a clean peak of the known sample from the chromatograph. The identifying peak should have a reasonable retention time and should be well separated from extraneous peaks at the detection levels. UV detectors are mostly utilized among all the detectors because they offer high sensitivity and also because maximum of naturally occurring compounds have some UV absorbance at low wavelengths (190–210 nm) (Cannell, 1998). The high sensitivity of UV detection is a plus point when a compound of interest is only present in small amounts within the sample. Besides UV, other detection methods are also being employed to detect phytochemicals, among which is the diode-array detector (DAD) coupled with mass spectrometry (MS) (Tsao and Deng, 2004). Liquid chromatography coupled with mass spectrometry (LC/MS) is also a powerful technique for the analysis of complex botanical extracts (Cai et al., 2002; Heinrich, 2000).

Tandem mass spectrometry (MS^n) is applied to provide abundant information for structural elucidation of the compounds. Therefore, the combination of HPLC and MS initiates fast and accurate identification of chemical compounds in medicinal herbs, especially when a pure standard is unavailable (Ye et al., 2007).

The fate of natural product isolation widely depends on the choice of solvent for sample reconstitution and the processing of a crude source material to provide a sample suitable for HPLC analysis. The source material, for example, dried powdered plant, firstly needs to be treated in such a way as to ensure that the compound of interest is efficiently liberated into solution. In the case of dried plant material, while an organic solvent (e.g., methanol and chloroform) may be utilized as the initial extractant followed by a period of maceration, solid material is then removed by decanting off the extract by filtration. The filtrate is then concentrated and injected into HPLC for separation. In the analysis of crude extract, the usage of guard columns is necessary. Guard columns will significantly ensure the protection of analytic columns as many natural product materials contain high levels of strongly binding components, such as chlorophyll and other endogenous materials that may in the long term harm and thus reduce the performance of analytic columns.

Nonchromatographic Techniques

Immunoassay

Immunoassays are nowadays an important tool in bioactive compound analyses. This technique uses monoclonal antibodies against drugs and low molecular weight natural bioactive compounds. This method is highly sensitive and specific for receptor binding analyses, enzyme assays, and qualitative and quantitative analytic procedures. Enzyme-linked immunosorbent assay (ELISA) based on MAbs is at times highly more sensitive than the conventional HPLC methods. Hybridoma technology, a specialized technique, is used to produce monoclonal antibodies (Shoyama et al., 1999). The following steps are involved in the production of monoclonal antibodies via hybridoma technology against plant drugs:

i. Specific antibodies are produced by repeatedly immunizing a rabbit with a specific plant drug, facilitated due to proliferation of the desired B cells.
ii. This in turn creates tumors in the rabbit.
iii. Spleen cells (these cells are rich in B cells and T cells) are cultured separately. The separately cultured spleen cells produce specific antibodies against the plant drug and against myeloma cells that produce tumors.
iv. Polyethylene glycol (PEG) is used for fusion of spleen cells to myeloma cells for production of hybridoma. The hybrid cells are grown in selective hypoxanthine-aminopterin-thymidine (HAT) medium.
v. The desired hybridoma is then selected for cloning and antibody production against a plant drug. This process is facilitated by preparing single cell colonies that will grow and can be used for screening of antibody producing hybridomas.
vi. The selected hybridoma cells are then cultured for the production of monoclonal antibodies on large scale against the specific plant drugs.
vii. The monoclonal antibodies are used to determine similar drugs in the plant extract mixture through enzyme-linked immunosorbent essay (ELISA).

Phytochemical Screening Assay

Chemical compounds obtained from plants are referred to as phytochemicals, and they encompass the plethora of plant secondary metabolites. For determining the phytochemicals in a mixture or for bioactive compound analyses, phytochemical screening assay is a basic, prompt, and cost-effective tool. A basic protocol for different phytochemical screening methods for detecting the secondary metabolites is shown in Table 4.2. The crude extract or active fraction from plant material is subjected to different phytochemical screening tests as shown in the table given.

Fourier-Transform Infrared Spectroscopy

FTIR has been considered to be a vital tool for the characterization and identification of compounds or functional groups (chemical bonds) present in an unknown mixture of plant extract (Hazra et al., 2007). What is more, FTIR spectra of pure compounds are basically so unique that they are like a molecular "fingerprint." In case of common plant compounds, the spectrum of an unknown compound can be identified by comparison with a library of known compounds. Various ways are used to prepare the samples for FTIR. One drop of sample placed between two plates of sodium chloride is the easy way for liquid samples. The drop forms a thin film between the plates. Potassium bromide (KBr) is used in case solid samples that are milled to it and then compressed into a thin pellet for analysis. Nevertheless, methylene chloride can be used to dissolve solid samples, and the solution is then placed onto a single salt plate. The solvent is then evaporated off, leaving a thin film of the original material on the plate.

Various chromatographic techniques and purification methods are utilized to isolate bioactive compounds from multicomponent mixture since bioactive compounds occur as such (Fig. 4.6).

4.1.5 Characterisation of Phytochemicals

Analytes ranging from large protein molecules to small inorganic ions can be separated through capillary electrophoresis (CE) and other related techniques that have rapidly developed from the last two decades.

Gas chromatography when coupled with mass spectrometry has been deemed as a powerful tool due to its sensitivity and selectivity. Moreover, liquid chromatography is the commonly utilized technique to separate and analyze phenolic compounds in plant and tissue samples. Liquid chromatography is carried out in columns. The regular columns are packed using reversed-phase C18-bonded silica gel as stationary phase. Elution systems are mainly binary, with one of the solvents being an acidified aqueous solvent. The second is an organic solvent (e.g., methanol), acidified with the same acid used in the aqueous solvent (Ibern-Gomez et al., 2002). Commonly, it covers gradient elution and, occasionally, isocratic elution.

TABLE 4.2 A Brief Summary of Phytochemical Screening of Secondary Metabolites (Sasidharan et al., 2011)

Secondary metabolite	Name of test	Methodology	Result(s)	Reference(s)
(1) Alkaloid	Dragendorff's test	Spot a drop of extract on a small piece of precoated TLC plate. Spray the plate with Dragendorff's reagent	Orange spot	Kumar et al. (2007)
	Wagner test	Add 2 mL filtrate with 1% HCl + steam. Then, add 1 mL of the solution with six drops of Wagner's reagent	Brownish-red precipitate	Chanda et al. (2006)
	TLC method 1	Solvent system: chloroform: methanol: 25% ammonia (8:2:0.5). Spots can be detected after spraying with Dragendorff's reagent	Orange spot	Tona et al. (1998)
	TLC method 2	Wet the powdered test samples with a half diluted NH_4OH and lixiviated with EtOAc for 24 h at room temperature. Separate the organic phase from the acidified filtrate and basify with NH_4OH (pH 11–12). Then, extract it with chloroform (3X), condense by evaporation, and use for chromatography Separate the alkaloid spots using the solvent mixture chloroform and methanol (15:1). Spray the spots with Dragendorff's reagent	Orange spot	Mallikharjuna et al. (2007)
(2) Anthraquinone	Borntrager's test	Heat about 50 mg of extract with 1 mL 10% ferric chloride solution and 1 mL of concentrated hydrochloric acid. Cool the extract and filter. Shake the filtrate with equal amount of diethyl ether. Further, extract the ether extract with strong ammonia	Pink or deep red coloration of aqueous layer	Kumar et al. (2007)
	Borntrager's test	Add 1 mL of dilute (10%) ammonia to 2 mL of chloroform extract	A pink-red color in the ammoniacal (lower) layer	Onwukaeme et al. (2007)

(Continued)

TABLE 4.2 A Brief Summary of Phytochemical Screening of Secondary Metabolites (Sasidharan et al., 2011)—cont'd

Secondary metabolite	Name of test	Methodology	Result(s)	Reference(s)
(3) Cardiac glycosides	Keller-Kiliani test	Add 2 mL filtrate with 1 mL of glacial acetic acid, 1 mL ferric chloride, and 1 mL concentrated sulfuric acid	Green-blue coloration of solution	Parekh and Chanda (2007)
	Keller-Kiliani test	Dissolve 50 mg of methanolic extract in 2 mL of chloroform. Add H_2SO_4 to form a layer	Brown ring at interphase	Onwukaeme et al. (2007)
	TLC method	Extract the powdered test samples with 70% EtOH on rotary shaker (180 thaws/min) for 10 h. Add 70% lead acetate to the filtrate and centrifuge at 5000 rpm/10 min. Further, centrifuge the supernatant by adding 6.3% Na_2CO_3 at 10000 rpm/10 min. Dry the retained supernatant, redissolve in chloroform, and use for chromatography. Separate the glycosides using EtOAc-MeOH-H_2O (80:10:10) solvent mixture	The color and hR_f values of these spots can be recorded under ultraviolet (UV 254 nm) light	Mallikharjuna et al. (2007)
(4) Flavonoid	Shinoda test	To 2–3 mL of methanolic extract, add a piece of magnesium ribbon and 1 mL of concentrated hydrochloric acid	Pink-red or red coloration of the solution	Kumar et al. (2007)
	TLC method	Extract 1 g powdered test samples with 10 mL methanol on water bath (60°C/5 min). Condense the filtrate by evaporation, add a mixture of water and EtOAc (10:1 mL), and mix thoroughly. Retain the EtOAc phase and use for chromatography. Separate the flavonoid spots using chloroform and methanol (19:1) solvent mixture	The color and hR_f values of these spots can be recorded under ultraviolet (UV 254nm) light	Mallikharjuna et al. (2007)
	NaOH test	Treat the extract with dilute NaOH, followed by addition of dilute HCl	A yellow solution with NaOH turns colorless with dilute HCl	Onwukaeme et al. (2007)

(5) Phenol	Phenol test	Spot the extract on a filter paper. Add a drop of phosphomolybdic acid reagent and expose to ammonia vapors.	Blue coloration of the spot	Kumar et al. (2007)
(6) Phlobatannin		2 mL extract was boiled with 2 mL of 1% hydrochloric acid HCl	Formation of red precipitates	Edeoga et al. (2005)
(7) Pyrrolizidine alkaloid		Prepare 1 mL of oxidizing agent, consisting of 0.01 mL hydrogen peroxide (30% w/v) stabilized with tetrasodium pyrophosphate (20 mg/mL) and made up of 20 mL with isoamyl acetate, and add to 1 mL of plant extract. Vortex the sample and add 0.25 mL acetic anhydride before heating the sample at 60°C for 50–70 s. Cool the samples to room temperature. Add 1 mL of Ehrlich reagent and place the test tubes in water bath (60°C) for 5 min. Measure the absorbance at 562 nm. The method of Holstege et al. (1995) should be used to confirm results of the screening method	Peaks were compared with the GC-MS library	McGaw et al. (2007), Mattocks (1967), and Holstege et al. (1995)
(8) Reducing sugar	Fehling test	Add 25 mL of diluted sulfuric acid (H_2SO_4) to 5 mL of water extract in a test tube and boil for 15 min. Then, cool it and neutralize with 10% sodium hydroxide to pH 7 and 5 mL of Fehling solution	Brick-red precipitate	Akinyemi et al. (2005)
(9) Saponin	Frothing test/Foam test	Add 0.5 mL of filtrate with 5 mL of distilled water and shake well	Persistence of frothing	Parekh and Chanda (2007)
	TLC method	Extract 2 g of powdered test samples with 10 mL 70% EtOH by refluxing for 10 min. Condense the filtrate, enrich with saturated *n*-BuOH, and mix thoroughly. Retain the butanol, condense, and use for chromatography. Separate the saponins using chloroform, glacial acetic acid, methanol, and water (64:34:12:8) solvent mixture. Expose the chromatogram to the iodine vapors	The color (yellow) and hR_f values of these spots were recorded by exposing chromatogram to the iodine vapors	Mallikharjuna et al. (2007)

(Continued)

TABLE 4.2 A Brief Summary of Phytochemical Screening of Secondary Metabolites (Sasidharan et al., 2011)—cont'd

Secondary metabolite	Name of test	Methodology	Result(s)	Reference(s)
(10) Steroid	Liebermann-Burchard test	To 1 mL of methanolic extract, add 1 mL of chloroform, 2–3 mL of acetic anhydride, and 1–2 drops of concentrated sulfuric acid	Dark green coloration	Kumar et al. (2007)
		To 1 mL of extract, add 2 mL acetic anhydride and 2 mL concentrated sulfuric acid H_2SO_4	Color change to blue or green	Edeoga et al. (2005)
	TLC method	Extract 2 g of powdered test samples with 10 mL methanol in water bath (80°C/15 min). Use the condensed filtrate for chromatography. The sterols can be separated using chloroform, glacial acetic acid, methanol, and water (64:34:12:8) solvent mixture. The color and hR_f values of these spots can be recorded under visible light after spraying the plates with an aldehyde sulfuric acid reagent and heating (100°C/6 min)	The color (greenish black to pinkish black) and hR_f values of these spots can be recorded under visible light	Mallikharjuna et al. (2007)
(11) Tannin	Braemer's test	10% alcoholic ferric chloride will be added to 2–3 mL of methanolic extract (1:1)	Dark blue or greenish grey coloration of the solution	Kumar et al. (2007) and Parekh and Chanda (2007)
(12) Terpenoid	Liebermann-Burchard test	To 1 mL of methanolic extract, add 1 mL of chloroform, 2–3 mL of acetic anhydride, and 1–2 drops of concentrated sulfuric acid	Pink or red coloration	Kumar et al. (2007)
	Salkowski test	5 mL extract was added with 2 mL of chloroform and 3 mL of concentrated sulfuric acid H2SO4	Reddish brown color of interface	Edeoga et al. (2005)
(13) Volatile oil		Add 2 mL extract with 0.1 mL dilute NaOH and small quantity of dilute HCl. Shake the solution	Formation of white precipitates	Dahiru et al. (2006)

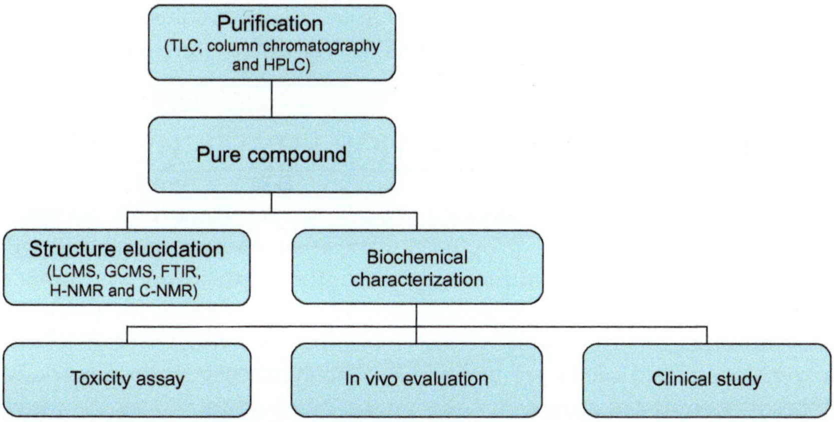

FIG. 4.6 The general approach involved in extraction, isolation, and characterization of bioactive compound from plants extract (Sasidharan et al., 2011).

pH in chromatography can harshly affect the separation of phenolic compounds so could buffer.

Mass Spectrometry Methods for Identification and Quantification

Mass spectrometry has proved to be of vital importance as a technique in identification and quantification of phenolics in fruit and vegetables. Scientific community has cited sensitivity and specificity as factors for explaining the acceptance of this method. Moreover, this technique might readily combine with different separation techniques such as CE, gas chromatography (GC), and liquid chromatography (LC), including HPLC and ultraperformance liquid chromatography (UPLC). Mass spectrometry applied to characterization of phenolic compounds has been widely reviewed (Fulcrand et al., 2008; Harnly et al., 2007; De Rijke et al., 2006; Prasain et al., 2004).

Other Methods for Identification and Quantification

Nuclear magnetic resonance (NMR) spectroscopy is a technique that plays a vital role in online identification of components of crude plant extracts (Wolfender et al., 2003). This individual technique does not provide enough spectroscopic information for a complete identification of natural products, and therefore, other techniques, such as LC-UV-DAD and LC-MS/MS, are needed for providing sufficient information. Furthermore, LC-NMR experiments are time-consuming and have to be performed on the LC peak of interest, identified by prescreening with LC-UV-MS. NMR applied to phenolic compounds includes 1H NMR, 13C NMR, correlation spectroscopy (COSY), heteronuclear chemical shift correlation NMR (C-H HECTOR), nuclear Overhauser effect in the laboratory frame (NOESY), rotating frame of reference (ROESY), total correlation spectroscopy (TOCSY) (Escribano-Bailon and

Santos-Buelga, 2003), heteronuclear multiple-quantum coherence (HMQC), and heteronuclear multiple-bond correlation (HMBC).

4.2 BIOSYNTHESIS OF MAJOR PHYTOCHEMICALS

The characterization of phytochemicals is very important because of the vital pharmacological properties. These secondary metabolites that are utmost exploited by humans are actually synthesized by plants for their own survival like defense compounds or as signals for pollinators or symbionts. All the plant organs including the roots, shoots, leaves, flowers, fruit, and seeds are utilized for the synthesis of these secondary metabolites and are stored in the specific compartments, which may be either whole organs or specialized cell types (Gutzeit and Ludwig-Müller, 2014). Because of the toxic nature of these secondary metabolites at high concentrations that provide an efficient defense against herbivores, these are compartmentalized in special organs. For example, the glandular trichomes are often rich in toxic compounds, and some pathways—an example being the synthesis of the artemisinin in *Artemisia annua* and labdane diterpene Zabienolin tobacco—occur exclusively in these trichomes (Maes et al., 2011). Similarly, the glandular trichomes of peppermint leave also contain high levels of aromatic oils (Turner et al., 2000).

The biosynthetic pathway of these phytochemicals is derived from several precursors of primary metabolism (Fig. 4.1). The precursor of a biosynthetic enzyme is defined as a molecule that is used as a substrate and converted to a product. The product can be an intermediate or final product in the pathway and in some cases can be used as precursor for the next biosynthetic enzyme.

$$\text{Substrate}(\text{precursor}) \rightarrow \text{Intermediate} \rightarrow \text{Intermediate} \rightarrow \text{Product}$$

$$\updownarrow$$

$$\text{Precursor} \rightarrow \text{Intermediate} \rightarrow \text{Product}$$

Thus, in the complex biosynthetic pathway, many junction points that act as an intermediate and are simultaneously used as a precursor in another pathway are found. For example, shikimic acid acts as an important intermediate for the amino acid metabolism and also acts as a precursor for the biosynthesis of aromatic secondary metabolites. In a similar manner, it has been found that the precursors that are used for the biosynthesis of one class of compounds may also act as precursors for a range of different metabolites too. From Fig. 4.7, it can be seen that three major classes of compounds act as major precursors for the biosynthesis of major metabolites. Hence, maximum precursors are derived from protein (amino acids), carbohydrate (sugars), and lipid (fatty acid) metabolism. It has been found that biosynthetic pathway for aromatic amino acids is one of the major sources for the synthesis of aromatic compounds such as phenols, flavonoids, and some alkaloids.

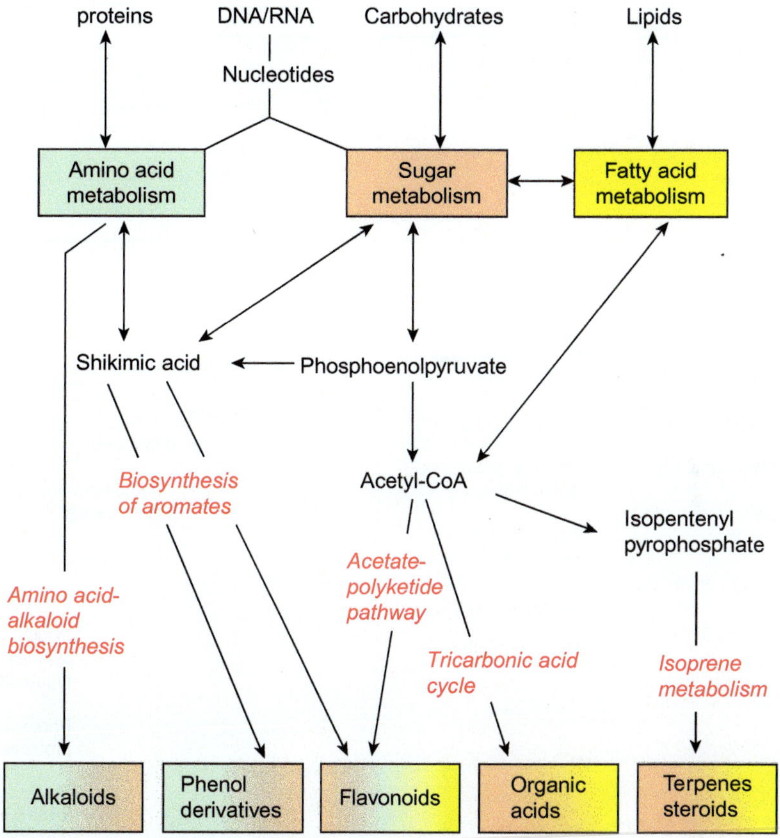

FIG. 4.7 General scheme of biosynthetic pathways and precursors for the major classes of secondary metabolites (Gutzeit and Ludwig-Müller, 2014).

Similarly, acetyl CoA, the central metabolite formed by glycolysis and β-oxidation of fatty acids and also utilized in the tricarboxylic acid cycle for synthesis of organic acids, also acts as precursor for secondary metabolites. Terpenes, one of the vital classes of secondary metabolites, are also derived from acetyl CoA. Thus, the detailed information regarding the biosynthetic pathways of these phytochemicals is essential for the targeted manipulation of these pathways in biotechnology. Besides the information regarding the involvement of genes and enzymes in the biosynthesis of these pathways, transcriptional regulation of these genes is equally important.

4.2.1 Terpenes

Terpenes can be classified as either essential or nonessential compounds depending upon the importance of these metabolites for the survival of plant. Essential terpenes play an important role in the basic and vital metabolic activities of plant, whereas nonessential

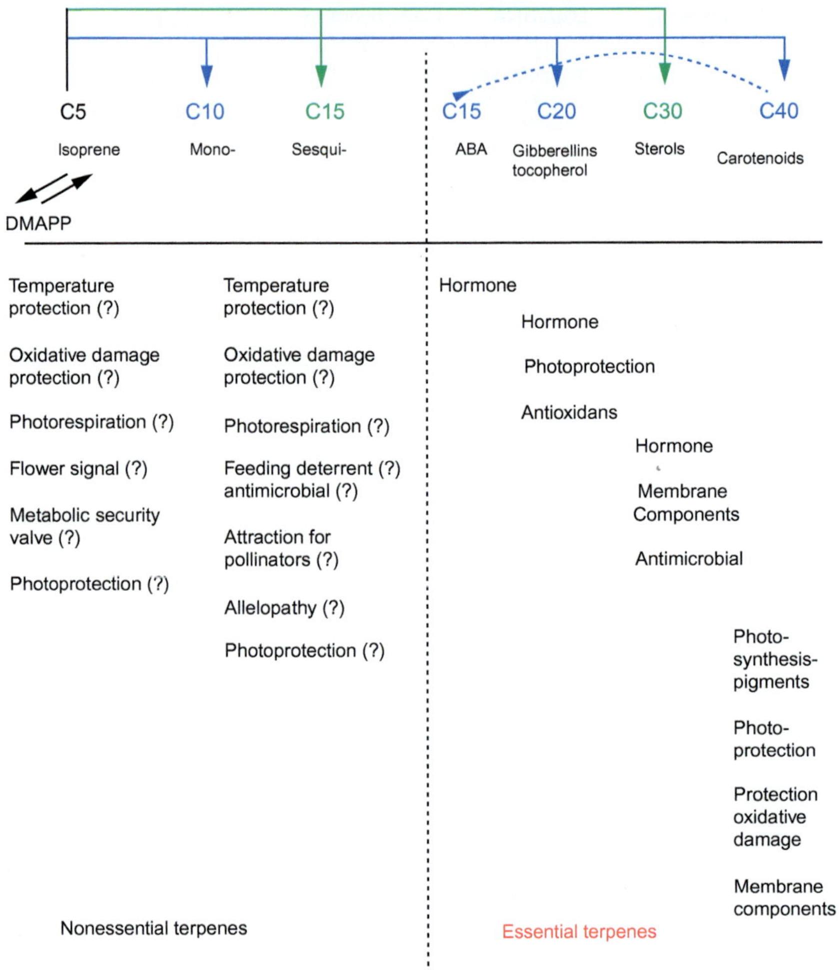

FIG. 4.8 Classification of terpenes from plants into essential and nonessential compounds (Gutzeit and Ludwig-Müller, 2014).

compounds are advantageous to the plant when produced under stress conditions; however, under normal conditions, these terpenes are not fatal for the plant and do not display any phenotype if the biosynthetic pathway is mutated (Fig. 4.8). For example, essential terpenes include carotenoids, which play an important role in photosynthesis as components of light-harvesting complexes and also as protective compounds against high light intensities; tocopherol also is an important terpenoid that has antioxidative role. Nonessential terpenes include many antimicrobial substances such as the monoterpenes (Gutzeit and Ludwig-Müller, 2014).

The biosynthesis of terpenes occurs through two different pathways:

(i) The mevalonate (MVA) pathway
(ii) The methylerythritol phosphate (MEP) pathway

Within the plant, both pathways can operate simultaneously, whereas in some organisms, only one pathway is involved for the biosynthesis of terpenes. It has been found that biosynthesis via MVA pathway takes place in cytosol and MEP pathway occurs in the plastids. The synthesis of mono-, di-, and tetraterpenes takes place in plastids; on the other hand, sesquiterpenes and triterpenes are usually synthesized in the cytosol (Gutzeit and Ludwig-Müller, 2014).The oxidation reaction involved in the biosynthesis of these terpenes takes place in ER where the cytochrome P450-dependent monooxygenases are present. The basic building blocks of terpene biosynthetic pathway, isopentenyl diphosphate (IPP), and dimethylallyl diphosphate (DMAPP) are produced by both the pathways that are ultimately used for the production of high-molecular-weight (complex) terpenes by head-to-tail addition by the action of enzyme prenyltransferase (Fig. 4.9). The key enzymes involved for the formation of mevalonic acid in the MVA pathway are hydroxymethylglutaryl CoA synthase and reductase. Similarly, enzymes

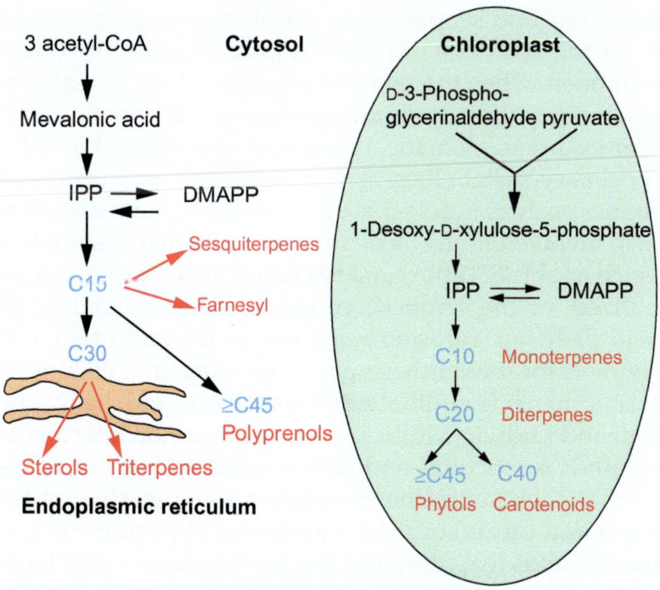

FIG. 4.9 Schematic view of terpene biosynthesis. The biosynthesis of terpenes takes place in two different organelles, the cytosol and the chloroplast, using different pathways. The mevalonate (MVA) pathway in the cytosol and MEP (methylerythritol phosphate) in plastids. The endoplasmic reticulum hosts cytochrome P450-dependent monooxygenases, which catalyze various oxidation reactions (Gutzeit and Ludwig-Müller, 2014).

involved for the synthesis of 2-C-methyl-D-erythritol-4-phosphate of the MEP pathway are 1-deoxy-D-xylulose-5-phosphate synthase and reductoisomerase. The future steps for both pathways are then catalyzed by terpene synthases that are encoded by multigene families (Tholl, 2006) and thus are able to synthesize various longer-chain terpenes from diphosphate molecules (Gutzeit and Ludwig-Müller, 2014).

4.2.2 Alkaloids

Biological precursors of most alkaloids are proteinogenic or nonproteinogenic amino acids, like lysine, phenylalanine, tyrosine, tryptophan, histidine, aspartic acid, ornithine, and anthranilic acid. Ornithine is an important precursor for various alkaloids and anthranilic acid, from which tryptophan is synthesized and is the precursor for acridine alkaloids. The shikimic acid pathway that is responsible for the biosynthesis of all aromatic amino acids is mostly involved in the biosynthesis of alkaloids. It has been found that for the biosynthesis of some alkaloids, more than one precursor is required. For example, in the biosynthesis of terpene indole alkaloids, tryptophan and monoterpenes act as a precursor for the indole moiety and for the synthesis of side chain, respectively (Gutzeit and Ludwig-Müller, 2014). Similarly, for the biosynthesis of tropane alkaloids hyoscyamine and scopolamine, ornithine and phenylalanine act as precursors for the synthesis of different parts of molecule (Fig. 4.10). As already mentioned, when the two pathways strive for the same precursor, the changes in one pathway should not hamper the operation of another pathway. For example, aromatic amino acids are utilized in the synthesis of various primary metabolites like proteins, cell wall components, and plant hormones such as indole-3-acetic acid (IAA). Similarly, these amino acids are also involved in the biosynthesis of phenolic alkaloids and plant pigments such as anthocyanins and betalains. Tryptophan and anthranilic acid are utilized for the synthesis of indole alkaloids and acridones via shikimic acid pathway. Tyrosine being one of the important amino acids acts as precursors for the synthesis of a large number of alkaloid types like protoberberine, benzylisoquinoline, morphinan, and benzophenanthridine (Gutzeit and Ludwig-Müller, 2014). The common intermediate for all these compounds is reticuline and is thus involved in the synthesis of isoquinoline, benzophenanthridine, morphines, sanguinarine, and berberine. Ornithine acts as a precursor for the synthesis of pyridine (e.g., nicotine) and tropane alkaloids (e.g., cocaine); the intermediate in this loop is tropolone with polyamine putrescine acting as common precursor. For the biosynthesis of hyoscyamine and scopolamine, ornithine along with amino acid phenylalanine is required for the formation of backbone molecule littorine. Also, for the synthesis of group of pyrrolizidine alkaloids, putrescine formed from the amino acids is converted into homospermidine and then metabolized to retronecine, the common precursor for pyrrolizidine

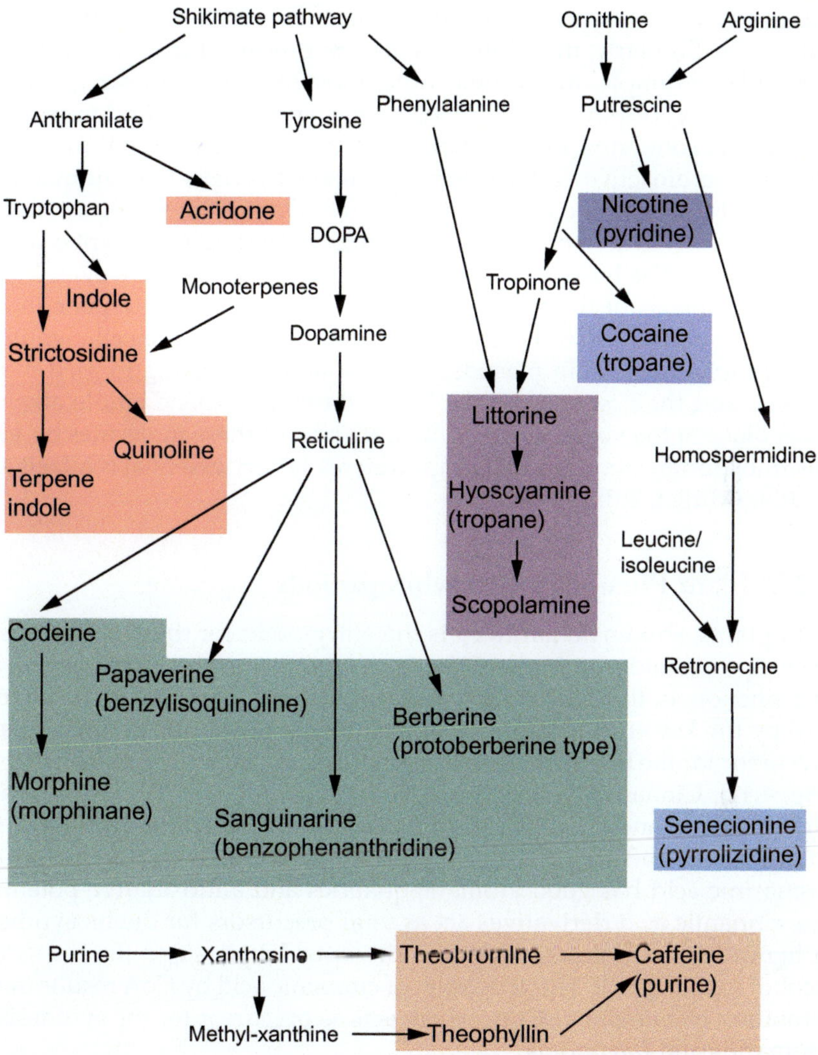

FIG. 4.10 Biosynthetic pathway of some of the alkaloids. The biosynthesis of different alkaloids by shikimate pathway. The different colors represent the different sources of origin (red, anthranilate/tryptophan, and green, tyrosine). In addition, ornithine and arginine via putrescine result in the synthesis of pyridine (dark blue), tropane (blue), and pyrrolizidine alkaloids (light blue). Phenylalanine together with ornithine is needed for the synthesis of a second group of tropane alkaloids (violet). Caffeine and related substances is derived from purine (brown) (Gutzeit and Ludwig-Müller, 2014).

alkaloids (Ober and Hartmann, 1999). Structurally, pyrrolizidine alkaloids are similar to the quinolizidine alkaloids and are synthesized from the amino acid lysine via polyamine intermediate cadaverine synthesized in the chloroplasts of mesophyll cells. Besides this, a large number of metabolites can be synthesized from different alternatives, like nicotine can be manufactured either from ornithine alone or by a fusion of ornithine with putrescine. Similarly, morphine can also be produced from two different routes; For example, in the first route, morphine is synthesized directly from thebaine, oripavine, and morphinone, and in the second route, first, thebaine is converted to neopinone and is then transformed into codeinone to the bioactive codeine (remedy against cough) that ultimately is converted to morphine (Novak et al., 2000). One of the important indole alkaloids such as serotonin is derived from tryptophan via tryptamine as intermediate. The biosynthesis of some of the alkaloids takes place in different organelles of the cell; for example, biosynthesis of terpene indole alkaloids takes place in the different compartments of the cell. Synthesis of tryptophan occurs in plastids, and tryptamine is being formed in the cytosol, and the first step for the biosynthesis of terpene indole alkaloid takes place in the vacuole. Thus, the transport of these molecules by various transporters between the compartments is very essential (Gutzeit and Ludwig-Müller, 2014).

4.2.3 Plant Phenolics: Phenylpropaniods

In plants, shikimate pathway is the entry point for the biosynthesis of phenylpropanoids, and one of the representative groups for these bioactive phenols is flavonoids. Phenylalanine that is converted to cinnamic acid by the key enzyme phenylalanine ammonia lyase (PAL) is the major precursor for the biosynthesis of flavonoids and cinnamic acid derivatives (Fig. 4.11). Cinnamic acid and rosmarinic acid are two important examples of simple and complicated phenolic molecules synthesized from the shikimic acid pathway (Huang et al., 2009). Among these two molecules, rosmarinic acid has good aromatic qualities and antioxidative potential. The cinnamic acid derivatives act as vital precursors for the biosynthesis of lignin that is produced from the cinnamaldehydes and the respective alcohol monolignols. The activation of cinnamic acid by CoA results in the formation of cinnamoyl-CoA, which acts as precursor for the synthesis of coumarins and flavonoids.

The flavonoid biosynthetic pathway has been very well studied in the model plant *Arabidopsis thaliana* (Winkel-Shirley, 2001) although recently much information regarding the biosynthesis of these flavonoids has been elucidated from the other plants also like maize and legumes. The schematic biosynthetic pathway for the flavonoids derived from various plant species is shown in Fig. 4.12. Chalcone synthase is one of the key enzymes involved in the biosynthesis of the first flavonoid naringenin chalcone. In *Arabidopsis*, the mutants of chalcone synthase have transparent seed coat (testa) and are therefore referred to as transparent testa (tt) mutants (Koornneef, 1990). The other mutant in *Arabidopsis* having the same phenotype was developed by mutating another important enzyme chalcone

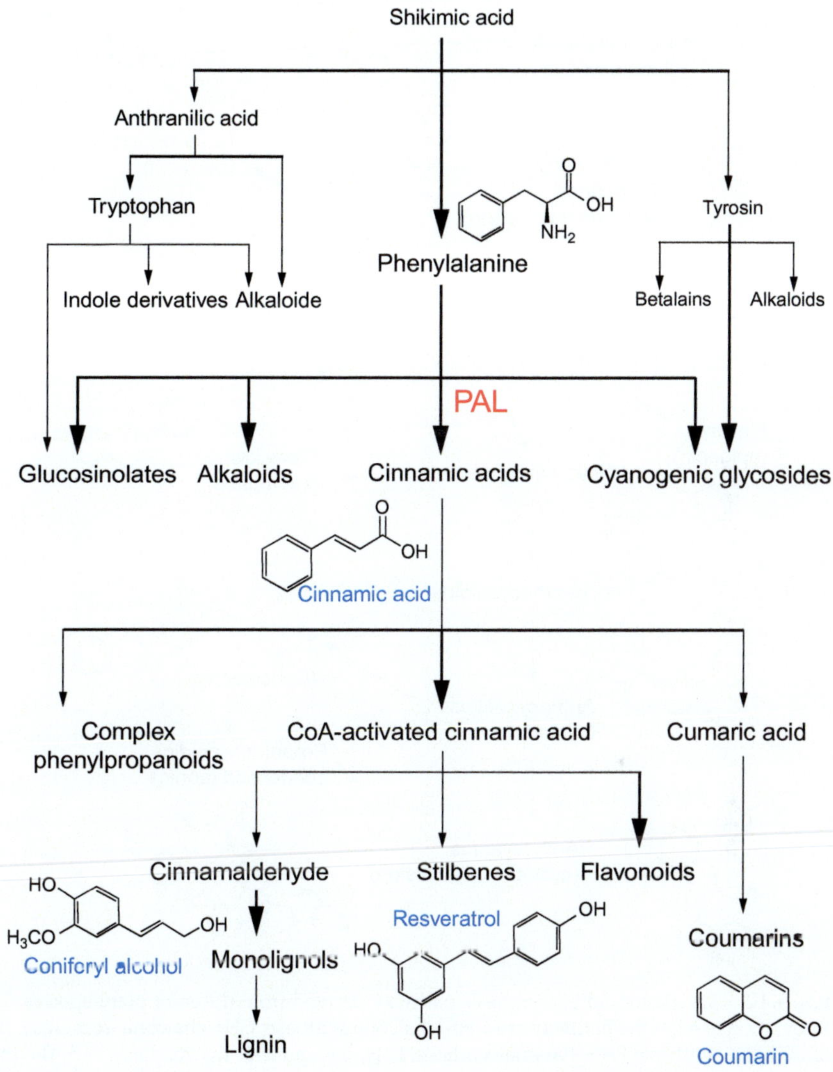

FIG. 4.11 The biosynthesis of phenols by using phenylalanine as precursor (Gutzeit and Ludwig-Müller, 2014).

isomerase that is responsible for the synthesis of naringenin. Naringenin in turn acts as a precursor for the biosynthesis of isoflavonoids of legumes, flavones, and flavonols, the largest group among the flavonoids. The key flavonols in model plant *A. thaliana* are kaempferol, quercetin, and their respective glycoside conjugates (Winkel-Shirley, 2001). The intermediates in the production of flavonols are dihydroflavonols that are precursors for the formation of anthocyanin pigments. Besides, they are also the

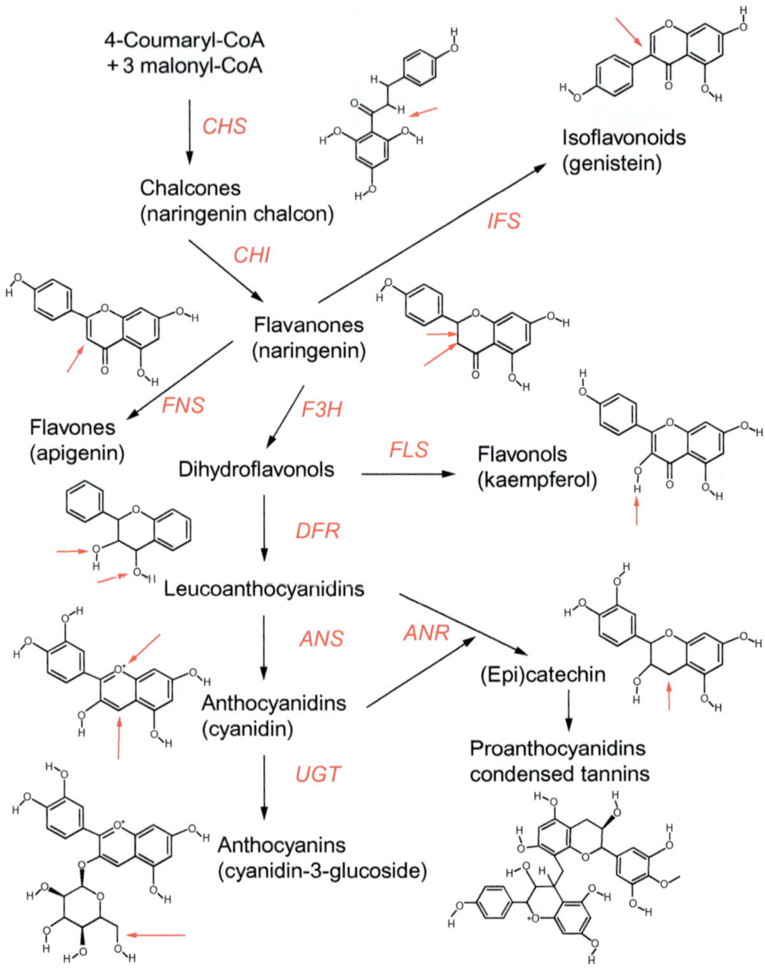

FIG. 4.12 The flavonoid biosynthesis pathway derived from different plant species. The abbreviation used in the pathway are CHS, chalcone synthase; CHI, chalcone isomerase; IFS, isoflavonoid synthase; FNS, flavone synthase; F3H, flavanone-3- hydroxylase; FLS, flavonol synthase; DFR, dihydroflavonol reductase; ANS, anthocyanidin synthase; UGT, glycosyl-transferase; and ANR, anthocyanidin reductase (Gutzeit and Ludwig-Müller, 2014).

precursors of condensed tannins also called proanthocyanidins that form the seed coat of many plants.

4.2.4 Carotenoids

Carotenoids are a group of isoprenoid molecules and because of their typical color in the range of yellow to red are commonly known as pigments. Their name is derived from the carrot (*Daucus carota*), the roots of which store the highest amount of the carotenoids. The characteristic color of flowers, fruits, and seeds is because of the carotenoids and is also exceptionally amassed in the underground organs of the plant like tubers and roots. For example, the yellow color of corn and bananas, the red color of tomato and watermelon, and the orange color of pumpkin and oranges is because of the carotenoids. Within the plants, carotenoids accumulate in the chromoplasts (specialized plastids) and thus act as visible signals for animals that contribute to pollination or seed dispersal. Plant carotenoids also take part in light harvesting and are vital for photoprotection against excess light. This physical property is due to a polyene chain with a large number of conjugated double bonds that acts as a chromophore. They also function as precursors for the production of apocarotenoid hormones such as abscisic acid and strigolactones. Carotenoids are synthesized not only by all photosynthetic organisms but also by some nonphotosynthetic bacteria and fungi. Animals cannot synthesize carotenoid with few exceptions but take them in their diets as an essential source of retinoids (including vitamin A) (Moran and Jarvik, 2010). Plant carotenoids are tetraterpenes derived from the 40-carbon structure isoprenoid phytoene. In general, carotenoids are grouped in two major classes: carotenes (hydrocarbons that can be cyclized at one or both ends of the molecule) and xanthophylls (oxygenated derivatives of carotenes). On the basis of the plants that accumulate large amount of carotenoids, the first major advancement in the field of plant carotenoids was accomplished using the plants that store large amounts of pigments in their chromoplasts like ripe fruit of *Solanum lycopersicum*, *Capsicum annuum*, *Tagetes* species, and *Narcissus* species (Cunningham and Gantt, 1998; Hirschberg, 2001; Fraser and Bramley, 2004). In the last half of the twentieth century, by using various biochemical and molecular approaches, the main carotenoid biosynthetic pathway was revealed. Before the end of the century, by the characterization and isolation of carotenoid-defective mutants from plants (*Arabidopsis*, tomato), algae (*Chlamydomonas* sp.), and cyanobacteria (*Synechocystis* sp.) and also by the development of metabolically engineered *Escherichia coli* strains, many carotenoid pathway genes have been elucidated (Cunningham and Gantt, 1998). Today, we have an almost-complete picture of the genes and enzymes that catalyze the core reactions of the carotenoid biosynthetic pathway in plants, including *Arabidopsis*.

Carotenoids being the major isoprenoid phytochemicals are synthesized from the five-carbon unit IPP and its isomer DMAPP. Within plants, two independent pathways like plastidial 2-C-methyl-D-erythritol-4-phosphate (MEP) and cytosolic mevalonic acid (MVA) exist for the production of these prenyl diphosphate precursors, but it has been found that the carotenoids are mainly produced from IPP and DMAPP from plastidial MEP pathway (Rodriguez-Concepcion, 2010), as shown in figure. The enzymatic addition of single molecule of DMAPP with the three molecules of

IPP results in the formation of geranylgeranyl diphosphate (GGPP), a key precursor for the biosynthesis of large number of plastidial isoprenoids including carotenoids (Lichtenthaler, 1999; Bouvier et al., 2005a,b). Two molecules of GGPP are condensed with the help of an important enzyme phytoene synthase (PSY) and result in the production of 40-carbon phytoene. This is the first committed step and is considered as the main bottleneck of carotenoid biosynthetic pathway (Hirschberg, 2001; Fraser and Bramley, 2004; Bouvier et al., 2005a; Sandmann et al., 2006; Cazzonelli and Pogson, 2010). Further, desaturation of phytoene results in the formation of chromophore-bearing chain of conjugated double bonds that forms the backbone of plant carotenoids and also controls their physical and biological activities (Britton, 1995). Phytoene is the uncolored product, and its desaturation and isomerization result in the synthesis of a red carotenoid, that is, lycopene. The end-to-end cyclization of lycopene polyene chain results in the production of carotenes and is the first branch point in the biosynthesis of complex carotenoids and xanthophylls. The carotenes are synthesized by the condensation of either one β ring with one ε ring to form α-carotene or with two β rings to form β-carotene. Carotenoids with two ε rings do not exist in *Arabidopsis* and are also uncommon in plants. Hydroxylation of the carotene rings produces xanthophylls such as lutein (from α-carotene) and zeaxanthin (from β-carotene). Zeaxanthin is epoxidated twice to make violaxanthin, which can be further modified to form neoxanthin (Fig. 4.13). The subsequent enzymatic modification of these carotenes and xanthophylls results in the generation of wide variety of carotenoids found in plants and other organisms (Walter and Strack, 2011). Additionally, carotenoids can be cleaved to generate apocarotenoids, a structurally diverse class of compounds and widely distributed in nature (Giuliano et al., 2003; Bouvier et al., 2005b; Auldridge et al., 2006; Walter and Strack, 2011). Plant apocarotenoids have the potent biological activities like they act as hormones ABA and strigolactones (Nambara and Marion-Poll, 2005; Van Norman and Sieburth, 2007; Xie et al., 2010). Other apocarotenoids are involved in the interaction of plants with their biological environment, acting as pigments and flavors that attract pollinators or seed-dispersing animals (Giuliano et al., 2003; Bouvier et al., 2005b; Auldridge et al., 2006; Walter and Strack, 2011).

4.3 CONCLUSION AND PERSPECTIVES

Although a lot of studies have been performed on the metabolic engineering strategies for the production of enormous phytochemicals in recombinant microbes, to date, only few studies have qualified in the production of commercially viable products because of the issues related to overall yields. From the studies, it is quite obvious that considerable improvements should be made making the microbial production

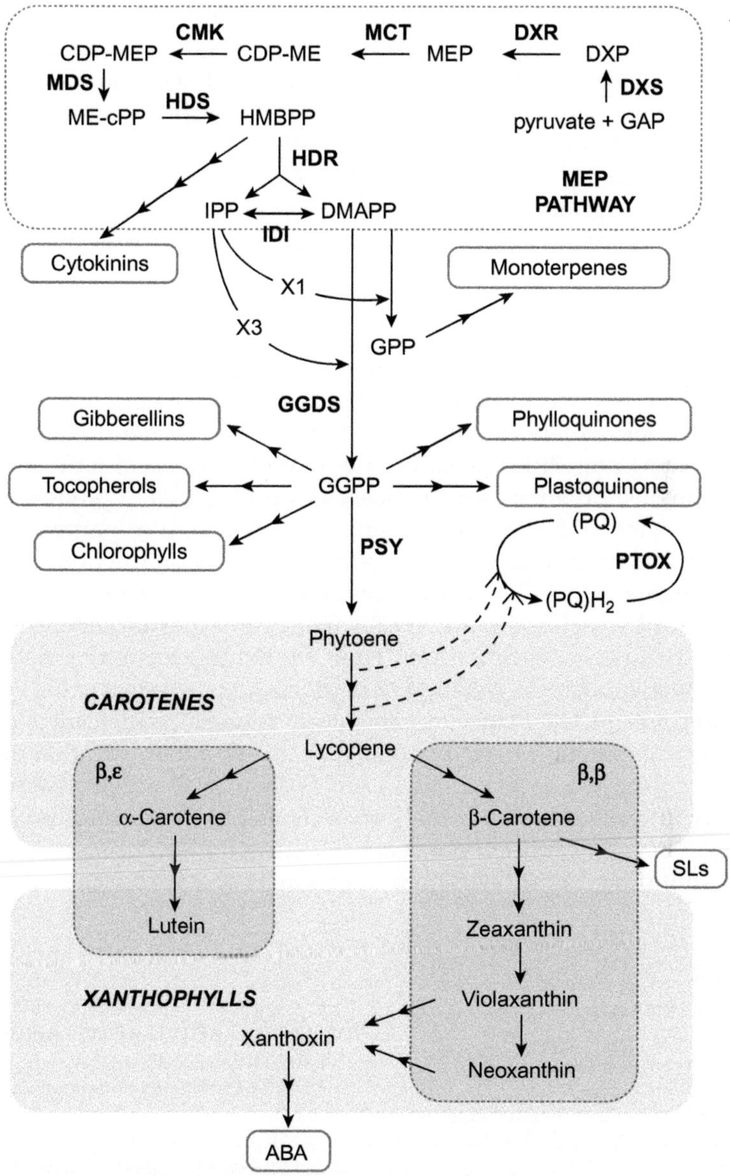

FIG. 4.13 Carotenoid biosynthetic pathway in *Arabidopsis* (M. Aguila Ruiz-Sola and Manuel Rodriguez-Concepcion).

of phytochemicals a complementary and competitive alternative to the current procedures. Further, detailed research is required for the effective understanding of cell metabolism, proficient expression of plant-specific enzymes, and alleviation of the toxicity displayed by many of the target compounds. Further, quite a few studies have also demonstrated the use of both in vivo and in vitro approaches for the diversification of plant natural products to generate libraries of compounds with potential pharmacological applications. Nonnatural analogs of secondary metabolites have also been generated by combining the metabolic engineering and in vitro biocatalysis. An important example is the nonnatural analogs of flavonoids, that is, halogens (Fowler et al., 2011), which can further be diversified using tailoring enzymes. The physiochemical and biological activity of the new compound being synthesized can be improved by the combination of more than one tailoring enzyme. Thus, the variety of chemical structure so formed can be tested as pharmaceutical candidates for a broad number of diseases. For the in vitro and in vivo production of natural and nonnatural compounds, the metabolic engineering of numeral variables have to be standardized like identification of ideal host organisms, fermentation operational variables, and also precursor selection (Jenkins et al., 2011). Besides, the total expression of pathway genes within the host organism for the proficient production of phytochemicals still poses a lot of drawbacks because reconstruction of pathway evokes a number of permutations and combinations. The production of diversified secondary metabolites within the cells brings a lot of stress to the cell as a result of which there are malformed prototypes of gene expression and metabolic enzyme inhibition. Thus, for the production of metabolites within the microbes, the detailed characterization of genes of biosynthetic pathways is required. Hence, the major question is the identification, production, and isolation of active enzymes and the identification and viability of cofactors needed for their catalytic activity. Also, generation of pathway intermediates and final products by in vitro biocatalysis should provide a better insight on the synergy between the constituent enzymes that permit the generation of natural products and analogs.

References

Aharoni, A., Giri, A.P., Deuerlein, S., Griepink, F., de Kogel, W.J., Verstappen, F., WVerhoeven, H.A., Jongsma, M.A., Schwab, W., Bouwmeester, H.J., 2003. Terpenoid metabolism in wild-type and transgenic *Arabidopsis* plants. Plant Cell 15 (12), 2866–2884.

Akula, R., Ravishankar, G.A., 2011. Influence of abiotic stress signals on secondary metabolites in plants. Plant Signal. Behav. 6 (11), 1720–1731.

Akinyemi, K.O., Oladapo, O., Okwara, C.E., Ibe, C.C., Fasure, K.A., 2005. Screening of crude extracts of six medicinal plants used in South-West Nigerian unorthodox medicine for anti-methicillin resistant *Staphylococcus aureus* activity. BMC Complement. Altern. Med. 5 (1), 6.

Alonso-Amelot, M.E., Oliveros-Bastidas, A., Calcagno-Pisarelli, M.P., 2007. Phenolics and condensed tannins of high altitude *Pteridium arachnoideum* in relation to sunlight exposure, elevation, and rain regime. Biochem. Syst. Ecol. 35 (1), 1–10.

Amme, S., Rutten, T., Melzer, M., Sonsmann, G., Vissers, J.P., Schlesier, B., Mock, H.P., 2005. A proteome approach defines protective functions of tobacco leaf trichomes. Proteomics 5 (10), 2508–2518.

Andersen, O.M., Markham, K.R. (Eds.), 2005. Flavonoids: Chemistry, Biochemistry and Applications. CRC Press, Boca Raton, FL.

Arbona, V., Manzi, M., Ollas, C.D., Gómez-Cadenas, A., 2013. Metabolomics as a tool to investigate abiotic stress tolerance in plants. Int. J. Mol. Sci. 14 (3), 4885–4911.

Auldridge, M.E., McCarty, D.R., Klee, H.J., 2006. Plant carotenoid cleavage oxygenases and their apocarotenoid products. Curr. Opin. Plant Biol. 9 (3), 315–321.

Battisti, V., Maders, L.D., Bagatini, M.D., Santos, K.F., Spanevello, R.M., Maldonado, P.A., Brule, A.O., do Carmo Araujo, M., Schetinger, M.R., Morsch, V.M., 2008. Measurement of oxidative stress and antioxidant status in acute lymphoblastic leukemia patients. Clin. Biochem. 41 (7), 511–518.

Bennett, R.N., Wallsgrove, R.M., 1994. Secondary metabolites in plant defence mechanisms. New Phytol. 127 (4), 617–633.

Bird, D.A., Franceschi, V.R., Facchini, P.J., 2003. A tale of three cell types: alkaloid biosynthesis is localized to sieve elements in opium poppy. Plant Cell 15 (11), 2626–2635.

Blytt, H.J., Guscar, T.K., Butler, L.G., 1988. Antinutritional effects and ecological significance of dietary condensed tannins may not be due to binding and inhibiting digestive enzymes. J. Chem. Ecol. 14 (6), 1455–1465.

Bouvier, F., Isner, J.C., Dogbo, O., Camara, B., 2005a. Oxidative tailoring of carotenoids: a prospect towards novel functions in plants. Trends Plant Sci. 10 (4), 187–194.

Bouvier, F., Rahier, A., Camara, B., 2005b. Biogenesis, molecular regulation and function of plant isoprenoids. Prog. Lipid Res. 44 (6), 357–429.

Britton, G., 1995. Structure and properties of carotenoids in relation to function. FASEB J. 9 (15), 1551–1558.

Brown, P.D., Tokuhisa, J.G., Reichelt, M., Gershenzon, J., 2003. Variation of glucosinolate accumulation among different organs and developmental stages of *Arabidopsis thaliana*. Phytochemistry 62 (3), 471–481.

Bundy, J.G., Davey, M.P., Viant, M.R., 2009. Environmental metabolomics: a critical review and future perspectives. Metabolomics 5 (1), 3.

Burchard, P., Bilger, W., Weissenböck, G., 2000. Contribution of hydroxycinnamates and flavonoids to epidermal shielding of UV-A and UV-B radiation in developing rye primary leaves as assessed by ultraviolet-induced chlorophyll fluorescence measurements. Plant Cell Environ. 23 (12), 1373–1380.

Cai, Z., Lee, F.S.C., Wang, X.R., Yu, W.J., 2002. A capsule review of recent studies on the application of mass spectrometry in the analysis of Chinese medicinal herbs. J. Mass Spectrom. 37 (10), 1013–1024.

Cannell, R.J. (Ed.), 1998. Natural Products Isolation. Vol. 4. Springer Science & Business Media, Totoa, NJ.

Cazzonelli, C.I., Pogson, B.J., 2010. Source to sink: regulation of carotenoid biosynthesis in plants. Trends Plant Sci. 15 (5), 266–274.

Chalker-Scott, L., 1999. Environmental significance of anthocyanins in plant stress responses. Photochem. Photobiol. 70 (1), 1–9.

Chanda, S.V., Parekh, J., Karathia, N., 2006. Evaluation of antibacterial activity and phytochemical analysis of *Bauhinia variegate* L. bark. Afr. J. Biomed. Res. 9, 53–56.

Chopra, A., Doiphode, V.V., 2002. Ayurvedic medicine: core concept, therapeutic principles, and current relevance. Med. Clin. 86 (1), 75–89.

Christie, P.J., Alfenito, M.R., Walbot, V., 1994. Impact of low-temperature stress on general phenylpropanoid and anthocyanin pathways: enhancement of transcript abundance and anthocyanin pigmentation in maize seedlings. Planta 194 (4), 541–549.

Clark, A.M., 1996. Natural products as a resource for new drugs. Pharm. Res. 13 (8), 1133–1141.

Coley, P.D., Bryant, J.P., Chapin III, F.S., 1985. Resource availability and plant antiherbivore defense. Science 230, 895–900.

Cos, P., Vlietinck, A.J., Berghe, D.V., Maes, L., 2006. Anti-infective potential of natural products: how to develop a stronger in vitro 'proof-of-concept'. J. Ethnopharmacol. 106 (3), 290–302.

Cunha, I., Sawaya, A.C., Caetano, F.M., Shimizu, M.T., Marcucci, M.C., Drezza, F.T., Carvalho, P.D.O., 2004. Factors that influence the yield and composition of Brazilian propolis extracts. J. Braz. Chem. Soc. 15 (6), 964–970.

Cunningham Jr., F.X., Gantt, E., 1998. Genes and enzymes of carotenoid biosynthesis in plants. Annu. Rev. Plant Biol. 49 (1), 557–583.

Dahiru, D., Onubiyi, J.A., Umaru, H.A., 2006. Phytochemical screening and antiulcerogenic effect of *Moringa oleifera* aqueous leaf extract. Afr. J. Tradit. Complement. Altern. Med. 3 (3), 70–75.

Dai, J., Mumper, R.J., 2010. Plant phenolics: extraction, analysis and their antioxidant and anticancer properties. Molecules 15 (10), 7313–7352.

Dash, B., Sharma, B.K., 2001. Charak Samhita. Chaukhamba Sanskrit Series, Varanasi.

De Bruyne, T., Pieters, L., Deelstra, H., Vlietinck, A., 1999. Condensed vegetable tannins: biodiversity in structure and biological activities. Biochem. Syst. Ecol. 27 (4), 445–459.

De Rijke, E., Out, P., Niessen, W.M., Ariese, F., Gooijer, C., Udo, A.T., 2006. Analytical separation and detection methods for flavonoids. J. Chromatogr. A 1112 (1), 31–63.

Debeaujon, I., Nesi, N., Perez, P., Devic, M., Grandjean, O., Caboche, M., Lepiniec, L., 2003. Proanthocyanidin-accumulating cells in *Arabidopsis testa*: regulation of differentiation and role in seed development. Plant Cell 15 (11), 2514–2531.

Dixon, R.A., Paiva, N.L., 1995. Stress-induced phenylpropanoid metabolism. Plant Cell 7 (7), 1085.

Dolara, P., Luceri, C., De Filippo, C., Femia, A.P., Giovannelli, L., Caderni, G., Cresci, A., 2005. Red wine polyphenols influence carcinogenesis, intestinal microflora, oxidative damage and gene expression profiles of colonic mucosa in F344 rats. Mutat. Res. Fundam. Mol. Mech. Mutagen. 591 (1), 237–246.

Edeoga, H.O., Okwu, D.E., Mbaebie, B.O., 2005. Phytochemical constituents of some Nigerian medicinal plants. Afr. J. Biotechnol. 4 (7), 685–688.

Edreva, A., Velikova, V., Tsonev, T., Dagnon, S., Gürel, A., Aktaş, L., Gesheva, E., 2008. Stress-protective role of secondary metabolites: diversity of functions and mechanisms. Gen. Appl. Plant Physiol. 34 (1–2), 67–78.

Escribano-Bailon, M.T., Santos-Buelga, C., 2003. Polyphenol Extraction From Foods. Methods in Polyphenol Analysis. In: Royal Society of Chemistry, Cambridge, pp. 1–16.

Fabricant, D.S., Farnsworth, N.R., 2001. The value of plants used in traditional medicine for drug discovery. Environ. Health Perspect. 109 (Suppl. 1), 69.

Fowler, Z.L., Shah, K., Panepinto, J.C., Jacobs, A., Koffas, M.A., 2011. Development of non-natural flavanones as antimicrobial agents. PLoS One 6 (10), e25681.

Fraser, P.D., Bramley, P.M., 2004. The biosynthesis and nutritional uses of carotenoids. Prog. Lipid Res. 43 (3), 228–265.

Fulcrand, H., Mané, C., Preys, S., Mazerolles, G., Bouchut, C., Mazauric, J.P., Cheynier, V., 2008. Direct mass spectrometry approaches to characterize polyphenol composition of complex samples. Phytochemistry 69 (18), 3131–3138.

Giner-Chavez, B.I., 1996. Condensed Tannins in Tropical Forages. Doctoral Dissertation Cornell University.

Giuliano, G., Al-Babili, S., Von Lintig, J., 2003. Carotenoid oxygenases: cleave it or leave it. Trends Plant Sci. 8 (4), 145–149.

Glynn, C., Rönnberg-Wästljung, A.C., Julkunen-Tiitto, R., Weih, M., 2004. Willow genotype, but not drought treatment, affects foliar phenolic concentrations and leaf-beetle resistance. Entomol. Exp. Appl. 113 (1), 1–14.

Gould, K.S., 2004. Nature's Swiss army knife: the diverse protective roles of anthocyanins in leaves. Biomed. Res. Int. 2004 (5), 314–320.

Grotewold, E. (Ed.), 2006. The Science of Flavonoids. Springer-Verlag, New York, NY.

Grubb, C.D., Abel, S., 2006. Glucosinolate metabolism and its control. Trends Plant Sci. 11 (2), 89–100.

Gutbrodt, B., Dorn, S., Unsicker, S.B., Mody, K., 2012. Species-specific responses of herbivores to within-plant and environmentally mediated between-plant variability in plant chemistry. Chemoecology 22 (2), 101–111.

Gutzeit, H.O., Ludwig-Müller, J., 2014. Plant Natural Products: Synthesis, Biological Functions and Practical Applications. John Wiley & Sons, Weinheim.

Hahlbrock, K., 1981. Flavonoids. Biochem. Plants 7, 425–456.

Hahn, N.I., 1998. Are phytoestrogens nature's cure for what ails us? A look at the research. J. Acad. Nutr. Diet. 98 (9), 974–976.

Halkier, B.A., Gershenzon, J., 2006. Biology and biochemistry of glucosinolates. Annu. Rev. Plant Biol. 57, 303–333.

Hamburger, M.O., Cordell, G.A., 1987. A direct bioautographic TLC assay for compounds possessing antibacterial activity. J. Nat. Prod. 50 (1), 19–22.

Harborne, J.B., 1994. The Flavonoids: Advances in Research Since 1986. Chapman & Hall, London.

Harborne, J.B., Williams, C.A., 1975. Flavone and flavonol glycosides. In: Harborne, J.B., Mabry, T.J., Mabry, H. (Eds.), The Flavonoids. Springer, Boston, MA.

Harborne, J.B., Williams, C.A., 2000. Advances in flavonoid research since 1992. Phytochemistry 55 (6), 481–504.

Harnly, J.M., Bhagwat, S., Lin, L.Z., 2007. Profiling methods for the determination of phenolic compounds in foods and dietary supplements. Anal. Bioanal. Chem. 389 (1), 47–61.

Hazra, K.M., Roy, R.N., Sen, S.K., Laskar, S., 2007. Isolation of antibacterial pentahydroxy flavones from the seeds of *Mimusops elengi* Linn. Afr. J. Biotechnol. 6 (12).

Heinrich, M., 2000. Ethnobotany and its role in drug development. Phytother. Res. 14 (7), 479–488.

Hirschberg, J., 2001. Carotenoid biosynthesis in flowering plants. Curr. Opin. Plant Biol. 4 (3), 210–218.

Holstege, D.M., Seiber, J.N., Galey, F.D., 1995. Rapid multiresidue screen for alkaloids in plant material and biological samples. J. Agric. Food Chem. 43, 691–699.

Homans, A.L., Fuchs, A., 1970. Direct bioautography on thin-layer chromatograms as a method for detecting fungitoxic substances. J. Chromatogr. A 51, 327–329.

Huang, W.Y., Cai, Y.Z., Zhang, Y., 2009. Natural phenolic compounds from medicinal herbs and dietary plants: potential use for cancer prevention. Nutr. Cancer 62 (1), 1–20.

Huie, C.W., 2002. A review of modern sample-preparation techniques for the extraction and analysis of medicinal plants. Anal. Bioanal. Chem. 373 (1), 23–30.

Ibern-Gomez, M., Andres-Lacueva, C., Lamuela-Raventos, R.M., Waterhouse, A.L., 2002. Rapid HPLC analysis of phenolic compounds in red wines. Am. J. Enol. Vitic. 53 (3), 218–221.

Jansen, M.A., Gaba, V., Greenberg, B.M., 1998. Higher plants and UV-B radiation: balancing damage, repair and acclimation. Trends Plant Sci. 3 (4), 131–135.

Jenkins, T., Bovi, A., Edwards, R., 2011. Plants: biofactories for a sustainable future? Philos. Trans. R. Soc. A Math. Phys. Eng. Sci. 369 (1942), 1826–1839.

Jones, O.A., Maguire, M.L., Griffin, J.L., Dias, D.A., Spurgeon, D.J., Svendsen, C., 2013. Metabolomics and its use in ecology. Aust. Ecol. 38 (6), 713–720.

Kazan, K., Manners, J.M., 2011. The interplay between light and jasmonate signalling during defence and development. J. Exp. Bot. 62 (12), 4087–4100.

Kim, H.K., Choi, Y.H., Verpoorte, R., 2010. NMR-based metabolomic analysis of plants. Nat. Protoc. 5 (3), 536.

Kliebenstein, D.J., 2004. Secondary metabolites and plant/environment interactions: a view through *Arabidopsis thaliana* tinged glasses. Plant Cell Environ. 27 (6), 675–684.

Koornneef, M., 1990. Mutations affecting the testa colour in *Arabidopsis*. Arabidopsis Inf. Serv. 27, 1–4.

Kouki, M., Manetas, Y., 2002. Resource availability affects differentially the levels of gallotannins and condensed tannins in *Ceratonia siliqua*. Biochem. Syst. Ecol. 30 (7), 631–639.

Krishnan, R., Chandravadana, M.V., Ramachander, P.R., Bharathkumar, H., 1983. Interrelationships between growth and alkaloid production in *Catharanthus roseus* G. Don. Herba Hungarica 22, 47–54.

Kumar, G.S., Jayaveera, K.N., Kumar, C.K., Sanjay, U.P., Swamy, B.M., Kumar, D.V., 2007. Antimicrobial effects of Indian medicinal plants against acne-inducing bacteria. Trop. J. Pharm. Res. 6 (2), 717–723.
Langenheim, J.H., 1994. Higher plant terpenoids: a phytocentric overview of their ecological roles. J. Chem. Ecol. 20 (6), 1223–1280.
Lattanzio, V., 2013. Phenolic compounds: introduction. In: Natural Products. Springer, Berlin Heidelberg. pp. 1543–1580.
Laughlin, J.C., 1993. Effect of agronomic practices on plant yield and antimalarial constituents of *Artemisia annua* L. Acta Hortic. (331)53–61.
Lepiniec, L., Debeaujon, I., Routaboul, J.M., Baudry, A., Pourcel, L., Nesi, N., Caboche, M., 2006. Genetics and biochemistry of seed flavonoids. Annu. Rev. Plant Biol. 57, 405–430.
Lichtenthaler, H.K., 1999. The 1-deoxy-D-xylulose-5-phosphate pathway of isoprenoid biosynthesis in plants. Annu. Rev. Plant Biol. 50 (1), 47–65.
Lommen, W.J.M., Bouwmeester, H.J., Schenk, E., Verstappen, F.W.A., Elzinga, S., Struik, P.C., 2008. Modelling processes determining and limiting the production of secondary metabolites during crop growth: the example of the antimalarial artemisinin produced in *Artemisia annua*. Acta Hortic. (765), 87–94.
Lukas, B., Schmiderer, C., Franz, C., Novak, J., 2009. Composition of essential oil compounds from different Syrian populations of *Origanum syriacum* L. (Lamiaceae). J. Agric. Food Chem. 57 (4), 1362–1365.
Maes, L., Van Nieuwerburgh, F.C.W., Zhang, Y., Reed, D.W., Pollier, J., Casteele, S.R.F.V., Inzé, D., Covello, P.S., Deforce, D.L.D., Goossens, A., 2011. Dissection of the phytohormonal regulation of trichome formation and biosynthesis of the antimalarial compound artemisinin in *Artemisia annua* plants. New Phytol. 189 (1), 176–189.
Mallikharjuna, P.B., Rajanna, L.N., Seetharam, Y.N., Sharanabasappa, G.K., 2007. Phytochemical studies of *Strychnos potatorum* Lf—a medicinal plant. J. Chem. 4 (4), 510–518.
Mandal, S.M., Chakraborty, D., Dey, S., 2010. Phenolic acids act as signaling molecules in plant-microbe symbioses. Plant Signal. Behav. 5 (4), 359–368.
Mangan, J.L., 1988. Nutritional effects of tannins in animal feeds. Nutr. Res. Rev. 1 (1), 209–231.
Mattocks, A.R., 1967. Spectrophotometric determination of unsaturated pyrrolizidine alkaloids. Anal. Chem. 39 (4), 443–447.
Maurya, R., Singh, G., Yadav, P.P., 2008. Antiosteoporotic agents from natural sources. Stud. Nat. Prod. Chem. 35, 517–548.
McCaskill, D., Croteau, R., 1998. Some caveats for bioengineering terpenoid metabolism in plants. Trends Biotechnol. 16 (8), 349–355.
McGaw, L.J., Steenkamp, V., Eloff, J.N., 2007. Evaluation of Athrixia bush tea for cytotoxicity, antioxidant activity, caffeine content and presence of pyrrolizidine alkaloids. J. Ethnopharmacol. 110 (1), 16–22.
McLeod, M.N., 1974. Plant tannins: their role in forage quality. Nutr. Abstr. Rev. 44, 803–812.
Ming, H.Q., Xia, G.C., Zhang, R.D., 1988. Advanced research on Rhodiola. Chin. Tradit. Herb. Drug 19 (5), 229–234.
Miranda, M.A., Varela, R.M., Torres, A., Molinillo, J.M., Gualtieri, S.C., Macías, F.A., 2015. Phytotoxins from *Tithonia diversifolia*. J. Nat. Prod. 78 (5), 1083–1092.
Mishra, S.N., 1989. In: Analytical methods for analysis of total alkaloids in root of *Withania* spp. Proc. All India Workshop on M&AP, pp. 492–495.
Modak, M., Dixit, P., Londhe, J., Ghaskadbi, S., Devasagayam, T.P.A., 2007. Recent advances in Indian Herbal Drug Research guest editor: Thomas Paul Asir Devasagayam Indian Herbs and herbal drugs used for the treatment of diabetes. J. Clin. Biochem. Nutr. 40 (3), 163–173.
Mole, S., Waterman, P.G., 1986. Tannic acid and proteolytic enzymes: enzyme inhibition or substrate deprivation? Phytochemistry 26 (1), 99–102.
Molyneux, R.J., Nash, R.J., Asano, N., 1996. In: Pelletier, S.W. (Ed.), Alkaloids: Chemical and Biological Perspectives. Vol. 11.

Moran, N.A., Jarvik, T., 2010. Lateral transfer of genes from fungi underlies carotenoid production in aphids. Science 328 (5978), 624–627.

Moteriya, P., Satasiya, R., Chanda, S., 2015. Screening of phytochemical constituents in some ornamental flowers of Saurashtra region. J. Phcog. Phytochem. 3 (5), 112–120.

Mueller-Harvey, I., McAllan, A.B., 1992. Tannins: their biochemistry and nutritional properties. Adv. Plant Cell Biochem. Biotechnol. 1, 151–217.

Nambara, E., Marion-Poll, A., 2005. Abscisic acid biosynthesis and catabolism. Annu. Rev. Plant Biol. 56, 165–185.

Ncube, B., Finnie, J.F., Van Staden, J., 2011. Seasonal variation in antimicrobial and phytochemical properties of frequently used medicinal bulbous plants from South Africa. S. Afr. J. Bot. 77 (2), 387–396.

Novak, B.H., Hudlicky, T., Reed, J.W., Mulzer, J., Trauner, D., 2000. Morphine synthesis and biosynthesis—an update. Curr. Org. Chem. 4 (3), 343–362.

Ober, D., Hartmann, T., 1999. Homospermidine synthase, the first pathway-specific enzyme of pyrrolizidine alkaloid biosynthesis, evolved from deoxyhypusine synthase. Proc. Natl. Acad. Sci. 96 (26), 14777–14782.

Oh, M.M., Trick, H.N., Rajashekar, C.B., 2009. Secondary metabolism and antioxidants are involved in environmental adaptation and stress tolerance in lettuce. J. Plant Physiol. 166 (2), 180–191.

Onwukaeme, D.N., Ikuegbvweha, T.B., Asonye, C.C., 2007. Evaluation of phytochemical constituents, antibacterial activities and effect of exudate of Pycanthus Angolensis weld warb (Myristicaceae) on corneal ulcers in rabbits. Trop. J. Pharm. Res. 6 (2), 725–730.

Parekh, J., Chanda, S., 2007. In vitro antimicrobial activity and phytochemical analysis of some Indian medicinal plants. Turk. J. Biol. 31 (1), 53–58.

Parker, J., 1977. Phenolics in black oak bark and leaves. J. Chem. Ecol. 3 (5), 489–496.

Patwardhan, B., Vaidya, A.D., Chorghade, M., 2004. Ayurveda and natural products drug discovery. Curr. Sci. 789–799.

Pavarini, D.P., Pavarini, S.P., Niehues, M., Lopes, N.P., 2012. Exogenous influences on plant secondary metabolite levels. Anim. Feed Sci. Technol. 176 (1), 5–16.

Pennycooke, J.C., Cox, S., Stushnoff, C., 2005. Relationship of cold acclimation, total phenolic content and antioxidant capacity with chilling tolerance in petunia (Petunia× hybrida). Environ. Exp. Bot. 53 (2), 225–232.

Perez-Estrada, L.B., Cano-Santana, Z., Oyama, K., 2000. Variation in leaf trichomes of *Wigandia urens*: environmental factors and physiological consequences. Tree Physiol. 20 (9), 629–632.

Pharmacopeia, U.S., 2002. United States Pharmacopeia. The Standard. US Pharmacopeia 1995, USA.

Pharmacopoeia, J., 2001. Japanese Pharmacopoeia. The Society of Japanese Pharmacopoeia, Tokyo.

Phrompittayarat, W., Putalun, W., Tanaka, H., Jetiyanon, K., Wittaya-areekul, S., Ingkaninan, K., 2007. Comparison of various extraction methods of *Bacopa monnieri*. Naresuan Univ. J. 15 (1), 29–34.

Pourcel, L., Routaboul, J.M., Cheynier, V., Lepiniec, L., Debeaujon, I., 2007. Flavonoid oxidation in plants: from biochemical properties to physiological functions. Trends Plant Sci. 12 (1), 29–36.

Prasain, J.K., Wang, C.C., Barnes, S., 2004. Mass spectrometric methods for the determination of flavonoids in biological samples. Free Radic. Biol. Med. 37 (9), 1324–1350.

Rahalison, L.L., Hamburger, M., Hostettmann, K., Monod, M., Frenk, E., 1991. A bioautographic agar overlay method for the detection of antifungal compounds from higher plants. Phytochem. Anal. 2 (5), 199–203.

Rahimmalek, M., Tabatabaei, B.E.S., Etemadi, N., Goli, S.A.H., Arzani, A., Zeinali, H., 2009. Essential oil variation among and within six *Achillea* species transferred from different ecological regions in Iran to the field conditions. Ind. Crop. Prod. 29 (2), 348–355.

Ram, J., Moteriya, P., Chanda, S., 2015. Phytochemical screening and reported biological activities of some medicinal plants of Gujarat region. J. Pharmacogn. Phytother. 4 (2).

Rao, R.K., Ali, N., Reddy, M.N., 1978. Occurrence of both sapogenins and alkaloid lycorine in *Curculigo orchioides*. Indian J. Pharm. Sci. 40, 104–105.

Rodriguez-Concepcion, M., 2010. Supply of precursors for carotenoid biosynthesis in plants. Arch. Biochem. Biophys. 504 (1), 118–122.

Rozema, J., van de Staaij, J., Björn, L.O., Caldwell, M., 1997. UV-B as an environmental factor in plant life: stress and regulation. Trends Ecol. Evol. 12 (1), 22–28.

Sandmann, G., Romer, S., Fraser, P.D., 2006. Understanding carotenoid metabolism as a necessity for genetic engineering of crop plants. Metab. Eng. 8 (4), 291–302.

Sasidharan, S., Chen, Y., Saravanan, D., Sundram, K.M., Latha, L.Y., 2011. Extraction, isolation and characterization of bioactive compounds from plants' extracts. Afr. J. Tradit. Complement. Altern. Med. 8 (1), 1–10.

Saxena, M., Saxena, J., Nema, R., Singh, D., Gupta, A., 2013. Phytochemistry of medicinal plants. J. Pharmacogn. Phytochem. 1 (6), 168–182.

Schofield, P., Mbugua, D.M., Pell, A.N., 2001. Analysis of condensed tannins: a review. Anim. Feed Sci. Technol. 91 (1), 21–40.

Shahverdi, A.R., Abdolpour, F., Monsef-Esfahani, H.R., Farsam, H., 2007. A TLC bioautographic assay for the detection of nitrofurantoin resistance reversal compound. J. Chromatogr. B 850 (1), 528–530.

Shakya, A.K., Shukla, S., 2011. Evaluation of hepatoprotective efficacy of Majoon-e-Dabeed-ul-ward against acetaminophen-induced liver damage: a Unani herbal formulation. Drug Dev. Res. 72 (4), 346–352.

Shakya, A.K., Sharma, N., Saxena, M., Shrivastava, S., Shukla, S., 2012. Evaluation of the antioxidant and hepatoprotective effect of Majoon-e-Dabeed-ul-ward against carbon tetrachloride induced liver injury. Exp. Toxicol. Pathol. 64 (7), 767–773.

Sharkey, T.D., Loreto, F., 1993. Water stress, temperature, and light effects on the capacity for isoprene emission and photosynthesis of kudzu leaves. Oecologia 95 (3), 328–333.

Sharkey, T.D., Yeh, S., 2001. Isoprene emission from plants. Annu. Rev. Plant Biol. 52 (1), 407–436.

Shoyama, Y., Tanaka, H., Fukuda, N., 1999. Monoclonal antibodies against naturally occurring bioactive compounds. Cytotechnology 31 (1–2), 9.

Solecka, D., Kacperska, A., 1995. Phenylalanine ammonia-lyase activity in leaves of winter oilseed rape plants as affected by acclimation of plants to low temperature. Plant Physiol. Biochem. Paris 33, 585.

Stashenko, E.E., Martínez, J.R., Ruíz, C.A., Arias, G., Durán, C., Salgar, W., Cala, M., 2010. Lippia origanoides chemotype differentiation based on essential oil GC-MS and principal component analysis. J. Sep. Sci. 33 (1), 93–103.

Szakiel, A., Pączkowski, C., Henry, M., 2011. Influence of environmental abiotic factors on the content of saponins in plants. Phytochem. Rev. 10 (4), 471–491.

Tapas, A.R., Sakarkar, D.M., Kakde, R.B., 2008. Flavonoids as nutraceuticals: a review. Trop. J. Pharm. Res. 7 (3), 1089–1099.

Taylor, L., 2000. Plant Based Drugs and Medicines. Raintree Nutrition Inc, Carson City NV.

Telascrea, M., de Araújo, C.C., Marques, M.O., Facanali, R., de Moraes, P.L., Cavalheiro, A.J., 2007. Essential oil from leaves of *Cryptocarya mandioccana Meisner* (Lauraceae): composition and intraspecific chemical variability. Biochem. Syst. Ecol. 35 (4), 222–232.

Theis, N., Lerdau, M., 2003. The evolution of function in plant secondary metabolites. Int. J. Plant Sci. 164 (S3), S93–S102.

Tholl, D., 2006. Terpene synthases and the regulation, diversity and biological roles of terpene metabolism. Curr. Opin. Plant Biol. 9 (3), 297–304.

Tona, L., Kambu, K., Ngimbi, N., Cimanga, K., Vlietinck, A.J., 1998. Antiamoebic and phytochemical screening of some Congolese medicinal plants. J. Ethnopharmacol. 61 (1), 57–65.

Treutter, D., 2005. Significance of flavonoids in plant resistance and enhancement of their biosynthesis. Plant Biol. 7 (06), 581–591.

Tsao, R., Deng, Z., 2004. Separation procedures for naturally occurring antioxidant phytochemicals. J. Chromatogr. B 812 (1), 85–99.

Turner, G.W., Gershenzon, J., Croteau, R.B., 2000. Development of peltate glandular trichomes of peppermint. Plant Physiol. 124 (2), 665–680.

Van Norman, J.M., Sieburth, L.E., 2007. Dissecting the biosynthetic pathway for the bypass1 root-derived signal. Plant J. 49 (4), 619–628.

Van Soest, P.J., 1994. Nutritional Ecology of the Ruminant. Cornell University Press, Ithaca, NY.

Veitch, N.C., 2007. Isoflavonoids of the Leguminosae. Nat. Prod. Rep. 24 (2), 417–464.

Veitch, N.C., 2010. Flavonoid chemistry of the Leguminosae. Recent Adv. Polyphenol Res. 2, 23–58.

Vilela, E.C., Duarte, A.R., Naves, R.V., Santos, S.C., Seraphin, J.C., Ferri, P.H., 2013. Spatial chemometric analyses of essential oil variability in *Eugenia dysenterica*. J. Braz. Chem. Soc. 24 (5), 873–879.

Walter, M.H., Strack, D., 2011. Carotenoids and their cleavage products: biosynthesis and functions. Nat. Prod. Rep. 28 (4), 663–692.

Walton, N.J., Mayer, M.J., Narbad, A., 2003. Vanillin. Phytochemistry 63 (5), 505–515.

Weid, M., Ziegler, J., Kutchan, T.M., 2004. The roles of latex and the vascular bundle in morphine biosynthesis in the opium poppy, *Papaver somniferum*. Proc. Natl. Acad. Sci. U. S. A. 101 (38), 13957–13962.

Wienkoop, S., Zoeller, D., Ebert, B., Simon-Rosin, U., Fisahn, J., Glinski, M., Weckwerth, W., 2004. Cell-specific protein profiling in *Arabidopsis thaliana* trichomes: identification of trichome-located proteins involved in sulfur metabolism and detoxification. Phytochemistry 65 (11), 1641–1649.

Wink, M., 1988. Plant breeding: importance of plant secondary metabolites for protection against pathogens and herbivores. TAG Theor. Appl. Genet. 75 (2), 225–233.

Winkel-Shirley, B., 2001. Flavonoid biosynthesis. A colorful model for genetics, biochemistry, cell biology, and biotechnology. Plant Physiol. 126 (2), 485–493.

Winkel-Shirley, B., 2002. Biosynthesis of flavonoids and effects of stress. Curr. Opin. Plant Biol. 5 (3), 218–223.

Woisky, R.G., Salatino, A., 1998. Analysis of propolis: some parameters and procedures for chemical quality control. J. Apic. Res. 37 (2), 99–105.

Wolfender, J.L., Ndjoko, K., Hostettmann, K., 2003. Liquid chromatography with ultraviolet absorbance-mass spectrometric detection and with nuclear magnetic resonance spectrometry: a powerful combination for the on-line structural investigation of plant metabolites. J. Chromatogr. A 1000 (1), 437–455.

World Health Organization, Research for International Tobacco Control, 2008. WHO Report on the Global Tobacco Epidemic, 2008: The MPOWER Package. World Health Organization.

Xie, X., Yoneyama, K., Yoneyama, K., 2010. The strigolactone story. Annu. Rev. Phytopathol. 48, .

Yadav, M., Chatterji, S., Gupta, S.K., Watal, G., 2014. Preliminary phytochemical screening of six medicinal plants used in traditional medicine. Int. J. Pharm. Pharm. Sci. 6 (5), 539–542.

Yan, X., Wu, S., Wang, Y., Shang, X., Dai, S., 2004. Soil nutrient factors related to salidroside production of *Rhodiola sachalinensis* distributed in Chang Bai Mountain. Environ. Exp. Bot. 52 (3), 267–276.

Ye, Z., Weinberg, H.S., Meyer, M.T., 2007. Trace analysis of trimethoprim and sulfonamide, macrolide, quinolone, and tetracycline antibiotics in chlorinated drinking water using liquid chromatography electrospray tandem mass spectrometry. Anal. Chem. 79 (3), 1135–1144.

Zhonghua Renmin Gongheguo wei sheng bu yao dian wei yuan hui, 2000. Pharmacopoeia of the People's Republic of China. Vol. 1. Chemical Industry Press, Beijing, China.

Zobayed, S.M.A., Afreen, F., Kozai, T., 2005. Temperature stress can alter the photosynthetic efficiency and secondary metabolite concentrations in St. John's wort. Plant Physiol. Biochem. 43 (10), 977–984.

Zygmunt, B., Namieśnik, J., 2003. Preparation of samples of plant material for chromatographic analysis. J. Chromatogr. Sci. 41 (3), 109–116.

Further Reading

De Brito, E.S., de Araújo, M.C.P., Lin, L.Z., Harnly, J., 2007. Determination of the flavonoid components of cashew apple (*Anacardium occidentale*) by LC-DAD-ESI/MS. Food Chem. 105 (3), 1112–1118.

Dudareva, N., Pichersky, E., Gershenzon, J., 2004. Biochemistry of plant volatiles. Plant Physiol. 135 (4), 1893–1902.

Fossen, T., Andersen, O.M., 2006. In: Andersen, O.M., Markham, K.R. (Eds.), Flavonoids: Chemistry, Biochemistry and Applications. CRC Press/Taylor & Francis, Boca Raton, pp. 52–68.

Ghidouche, S., Es-Safi, N.E., Ducrot, P.H., 2008. Mechanistic study on the enzymatic oxidation of flavonols. Tetrahedron Lett. 49 (4), 619–623. f.

Lin, L.Z., He, X.G., Lindenmaier, M., Nolan, G., Yang, J., Cleary, M., Cordell, G.A., 2000. Liquid chromatography-electrospray ionization mass spectrometry study of the flavonoids of the roots of *Astragalus mongholicus* and *A. membranaceus*. J. Chromatogr. A 876 (1), 87–95.

Morinaga, O., Tanaka, H., Shoyama, Y., 2006. Detection and quantification of ginsenoside Re in ginseng samples by a chromatographic immunostaining method using monoclonal antibody against ginsenoside Re. J. Chromatogr. B 830 (1), 100–104.

Zhang, W., Cai, Y., Tu, C., Ma, L.Q., 2002. Arsenic speciation and distribution in an arsenic hyperaccumulating plant. Sci. Total Environ. 300 (1), 167–177.

CHAPTER

5

Biotechnological Strategies for Improvement of Aromatic and Medicinal Herbs

OUTLINE

Himalayan Phytochemicals
https://doi.org/10.1016/B978-0-08-102227-6.00005-X

5.1 INTRODUCTION

Medicinal plants are the important reservoirs for the formulation of new drugs (Tsay et al., 2003; Leon-Diaz et al., 2010; Nishigaki et al., 2010; Hamilton, 2004). In Europe, around 1300 herbal plants are used as medicine among which 90% are harvested from wild resources in the United States. Moreover, about 118 of the peak 150 prescription drugs are based on the natural sources (Balunas and Kinghorn, 2005). It has been estimated that in developing countries, about 80% of people are totally dependent on herbal drugs for their primary health care, and in developed countries, 25% of prescribed medicines are derived from wild medicinal plants (Hamilton, 2004). So with the escalating demand for these medicinal herbs, their use is alarmingly increasing throughout the world (Tsay et al., 2003; Murch et al., 2007). Thus, it has been seen that genetic diversity of these important medicinal plants is constantly under the threat of extinction as a result of habitat abolition and unmonitored trade of medicinal plants. It has been estimated by the International Union for Conservation of Nature and the World Wildlife Fund that worldwide, around 50,000–80,000 flowering plant species are used for medicinal purpose, among which 15,000 species are under the threat of extinction as a result of overharvesting and habitat destruction (Bentley, 2000). In order to deal with this alarming situation, the major biotechnological technique, that is, plant tissue culture, serves as a boon to acquire genetically elite populations under in vitro conditions. One of the main problems with most medicinal plants is either they don't set seeds or seeds are too small that they don't even germinate. In this case, micropropagation holds important promise for rapid and mass multiplication under disease-free conditions. Besides, the callus-derived plants show enormous genetic variation that could be exploited for developing superior clones/varieties predominantly in vegetatively propagated plant species (Thangavel et al., 2014).

5.2 APPLICATIONS OF IN VITRO PLANT REGENERATION AND MICRO PROPAGATION IN MAPs IMPROVEMENT

In vitro plant regeneration and micropropagation are vital biotechnological tools to select, multiply, and safeguard the crucial genotypes of medicinal plants. As compared with conventional methods of vegetative propagation, micropropagation has a lot of advantages (Nehra and Kartha, 1994). With the help of micropropagation, the multiplication rate is significantly increased and also results in the production of pathogen-free plants. The propagula derived from existing meristem yield plants that are genetically identical with the donor plants (Hu and Wang, 1983). The plant regeneration from shoot and stem meristems of medicinal plants like *Catharanthus roseus, Cinchona ledgeriana, Digitalis* spp., *Rehmannia glutinosa, Rauvolfia serpentina,* and *Isoplexis canariensis* has yielded encouraging results (Paek et al., 1995; Roy et al., 1994; Pérez-Bermúdez et al., 2002). Thus, the appropriate selection of explants is the critical step for in vitro propagation, and it has been seen that nodal and axillary bud explants have higher rate of multiplication and are extensively used in vitro propagation purposes.

5.2.1 Regeneration by Axillary Buds/Apical Shoots

Regeneration by axillary or apical buds is an important technique of micropropagation in which nodal segments bearing axillary or apical buds are directly involved in the regeneration of multiple shoots without the intervention of callus phase (Pati et al., 2006). This micropropagation technique plays a vital role in the regeneration of virus-free seeds from different plant species (Grout, 1999; Faccioli, 2001; Rout et al., 2006). Another significant role of this type of regeneration is the conservation of most vulnerable and critically endangered medicinal and aromatic plants. By this technique, these plants can be grown and multiplied on a large scale without depending on their natural habitats, thus preserving the natural resources that are otherwise under the constant threat of exploitation (Prevalek-Kozlina et al., 1997; Cuenca et al., 1999). One of the prerequisites for the mass propagation of medicinal plants by micropropagation depends upon the media composition and appropriate concentration of phytohormones. The various medicinal and aromatic plants that have been successfully raised and transferred to the fields by micropropagation are discussed in detail. For example, in *Solanum surattense,* multiple shoots were regenerated from the inoculation of nodal segments. The plantlets raised from the nodal segments were acclimatized and successfully transferred to the fields (Pawar et al., 2002). Similarly, utmost shootlet formation was observed in some of the medicinal plants of economic importance like *Withania somnifera, R. serpentina, Ocimum basilicum,* and *Enicostemma littorale* when cultured in different composition of basal media with various combinations of phytohormones (Begum et al., 2002; Jain et al., 2003; Autade et al., 2016; Nagarathnamma et al., 2010). It has also been reported that in peppermint (*Mentha* x *piperita* L.), multiple shoots were regenerated form the nodal explants without the intervention of callus phase when grown in MS media supplemented with BAP, kinetin, and zeatin. Thus, by using the nodal and shoot tip for regeneration process, enormous planting material can be raised without the intervention of seasonal effect and is also the source of superior quantities of biologically important compound (Kiran et al., 2014). The important medicinal plants that have been raised via micropropagation from the shoot apical meristem have been reported in different studies like *C. roseus*, *C. ledgeriana*, *Digitalis* spp., *R. glutinosa*, *R. serpentina*, *I. canariensis*, *Citrullus colocynthis*, *Zephyranthes bulbous*, *Plectranthus vettiveroids*, and *Glossocardia bosvallea* (Roy et al., 1994; Paek et al., 1995; Sivasubramanian et al., 2002; Geetha and Gopal, 2007; Gayathri and Ramagopal, 2007; Meena and Patni, 2007). In 1987, Nadgauda et al. have used the bud explants of rhizome from the micropropagation and mass cultivation of various turmeric varieties like Duggirala and tekkurpetta. Some of the important Himalayan medicinal plants like *Potentilla fulgens* were also regenerated by the tissue culture from the axillary bud explants (Sambyal et al., 2006). Similarly, the mass multiplication of number of Himalayan threatened and endemic medicinal plants was performed by this important technique of micropropagation, for example, *Gentiana kurroo*, *Picrorhiza kurroa*, *Coleus forskohlii*, *Saussurea lappa*, *Atropa acuminata*, *Salvadora persica*, *Paederia foetida*, *Tylophora indica*, *Ginkgo biloba*, *Bacopa monnieri*, *Eclipta alba*, *Decalepis hamiltonii*, and *Janakia arayalpathra* (Ahuja, 1987; Sood and Chauhan, 2009; Sharma and Singh 1995; Mathur et al., 2002, 2003; Srivastava and Srivastava, 2004; Thangavel, 2008).

5.2.2 Regeneration by Adventitious Shoots

In the present era, the demand of the herbal medicine is going up at the alarming speed as a result of which there is increasing requirement of huge raw material. For the consistency of these herbal formulations, these plants should be harvested at the particular stage when these active compounds are available in optimum quantities. So, the intrusion of plant tissue culture is of great advantage for the rapid multiplication of desired clones and high-yielding medicinal plants. Micropropagation of medicinal and aromatic plants by using the adventitious shoots and their conservation by establishing gene or tissue banks thus plays an important role for the rapid supply of good quality planting material. Thus, in order to retain the genetic makeup of high-yielding medicinal plants under the controlled environmental conditions, micropropagation plays a sole role for at least those plants where vegetative propagation by conventional methods is unsatisfactory (Chaturvedi et al., 2007). It has been assumed that by micropropagation, a single node can produce a multibranched plant after the interval of 60 days that can in turn produce 40 single-node leaf cuttings. Thus after the incubation of 60 days, 40 cultures of multibranched plants can be raised, which in 1 year can produce at least 2,560,000 plants. So practically, it has been seen in the case of *Dioscorea floribunda* that by the micropropagation via nodal segments and shoot-to-shoot proliferation, there has been a million-fold increase in the rate of its multiplication over the conventional methods (Chaturvedi, 1975; Chaturvedi and Sinha, 1979) In the case of many medicinal plants, the various combinations of different hormones in the basal media have resulted in the direct organogenesis without the intermediate callus phase. For example, in *C. colocynthis*, direct organogenic response was observed from the cotyledon explants in the presence of zeatin and 2,4-D. Similarly, in *C. lanatus* (Adelberg et al., 1997; Srivastava et al., 1989) and *Coccinia indica* (Josekutty et al., 1993), the direct organogenesis was seen when these plants were grown on basal media supplemented with 11 μM IBA and 2.32 μM kinetin. Various investigations have proved that shoot buds were having inhibitory effect when zeatin was only used, but this negative effect was nullified in combination with 2,4-D. In the case of *Brassica* spp., the shoot bud formation was having stimulatory effect when BAP or kinetin was used in the presence of 2,4-D (Wong and Loh, 1987). It has been shown by various studies that IAA alone was not efficient enough to promote growth, but when used in combination with BAP, it was able to replace NAA. Such observation was proved in the case of *Trichosanthes dioica* (Hossain et al., 1997), *Sesbania speciosa* (Narmatha Bai et al., 1997), and *Momordica dioica* (Islam et al., 1994). Similarly, in the case of *Trichosanthes anguina* and *Luffa acutangula*, shoot buds were differentiated when the explants of these plants were grown on media supplemented with 4 mg/L BAP + 1 mg/L NAA and 3 mg/L BAP + 1 mg/L IAA, respectively, in two subsequent culturing. Thus, plantlets were regenerated from these shoot buds after growing these plants in MS media without any hormonal combination (Mustafa and Mallaiah, 1991).

5.2.3 Regeneration Through Somatic Embryogenesis

Plant regeneration by somatic embryogenesis is an important micropropagation process by which a zygotic embryo like structure is developed from the either haploid or diploid somatic cells without the fusion of gametes. Thus, by the series of embryological stages, a viable plant can be generated from the somatic embryo. This technique is very important and can be used for the conservation of rare and endangered plant species that are the storehouses of precious active constituents that are present in lesser quantities but have the high industrial demand like *C. roseus* and *Taxus brevifolia.* Similarly, speedy micropropagation is vital for endangered plants like *Dioscorea deltoidea* that has the highest diosgenin content among *Dioscorea* species but has a regeneration cycle of 10 years and is thus slow-propagating plant. Some of the medicinal plants that have been regenerated by somatic embryogenesis are as follows:

(I) *Podophyllum hexandrum*

Somatic embryogenesis was successfully used for the regeneration of one of the important medicinal plants, that is, *P. hexandrum* of the Himalayan region in 1990 that has a dormancy period of about 10 months (Arumugam and Bhojwani, 1990;

Badhwarr and Sharmab, 1963). The podophyllotoxin derived from the rhizome of this plant has anticancerous properties and thus has a high market demand (Kamil and Dewick, 1986). For the somatic embryogenesis, the excised embryos were germinated within only 7 days on the basal media in the presence of 2,4-D, suggesting that that endosperm and seed coat were responsible for the prolonged dormancy of this plant. The maturation of somatic embryos was only possible when transferred to the media supplemented with NAA.

(II) *Aconitum heterophyllum*

A. heterophyllum, an important perennial medicinal herb of the Himalayan region, was cultured through somatic embryogenesis via callus derived from leaf and petiole explants. Although the plant can be multiplied by either seed or root tubers, there is a huge loss of seed viability. So, *A. heterophyllum* was successfully raised from the callus that produced somatic embryos after 2–3 months and was later on transformed in the plantlet after the proper administration of different combinations of phytohormones (Giri et al., 1993). Most of the bioactive molecules in the medicinal herbs are derived from roots that result in the rapid depletion of these resources from their natural habitats. Thus, micropropagation provides the best opportunity for the rapid multiplication of these elite medicinal plants in the laboratory.

(III) *Rauvolfia serpentina*

R. serpentina L. is also an important and endangered medicinal shrub because its roots are a rich source of group of around 50 alkaloids and large number of bioactive chemicals (Nathan Kline, 1954; Anonymous, 1969; Lewis and Elvin-Lewis, 2003). The seeds of this plant have poor viability and germination rate, and only propagation by root cuttings is also a limiting factor. Thus, micropropagation holds an essential proposition in order to cater the roaring demand of pharmaceutical industry. These plants have been regenerated from somatic tissue via four morphogenic differentiation patterns like somatic embryogenesis, caulogenesis, direct plantlet formation, and regeneration of shoot buds from roots differentiated from somatic tissue. From each organized structures, the normal plants were regenerated (Chaturvedi, 1968; Chaturvedi et al., 2007). In vitro multiplication of *R. serpentina* by using leaf explants was developed by Singh et al. (2009). The callus was induced when these leaf explants were supplemented with 2.5 mg/L BAP + 2.0 mg/L IAA and 1.0 mg/L BAP + 0.5 mg/L IAA. After proper shooting and rooting, the plantlets were transferred to hardening unit for acclimatization. There was 67% survival rate of these plants after hardening (Singh et al., 2009).

(IV) *Angelica glauca*
A. glauca is a perennial, medicinal, and aromatic herb, with wide distribution across the Indian Himalaya (Anonymous, 1985). The plant is the source of many metabolites of pharmaceutical importance (Wang et al., 1996). As a result of which, the species has reached to endangered status and has been included in Indian Red Data Book (Anonymous, 1997; Jain and Sastry, 1984). The leaf explants raised from newly sprouted rhizomes were differentiated into somatic embryos and later on developed into plantlet when supplemented with various combinations of phytohormones (Bisht et al., 2015).

(V) *Picrorhiza kurroa*
In *P. kurroa*, germplasm conservation was performed by somatic embryogenesis from the leaf explants of newly sprouted rhizomes. Thus, the multiplication and regeneration of large number of plants were possible as compared with the conventional methods of propagation via rhizomes and seeds (Bisht et al., 2017).

(VI) *Withania somnifera* (Indian ginseng, Ashwagandha)
W. somnifera, the Indian ginseng, is also listed as endangered species because of the overexploitation as a result of immense medicinal properties like antiinflammatory, antiarthritic, curative for nervous and gynec disorders, and immunomodulatory to antitumour. Its medicinal properties are mainly due to alkaloids, like somniferine and withasomnine and in particular withanolide, namely, withaferin A. This plant has a poor germination rate, and the estimated production via root is about 1500 t annually, against the annual requirement of more than 7000 t. Thus, micropropagation of *W. somnifera* plays an important role for the rapid multiplication of this plant. Several workers have used nodal and internodal explants from seedlings for its micropropagation (Kulkarni et al., 2000). The regeneration from somatic callus and leaf segments taken from field-grown plants has also been practiced for its micropropagation (Rani and Grover, 1999; Sivanesan and Murugesan, 2005). The multiple shoots regenerated from leaf explants after 100% rooting have shown 87% plantlet survival under field conditions (Chaturvedi et al., 2007).

(VII) *Dioscorea deltoidea*
D. deltoidea is the richest source for commercial production of steroidal drug diosgenin. *D. deltoidea* being enlisted as endangered medicinal plant and the large-scale propagation by in vitro technique is essential for supplying its alarming demand. This plant has been clonally multiplied through tissue culture of shoot

tips and nodal stem cuttings but at a much lower rate than *D. floribunda* (Chaturvedi and Sinha, 1979; Chaturvedi et al., 1977). Similarly, *Dioscorea composita* has been multiplied through nodal stem segments at a moderate rate (Datta et al., 1982). Plants of *D. deltoidea* have been propagated under glasshouse conditions and those of *D. composite* under field conditions.

In addition to medicinal plants, a few aromatic plants with medicinal value have also been micropropagated to a rational extent.

(VIII) *Rosmarinus officinalis* L. and *Thymus vulgaris* L.
R. officinalis, a woody perennial herb, is an important source of rosemary oil that is commercially used in cosmetic industry due to its antiwrinkle and anabolic properties. This plant is very difficult to multiply via seeds due to less germination rate and is thus vegetatively propagated. A protocol for propagation of cloned explants of *R. officinalis* was developed by employing nodal stem segments and leaf segments. The plants were successfully cloned by micropropagation and were acclimatized in the soil too. It has been estimated that around 5000 plants were produced from one explant in a year (Purohit et al., 1994; Dave et al., 2003). Similarly, *T. vulgaris* is a commercial source of thymol and thyme oil, which are used as disinfectants externally and internally and are the main ingredients of cough expectorants. *T. vulgaris* has been micropropagated by shoot tips and nodal stem segments, producing about 22,000 plants from one explant just in 5 months, which have been satisfactorily grown in soil (Koblitz et al., 1983).

Although there are a number of papers on micropropagation of medicinal plants and in the Table 5.1 are given, the number of the medicinal plants has been successfully propagated via micropropagation and acclimatized to soil conditions.

5.2.4 Production of Secondary Metabolites From Medicinal Plants

Due to urbanization, a huge number of natural habitats for an enormous range of plants are swiftly destroyed resulting in extinction of many precious or even endemic species. Micropropagation of plants upholds superb promise for meticulous production of countless beneficial secondary metabolites. In order to cater the increasing industrial demand for the production of plant metabolites, biotechnological approaches, mainly plant tissue culture, own a potential to complement traditional agricultural practices (Vijaya et al., 2010). Some of medicinally vital alkaloids, anticancer drugs, recombinant proteins, and food components are created

TABLE 5.1 In Vitro Cloned Important Medicinal Plants Grown Ex Vitro

S. No.	Plant species	(Common name)	Explants	Ex vitro cultivation	Reference
1.	*Allium sativum* L.	(Garlic)	Apical meristem Meristem culture	Field cultivation Mass propagation	Ayabe and Sumi, 1998 Walkey et al., 1987
2.	*Aloe vera* L. *Atropa belladonna* L.	(*Aloe*) (Belladonna)	Shoot apex Various parts of plant Leaf, root segment	Field cultivation	Cavallini et al., 1991 Chaturvedi et al., 1982 Zenkteler, 1971
3.	*Artemisia annua* L.	(Sweet wormwood, qing hao)	Nodal stem segment		Gulati et al., 1996
4.	*Azadirachta indica A.* Juss	(Neem)	Nodal stem segment		Sharma et al., 2002; Chaturvedi et al., 2004
5.	*Bacopa monnieri* (L.) Pennell	(Waterhyssop)	Leaf, internode, and nodal stem segment		Tiwari et al., 2000, 2001; Ali et al., 1996
6.	*Carica papaya* L.	(Papaya)	Shoot tip and Inflorescence tip		Agnihotri et al., 2004
7.	*Cassia fistula* L.	(Golden shower)	Embryo Axillary shoot bud	Field cultivation	Bajaj et al., 1988 Gharyal and Maheshwari, 1990
8.	*Catharanthus roseus* (L.) G. Don	(Madagascar periwinkle)	Shoot tip		Bajaj et al., 1988
9.	*Chlorophytum borvilianum* Sant. et Fernand	(Chlorophytum)	Shoot tip	Field cultivation	Purohit et al., 1994 Dave et al., 2003

(Continued)

TABLE 5.1 In Vitro Cloned Important Medicinal Plants Grown Ex Vitro—cont'd

S. No.	Plant species	(Common name)	Explants	Ex vitro cultivation	Reference
10.	*Cinchona ledgeriana*	*Moens ex Trimen* (quinine)	Shoot meristem Shoot apex	Mass propagation Mass propagation	Koblitz et al., 1983 Hunter, 1988
11.	*Coleus forskohlii*	Briq.	Axillary bud		Sharma et al., 1991
12.	*Commiphora wightii* (Arnott) Bhandari.	(Indian bdellium tree)	Nodal explant		Barve and Mehta, 1993
13.	*Costus speciosus* (Koen. Ex. Retz) Sm	(Spiral ginger)	Rhizome Stem segments Shoot apex	Mass propagation Mass propagation	Chaturvedi et al., 1984 Indrayanto et al., 1999 Balachandran et al., 1990
14.	*Curcuma domestica* Val.	(Turmeric)	Rhizome bud Rhizome bud	Large scale Large scale	Nadgauda et al., 1978 Salvi et al., 2002
15.	*Digitalis lanata* Ehrh.	(Grecian foxglove)	Shoot tip Meristem		Erdei et al., 1981 Schöner and Reinhard, 1982
16.	*Dioscorea composita* Hemsl.	(Medicinal yam)	Stem and nodal tissue		Datta et al., 1982
17.	*Dioscorea deltoidea* Wall. ex Kunth	(Medicinal yam)	Tuber		Mascarenhas et al., 1976
18.	*Dioscorea floribunda* Mart. and Gal.	(Medicinal yam)	Single-node stem segment	Field cultivation	Chaturvedi 1975; Chaturvedi and Sinha, 1979

19.	*Duboisia myoporoides* R. Br.	(Corkwood)	Nodal stem segment		Kukreja and Mathur, 1985
20.	*Glycyrrhiza glabra* L.	(Licorice)	Shoot bud Shoot tip and nodal stem segment		Shah and Dalal, 1982 Thengane et al., 1998
21.	*Hyoscyamus niger* L.	(Henbane)	Petiole		Cheng and Raghavan, 1985
22.	*Mentha arvensis* L.	(Mint)	Axillary bud		Rech and Pires, 1986 Dhawan et al., 2003
23.	*Mentha piperita* L.	(Peppermint)	Axillary bud		Rech and Pires, 1986
24.	*Ocimum basilicum* L.	(Sweet basil)			Ahuja et al., 1982
25.	*Panax ginseng* Mey.	(Ginseng)	Cotyledon		Choi et al., 1998
26.	*Papaver somniferum* L.	(Opium poppy)	Hypocotyl		Nessler, 1982 Wakhlu and Bajwa, 1986
27.	*Picrorhiza kurroa* Royle. ex. Benth	(Picrorhiza)	Shoot tip Nodal stem segments	Field cultivation	Lal et al., 1988 Chandra et al. 2006
28.	*Rauvolfia serpentina* (L.) Benth. ex Kurz	(Serpentine)	Nodal stem segment	Mass propagation	Chaturvedi and Sinha, 1979
29.	*Rosmarinus officinalis* L.	(Rosemary)	Nodal stem segment	Field cultivation	Chaturvedi et al., 1984
30.	*Saussurea lappa* C. B. Clarke	(Costus)	Shoot tip		Arora and Bhojwani, 1989

(Continued)

TABLE 5.1 In Vitro Cloned Important Medicinal Plants Grown Ex Vitro—cont'd

S. No.	Plant species	(Common name)	Explants	Ex vitro cultivation	Reference
31.	*Solanum khasianum* Clarke		Nodal stem segment Excised-root segment	Field cultivation Field cultivation	Chaturvedi and Sinha, 1979 Chaturvedi et al., 2004
32.	*Solanum surattense* Burm. f.	(Yellow-berried nightshade)	Leaf, stem, and root segment		Prasad and Chaturvedi, 1978
33.	*Tylophora indica* (Burm. F.) Merrill	(Emetic swallow wort)	Nodal stem segment		Sharma and Chandel, 1992
34.	*Withania somnifera* Dunal	(Indian ginseng)	Shoot tip leaf segment		Kulkarni et al., 2000 Sivanesan and Murugesan, 2005
35.	*Zingiber officinale* Rosc.	(Ginger) excised	Shoot bud Meristem culture	Mass propagation Mass propagation	Nadgauda et al., 1980 Bhagyalakshmi and Singh, 1988

by numerous cultures of plant cells and tissues. A number of these compounds at the moment are commercially made available in the market, for example, shikonin and paclitaxel (taxol). To date, 20 diverse recombinant proteins like antibodies, enzymes, edible vaccines, growth factors, and cytokines have been produced in plant cell culture. Recent advances in scale-up methodologies and immobilization procedures have resulted in the substantial increase in the production of compounds with a high-added value (Sajc et al., 2000). Cell suspension cultures of various medicinal and aromatic plants for the mass production of plant secondary metabolites are nowadays widely practiced. Cell cultures cannot solely yield outlined characterized standard phytochemicals in substantial volumes but can additionally wipe out the presence of interfering molecules that occur in naturally grown plants. The advantage of this methodology is that it will ultimately give an eternal, reliable supply of natural products. The variety of various forms of bioreactors is used for mass cultivation of plant cells. The primary industrial application of huge-scale cultivation of plant cells was carried out in stirred tank reactors to provide shikonin by cell culture of *Lithospermum erythrorhizon*. Similarly, cells of *C. roseus*, *D. deltoidea*, *Digitalis lanata*, *Hypericum perforatum*, *Maackia amurensis*, *Panaxnoto ginseng*, *Taxus wallichiana*, and *Sophora flavescens* have been cultured in various bioreactors for the production of secondary plant products (Davioud et al., 1989). Thus, plant tissue culture signifies the most encouraging areas of application at present time and gives an outlook into the future. The effective attempts to produce some of these valuable pharmaceuticals in relatively large quantities by cell cultures are enlisted as (Jordan and Wilson, 1998; Abdin et al., 2003).

Taxol

Taxol (paclitaxel) is one of the WHO's model list of essential cytotoxic drug that kills cancerous cells. It is a fancy diterpene organic compound found within the bark of the genus *Taxus* and is one among the foremost promising antitumor agents known because of its distinctive mode of action on the microtubular cell system (Cragg et al., 1993). It is used to treat breast cancer, ovarian cancer, nonsmall-cell lung cancer, pancreatic cancer, and AIDS-related Kaposi sarcoma. In order to overcome the overexploitation of *Taxus* tree and the hefty amount required for the chemical synthesis of taxol, plant cell culture of various *Taxus* species for the mass production of taxol has been extensively explored (Suffness, 1995; Fett-Neto et al., 1994). To increase the production of taxol in these cultures, the effect of various amino acids to the culture was studied, and one of the essential amino acid, phenylalanine, was found to help in most taxol production in *T. cuspidata* cultures (Ciddi et al., 1995). The impact of various biotic and abiotic elicitors was conjointly studied to enhance the production and accumulation of taxol through tissue cultures (Strobel et al., 1992; Yukimune et al., 1996; Tam et al., 1980).

Morphine and Codeine

Papaver somniferum (opium poppy) is the industrial source of the analgesics, morphine, and codeine. For the large-scale production of these analgesics, callus and suspension cultures of *P. somniferum* have been investigated by various groups. Thus, the morphologically undifferentiated cultures were able to produce morphine and codeine (Yoshikawa and Furuya, 1985; Furuya et al., 1972). It has been found that the production of morphine and codeine was higher in the presence of hormones; for example, in the absence of hormones, maximum concentration of codeine and morphine was 3.0 and 2.5 mg/g dry weight, respectively, which was about three times higher in cultures supplied with hormones. Biotransformation of codeinone to codeine has also been reported by Furuya et al. 1972 by the immobilization of cells of *P. somniferum* (Guggenheim, 1913). The conversion yield by this technique was about 70%, and about 88% of the converted codeine was emitted into the medium.

L-DOPA

L-3,4-dihydroxyphenylalanine (L-DOPA) is an imperative secondary metabolism intermediate in higher plants and is a precursor of alkaloids, betalains, and melanins. L-DOPA has been isolated from *Vicia faba* (Daxenbichler et al., 1971) *Mucuna, Baptisia,* and *Lupinus* (Brain and Lockwood, 1976). It is also a precursor of catecholamines in animals and is used as an effective drug against Parkinson's disease, a progressive disabling disorder associated with a deficiency of dopamine in the brain. As a result of extensive appliance of this therapy, there has been increasing demand for the large-scale production of the L-DOPA at an economical price. Thus, the introduction of cell cultures has become an important means for the mass production of L-DOPA under in vitro conditions. It has been observed by Brain and Lockwood (1976) that the callus culture of *Mucuna pruriens* accumulated 25 mg/L L-DOPA in the medium fortified with enriched amount of auxin, 2,4-D. Later on, Teramoto et al. (1988) induced callus tissues in many other species of *Mucuna* like *M. hassjoo, M. pruriens,* and *M. deeringiana* and optimized the culture conditions. The maximum concentration of L-DOPA was attained when *M. hassjoo* cells were cultivated in MS medium supplemented with 0.025 mg/L 2,4-D and 10 mg/L kinetin.

Diosgenin

Diosgenin, a steroidal sapogenin, is an important precursor for the chemical synthesis of the steroidal drugs that occurs abundantly in plants. This bioactive phytochemical is tremendously significant to the pharmaceutical industry (Tal et al., 1983). Besides being used as steroidal drug, it has a high potential and interest in the treatment of various types of disorders like cancer, hypercholesterolemia, inflammation, and several types of infections (Jesus et al., 2016). In 1983, Tal et al. first time reported the use of cell cultures of *D. deltoidea* for the mass production of diosgenin. It has been found that the prominent levels of carbon and nitrogen significantly powered diosgenin amassing in one cell line. Later on, Ishida (1988) established immobilized cell cultures of *Dioscorea*, in which reticulated polyurethane foam was revealed to stimulate diosgenin production, escalating the cellular concentration by 40% and total yield by 25%. Tal et al. have been able to achieve diosgenin levels as high as 8% in batch-grown *D. deltoidea* cell suspensions.

Camptothecin

Plants are believed as a treasure trove of medicinal compounds to lessen the human ailments (Briskin, 2000). Camptothecin, a potent cytotoxic monoterpenoid indole alkaloid, was first time isolated from *Camptotheca acuminata* (Sakato and Misawa, 1974). The production of this drug from the natural sources in fact threatens the very existence of this plant, as a result of which it is very important to find some other alternatives. In vitro micropropagation has been utilized as a sustainable source of the production of this leading drug. Thus, biotechnological methods like in vitro culture along with bioreactors play an important role for the mass production of these natural products for the preparation of pharmaceuticals. Initially, in 1974, Sakata has first time produced camptothecin from cell suspension cultures of *C. acuminata* (Sakato and Misawa, 1974). However, a reasonable level of camptothecin was not attained by this method. Later on, the shoot cultures of *C. acuminata* were found to produce higher amounts of camptothecin as compared with cell suspension (Sudo et al., 1991).

Berberine

Berberine is an important isoquinoline alkaloid isolated from the roots of *Coptis japonica* and also from the cortex of *Phellodendron amurense*. This antibacterial alkaloid has also been identified from a number of cell cultures, remarkably from those of *C. japonica*, *Thalictrum* spp., and *Berberis* spp. (Vanisree et al., 2004; Breuling et al., 1985; Sato and Yamada, 1984; Nakagawa et al., 1984; Suzuki et al., 1988). The production of this alkaloid was enhanced in the cell cultures by the supplementation of various hormones and fortification of nutrient medium (Sato and Yamada, 1984; Nakagawa et al., 1986; Morimoto et al., 1988). Mitsui et al. produced berberine on a large scale from the high-yielding cell lines, which reached the productivity of 1.4g/L over 2 weeks. Berberine has also been produced from the cell cultures of *Thalictrum rugosum* culture upon elicitation of cultures with a yeast polysaccharide elicitor (Funk et al., 1987). Similarly, the inclusion of spermidine in the cell cultures of *Thalictrum minus* has also enhanced the berberine production as reported by Hara et al. (1991).

5.2.5 Conservation Through Cryopreservation

Cryopreservation, established from the last 25 years, is a crucial and most worthy technique for long-term preservation of biological resources. Cryopreservation is the conservation of living cells, tissues organs, and microorganisms at ultralow temperature (usually that of liquid nitrogen, −196°C). As compared with serial subculturing, under cryogenic storage at liquid nitrogen temperature, cell division and all other metabolic activities are halted, as a result of which the stored cells will not undergo via any genetic changes and can be conserved for extended durations. One of the most important features of cryopreservation is that the cells can be stored in the small space, require very minimal maintenance, and have the less risk of contamination and operator errors (Engelmann, 2004). Thus, by this safe and cost-efficient technique, the various categories of plants like vegetatively propagated plants, rare and endangered species, and non-orthodox seed can be preserved for longer durations. In order to prevent cryopreserved materials from the freezing injury and to maintain the post-thaw viability of these samples, the water should be removed from these samples. On the basis of formation of ice within the plant material, there are two different mechanisms of cryopreservation: the classical cryopreservation technique, in which cooling is performed in the presence of ice, and newer method based on vitrification, in which cooling takes place without ice formation (Gonzalez-Arnao et al., 2008). The new cryopreservation methods include encapsulation-dehydration, vitrification, encapsulation-vitrification, desiccation, pregrowth, pregrowth-desiccation, and droplet freezing (Engelmann, 1997). The various cryopreservation methods are as follows:

1. Desiccation method (air drying)
   Desiccation is the least difficult strategy and composed of hydrating explants and freezing them quickly by drenching in LN. The stream of sterile air under the laminar flow is used for the drying of the samples. The air drying in the closed tube in the presence of silica gel is also one of the efficient methods of drying (Panis et al., 2001). The strategy is especially applied to commonest agricultural and horticultural species, orthodox seeds, zygotic embryos, embryogenic axes, and pollen grains (Uragami et al., 1990; Engelmann, 2004). In 1999, Changrum et al. have proposed that the rate of water loss among tissues and various tissues of different species is highly variable. It has been found that some of orthodox seeds are very resistant to drying, and there has been no reduction in viability when the moisture content was even below 3% (Uragami et al., 1990; Panis et al., 2001; Engelmann, 2004). In some cases, drying caused expulsion of water from the cell, which resulted in various solute impacts, for example, pH changes, increasing electrolyte concentrations, protein denaturation, macromolecular interactions, and

ultimately severe damage to cell (Dumet and Benson, 2000). Therefore, drying may not be fundamentally advantageous for cryopreservation, if uneven circulation of water results in changed freezing responses in different cells of the tissue.

2. Slow freezing (classical method)
It is also known as a two-step freezing and slow-controlled freezing method. In this method, the tissues are cooled down gradually at a controlled rate from 0.1 to −40°C and are afterward rapidly immersed in LN. Among the various cryopreservation techniques, slow-freezing method is the most common and requires a programmed freezing apparatus (Zhao et al., 2005). The cold acclimatization causes complex physical and biochemical changes within a cell, for example, alters the membrane composition that results in increasing the dehydration tolerance of cell (Sugawara and Steponkus, 1990; Hannah et al., 2005). The various tropical and subtropical species that are not acclimatized to the cold temperature have shown improved cryopreservation results as a result of cold precultures (Chang et al., 2000; Leunufna and Keller, 2005). The cold-acclimated plants accumulate the higher concentration of soluble sugars (Levitt, 1972). These soluble sugars protect the cells from the frost injury and play an important role in osmoprotection, cryoprotectant, and metabolization of other protective substances during cryopreservation (Hincha, 1990; Hitmi et al., 1999). During freeze-induced dehydration, free water molecules are drawn out from the cytoplasm to intercellular spaces where it crystallizes via osmotic gradient (Benson, 2004). As a result of this dehydration, the solute concentration within the cells rises that prevents the formation of ice crystals within the cells. Thus, the cells solidify without the formation of crystals that give the glassy status to the cells. Various cryoprotective solutes like DMSO, sucrose, and PVS2 decrease the free water content within the cells (Sakai et al., 1990). In 2008, Benson has found that DMSO and glycerol prevented ice formation within the cells and increased the cellular osmolality of the cells by penetrating the call wall and membranes. Also it has been shown that sucrose was not able to penetrate the plasma membrane (Tao and Li, 1986). Thus when the cells were frozen, sucrose was accumulated in the cell-wall space and formed the buffer layer between the cell wall and protoplasm, as a result of which the cells were protected from freeze-induced dehydration (Tao and Li, 1986). During the cold acclimation, the plant cells accumulated higher content of essential amino acids. In potato shoot tips, there was a higher concentration of amino acids like glycine, glutamine, asparagines, and proline during cold acclimatization (Stewart and Larher, 1980). Levitt, in 1972, has found that the nitrogen content was increased during the growth of the plants at

lower temperatures. Cooling speed was also found to be a main factor responsible for slow-freezing method of cryopreservation; there were higher survival rates in plants that were slowly cooled down (Benson et al., 1996; Chang and Reed, 2000). A combination of slow-freezing and vitrification methods has shown 61% regrowth rate as compared with only vitrification that showed 43% regrowth rate only in 28 *Pyrus* genotypes (Luo et al., 1995). In addition, classical cryopreservation methods have also been successfully applied to cell suspensions and callus cultures (Kartha and Engelmann, 1994). Fukai (1990) has successfully applied a controlled-rate freezing method of cryopreservation for *Chrysanthemum morifolium* and other *Chrysanthemum* species. In potato, survival rate was increased by slow-freezing method as compared with that of the basic cryopreservation method of vitrification alone (Zhao et al., 2005).

3. Preculture and preculture/dehydration
This method is mostly used for the cryopreservation of zygotic and somatic embryos of some of the plant species. This method involves preculture or pregrowth of germplasm on a medium supplemented with various cryoprotectants like sucrose or glucose and then immersion in LN (Dumet et al., 1993; Engelmann, 1997). This technique was used for the cryopreservation of highly proliferating meristems of banana. The meristem cultures were grown on proliferating medium supplemented with 0.4M sucrose for 2 weeks; later on, these meristem clumps were excised, transferred to cryovials, and frozen rapidly. The regeneration rate of postthaw cultures was varying between 0% and 69% depending on the cultivar (Panis et al., 1996). In the case of asparagus, the nodal explants were precultured on 0.7M sucrose for 2 days, after which the nodal segments were desiccated to 20% water content with silica gel. The survival rate was raised up to 63% after cooling in LN (Uragami et al., 1990). In the case of *Lilium ledebourii*, the seeds pretreated with sucrose and dehydration have shown survival rate of 75% (Kaviani et al., 2009). Similarly in *Pyrus*, 80% recovery was obtained on 0.75M sucrose preculture with 4h dehydration (Scottez et al., 1992). For the major cryopreservation techniques, sucrose increases the desiccation tolerance and thus is an important pregrowth additive during cryopreservation. Amid diverse types of sugars (fructose, glucose, sorbitol, and sucrose) used as somatic agents in preculture medium, sucrose was the best for the survival of cryopreserved date palm tissue culture (Bekheet et al., 2007). The survival rates of various plants like asparagus (Uragami et al., 1990), carrot (Dereuddre et al., 1991), and date palm (Bagniol and Engelmann, 1992) have been increased by preculturing these tissues on media containing high concentration of sugar (Bekheet et al., 2007).

4. Encapsulation-dehydration

The encapsulation-based techniques are based on the production of high-quality "synthetic seeds." Initially, this technique was used to encapsulate single somatic embryos surrounded by artificial seed coat (Murashige, 1977), but nowadays, various explants like shoot tips, nodal segments, bulblets, and even callus samples are used to produce synthetic seeds (Lambardi et al., 2006). The most widely used encapsulation method to produce synthetic seeds was developed by Redenbaugh et al. (1991). For the production of synthetic seeds, the explants are first incubated in a Na-alginate solution (1%–5% and 3%) to form the alginate drops that are subsequently released into a complexity agent (50–100 mM $CaCl_2$ solution) where bead hardening occurs in 20–30 min. Encapsulation-dehydration (ED) is an important new cryopreservation technique and is based on the artificial seed technology. Fabre and Dereuddre (1990) have developed this method, and it was used for *Solanum*. The artificial seeds of *Solanum* explants were prepared in alginate beads followed by subsequent culture in a highly concentrated sucrose solution (0.7–1.5 M) and finally physical dehydration or air drying till the moisture content decreases to 20%–30% and lastly direct immersion in LN. It has been found that the presence of a nutritive matrix (the bead) around the explants can promote the regrowth of cryopreserved explants after thawing. Although the ED method is very simple, some species cannot tolerate the high concentration of sucrose used, and proper handling of alginate beads is also required. Physical desiccation used in this method is carried out either with silica gel or in the airflow of the laminar flow cabinet (Paulet et al., 1993). The ED method was used for cryopreservation of *Cynodon*, in which 1- to 4-week cold-acclimation period was combined with dehydration up to 19%–23% moisture before exposure to LN (Reed et al. 2005). The ED method has been used for the formation of synthetic beads from embryonic axes of *Melia azedarach* L. in the presence of sucrose (0.75 M) and various concentration of salicylic acid, which enhanced the viability rate of cryopreserved explants (Bernard et al., 2002; Kaviani, 2007). Similarly the somatic embryo of Citrus was cryopreserved by ED method, and the survival rate of 100% was obtained after the beads in enclosed somatic embryos were pretreated on media with high sucrose concentration, dehydrated below 25% moisture content, and directly immersed in LN (Duran-Vila, 1995).

5. Vitrification

Vitrification is the important technique and can be defined as "the solidification of a liquid brought about not by crystallization but by an extreme elevation in viscosity during cooling" (Fahy et al., 1984). It is a simple method of cryopreservation and prevents the cells from the mechanical injury from ice. Thus, vitrification-based methods result in the elimination of most or all freezable water from cells by physical or osmotic dehydration of explants, followed by ultrarapid freezing that finally results in vitrification of intracellular solutes, that is, formation of an amorphous glassy structure without occurrence of ice crystals that are detrimental to cellular structure integrity. The planting material revived via vitrification has a higher recovery rate and can be used for the cryopreservation of those species that are even sensitive to low temperature (Takagi et al., 1997; Thinh et al., 1999). The procedure for vitrification involves exposure of planting material to the highly concentrated cryoprotectant solution for the erratic time period of about 15 min up to 2 h, followed by the direct immersion in the LN. During vitrification, there is the enormous increase in the viscosity of cells, as a result of which plants are able to withstand deadly osmotic and evaporative dehydration stresses (Reed et al., 2005). By this method, a broad range of plant species have been cryopreserved that involve the complex organs like shoot apices or embryos (Sakai, 1993; Huang et al., 1995; Takagi, 2000; Vidal et al., 2005; Wang et al., 2005). In order to tolerate the severe effect of dehydration, the osmoprotectants like 2 M glycerol and 0.4 M sucrose along with plant vitrification solution 2 (PVS2) were widely used by various studies (Hirai and Sakai, 2003; Matsumoto and Sakai, 2003; Kobayashi et al., 2006). Plant vitrification solution 2 (PVS2) was developed by Sakai et al. in 1990 and was composed of 30% (w/v) glycerol, 15% (w/v) ethylene glycol, and 15% (w/v) DMSO in liquid medium with 0.4 M sucrose. Thus, vitrification is much simple than ED as it requires lower handling, but the only disadvantage is the toxicity of the concentrated vitrification solutions. The toxicity of vitrification solution can be nullified by cold and sugar hardening, a loading phase, and the incubation of planting material in the vitrification solution at 0°C in place of RT. The vitrification solution causes the chemical toxicity or osmotic stress that results in the severe damage to planting material (Sakai, 2000). Therefore, permeation of some components of vitrification solution should be delayed, and adequate dehydration should be critically allowed to avoid the toxicity effect (Charoensub et al., 1999; Tsukazaki et al., 2000). Also it has been found that direct exposure of germplasm to PVS2 reduces viability and increases toxicity. Thus, it is recommended that there should be stepwise increase in PVS2 concentration that diminishes this toxic effect (Kobayashi et al., 2006). It has been shown that temperate species that amass sugars and protective proteins during cold season are better able to withstand as compared with tropical species (Chang et al., 2000; Reed et al., 2005). Vitrification has also been used for tropical and warm-temperature plant germplasm by increasing the tolerance of plants by pretreatment with high sucrose (Benson, 2004; Reed et al., 2004; Sakai, 2004). The slow-cooling and innovative (vitrification/one-step freezing) methods have all been successfully applied to *Citrus* cryopreservation (Wang and Deng, 2004; De Carlo and Lambardi, 2005). In the case of *Carica papaya* L., Wang et al. (2005) have found that the recovery percentage of six cultivars of shoot tips of this plant was between 48% and 93%.

5.3 GENETIC TRANSFORMATION FOR IMPROVEMENT OF YIELD AND QUALITY OF MAPs

From the past two decades, there has been a significant upsurge in the number of reports regarding the successful genetic transformation of numerous plant species, variants, and cultivars mediated by the plant genetic engineer, *Agrobacterium tumefaciens* (Herrera-Estrella et al., 2004). It is a well-studied soil bacterium that has a natural ability to infect plants at the site of injuries and results in the formation of crown gall at the site of infection. During *Agrobacterium*-mediated transformation, a fragment of Ti plasmid carrying the gene of interest gets transferred and integrated into the plant genome. The genetic transformation of medicinal plants is less efficient and depends upon the number of factors like *Agrobacterium* strain, methods, and procedures of transformation and on plant species, type, and age of the explants and regeneration conditions. The genetic transformation of medicinal plants is very important to increase the production of bioactive compounds and biopharmaceuticals that are otherwise available in limited amounts. One of the prerequisites for the stable genetic transformation of medicinal plants is the ability of plants to regenerate. To date, various attempts have been successfully tried for the stable *Agrobacterium*-mediated genetic transformation of medicinal plants belonging to the families like Araceae, Apocynaceae, Asteraceae, Araliaceae, Asphodelaceae, Begoniaceae, Crassulaceae, Fabaceae, Linaceae, Lamiaceae, Plantaginaceae, Papaveraceae, Scrophulariaceae, and Solanaceae.

5.3.1 *Agrobacterium tumefaciens* Mediated Transformation

Agrobacterium hereditarily changes its host by transmitting its well-characterized DNA segment from its tumor-inducing (Ti) plasmid to the host-cell genome (Gelvin, 1998). The transferred DNA (T-DNA) of *A. tumefaciens* carries an arrangement of oncogenes (Gaudin et al., 1994) and genes for opine catabolism, the expression of which within plants prompts neoplastic growth of the transformed tissue, the synthesis of opines, and the amino acid by-products solely utilized by bacterium as a nitrogen source. Thus, T-DNA of recombinant *Agrobacterium* strains in which some native insignificant genes have been replaced by the particular gene of interest is the most proficient vehicle utilized nowadays for the introduction of foreign genes into plants and further generation of transgenic plant species (Draper et al., 2001) although the main aim of the genetic transformation and manipulation of genes in medicinal plants is the ample production of bioactive compounds. Besides, other traits related to the plants like uniformity, stability, growth and development, and resistance to biotic and abiotic stresses (Bajaj et al., 1988; Dubey and Guerra, 2002; Bernard et al., 2002). Nowadays, a lot of research is going on in manipulating plant biosynthetic pathways involved in the secondary metabolism to produce drug precursors, food components, or pesticides (Stevenson, 2004). Biosynthetic pathways of *Mentha* spp. (mints) have been engineered in order to vary essential oil production in trichomes and also to improve the resistance of the plant to fungal infection and other abiotic stresses (Veronese et al., 2001). *Agrobacterium*-mediated genetic transformation has been widely employed to transfer genes into the cultured plant DNA in order to improve product output in plant systems. For example, hairy-root cultures, transformed by *Agrobacterium* rhizogenes, often withstand stable and high productivity in hormone-free culture conditions (Sevón and Oksman-Caldentey, 2002). The proficient transgene delivery system has already been stabilized for several important medicinal plants, like *P. somniferum* (Park and Facchini, 2000), *Artemisia* spp. (wormwoods) (Chen et al., 1999), members of the family Solanaceae (Moyano et al., 1999), and *Taraxacum platycarpum* (Lee et al., 2004). Transformation systems based on *A. tumefaciens* are well established for *Taxus* (yew) (Han et al., 1994), *Echinacea* (Wang and To, 2004), *Scrophularia* (figwort) (Park et al., 2003b), *Digitalis* (foxglove) (Sales et al., 2003), *Thalictrum* (meadow rues) (Samanani et al., 2002), and *Artemisia* (Chen et al., 2000). Although the problems for the regenerating of whole plants from cultures have been overcome for many plants, still, some important species remain recalcitrant, notably *G. biloba* (Balz et al., 1999). The major challenge for the effective engineering of pathways in medicinal plants is the poor understanding of the metabolic pathways involved in the biosynthesis of active compounds. Recently, in hairy root cultures of *Hyoscyamus niger* (black henbane), simultaneous overexpressing of two genes encoding the rate-limiting upstream and downstream biosynthetic enzymes has resulted in the ninefold increase in production of the sedative compound scopolamine (Zhang et al., 2004). Yun et al. have increased the production of scopolamine in *A. belladonna*, by transforming with the enzyme hyoscyamine 6-b-hydroxylase from *Hyoscyamus* (Yun et al., 1992). Similarly, studies have been made toward engineering of alkaloid production in *P. somniferum* (Facchini et al., 2000). A threefold improvement in the production of the putative antimalarial and anticancer drug artemisinin has been described in transgenic *Artemisia* plants overexpressing farnesyl diphosphate synthase, the enzyme immediately preceding the first committed biosynthetic step (Chen et al., 1999, 2000). Regulation of transcription factors that turn whole secondary pathways on or off shows great promise in engineering metabolic pathways as compared with targeting an individual rate-limiting enzyme reaction (Yun et al., 1992). Herbicide-resistant *Atropa* and *Panax ginseng* transgenic plants have been developed that are resistant to bialaphos, glufosinate and Basta by transformation with the enzyme phosphinothricin acetyltransferase (Saito et al. 1992; Choi et al. 2003). The herbicide-tolerant *Solanum nigrum* (black nightshade) was developed by somatic hybridization, and also *Panax quinquefolius* was developed that was resistant to fungal diseases by transforming with antifungal gene chitinase or a thaumatin; both of these transgenic plants were successfully regenerated (de Donato et al., 1989; Chen and Punja, 2002; Punja and Chen, 2003). In 2014, Li et al. have established the *A. tumefaciens*-mediated transformation of in vitro raised plants of *D. purpurea* L. by using mature leaf explants.

5.3.2 RNAi Technology

To date, humans have used over 50,000 plant species as medicines for treating various types of diseases. Some of the irreplaceable medicinal products derived from plants are as antihypertensive (e.g., reserpine), analgesic (e.g., morphine), antitussive (e.g., codeine), antineoplastic (e.g., vinblastine and paclitaxel), cardiotonic (e.g., digoxin), or antimalarial (e.g., artemisinin) (Nessler, 1982). From the past 30 years, over 25% of the new drugs have been approved that are derived from plants, and around 50% of the top-selling chemicals in the market are derived from the knowledge based on plant secondary metabolism (Terryn et al., 2006). Recently, FDA has approved seven plant-derived compounds with anticancer activity for clinical use: taxol/paclitaxel (*T. brevifolia*), topotecan and irinotecan (*C. acuminata*), vinblastine and vincristine (*C. roseus*), and etoposide and teniposide (*Podophyllum peltatum*). In order to increase the beneficial or desirable components in food crops and to enhance the levels of target compounds that might be present in trace amounts in plants, various characteristic features of the plant can be changed like improving agronomic traits through the expression of a metabolite that might protect the plant against insect or pathogen attack and lowering levels of noxious or antinutritional factors in food using intact plants, plant organs, or cells as production vehicles for novel compounds. Thus, various approaches used to increase levels of a valuable molecule in a target plant include increasing flux to the target molecule, reducing flux through competing pathways, and overcoming rate-limiting steps and regulatory genes or transcription factors that induce the pathway, inhibit, or limit catabolism of the molecule or increase the number of specialized cells producing the compound in question. Successful RNAi-mediated gene-silencing efforts in medicinal plants are listed in Table 5.2. Further, for the synthesis of novel

TABLE 5.2 RNAi Mediated Gene-Silencing in Various Medicinal Plants

S. No.	Enzyme (gene)	Host species	Product	References
1	Cytochrome P450 (+) menthofuran synthase	*Mentha x piperita*	Menthofuran	Mahmoud and Croteau, 2001
2	Berberine bridge enzyme (BBE)*	*Eschscholzia californica*	Benzophenanthridine alkaloids	Park et al., 2002
3	BBE*	*Eschscholzia californica*	(*S*)-reticuline	Park et al., 2003a
4	Putrescine *N*-methyltransferase (PMT)	*Nicotiana tabacum*	Pyridine and tropane alkaloids	Chintapakorn and Hamill, 2003
5	Codeinone reductase (COR)*	*Papaver somniferum*	Codeine and morphine	Allen et al., 2004
6	Limonene-3-hydroxylase gene	*Mentha x piperita*	Limonene	Mahmoud et al., 2004
7	Berberine bridge enzyme (BBE) and the *N*-melhylcoclaurine 3′-hydroxylase (CYP80B1)	*Papaver somniferum*	Morphine, codeine, sanguinarine	Frick et al., 2004
8	De-etiolated1 DET1	*Solanum lycopersicum*	Carotenoid and flavonoid contents of fruits	Davuluri et al., 2005
9	Benzoic acid/salicylic acid carboxyl methyltransferase PhBSMT1	*Petunia x hybrida*	Methylbenzoate	Underwood et al., 2005
10	R2R3 MYB-type transcription factor ODORANT1	*Petunia x hybrida*	Fragrance	Verdonk et al., 2005
11	Glutatione S-transferase 1 and 2, ubiquitin conjugating enzyme, putative cullin anthocyanidin synthase, putative flowering-time gene, GPT and the glutathione conjugate transporter (Anthocyanin Nagira et al., 2006)	*Torenia fournieri*	Anthocyanin	Nagira et al., 2006

12	Dammarenediol synthase gene	*Panax ginseng*	Ginsenoside	Han et al., 2006
13	Phenylacetaldehyde synthase (PAAS)	*Petunia x hybrida*	Complete suppression of phenylacetaldehyde and 2-phenylethanol emission	Kaminaga et al., 2006
14	Benzoyl-CoA/benzyl alcohol/2-phenylethanol benzyltransferase (BPBT)	*Petunia* x *hybrida*	Benzaldehyde	Orlova et al., 2006
15	Cinnamoyl-CoA reductase	*Solanum lycopersicum*	Phenolics	Van der Rest et al., 2006
16	Codeinone reductase	*Papaver somniferum*	Morphinan alkaloid	Larkin et al., 2007
17	BBE	*Eschscholzia californica* (*S*)-	Reticuline	Fujii et al., 2007
18	Norcoclaurine 6-O-methyltransferase	*Coptis japonica*	Benzylisoquinoline alkaloid	Inui et al., 2007
19	Coniferyl alcohol acyltransferase (PhCFAT)	*Petunia x hybrida*	Coniferyl aldehyde and homovanillic acid	Dexter et al., 2007
20	Cinnamyl alcohol dehydrogenase	*Linum usitatissimum*	Lignin reduction	Wróbel-Kwiatkowska et al., 2007
21	Endogenous dihydroflavonol 4-reductase (DFR) and overexpression of both iris DFR and viola flavonoid 3′5′-hydroxylase	*Rosa hybrida*	Delphinidin	Katsumoto et al., 2007
22	VEGF-R2 protein	*Commiphora mukul*	*z*-Guggulsterone	Xiao and Singh, 2008
23	Salutaridinol 7-*O*-acetyltransferase	*Papaver somniferum*	Morphinan alkaloids	Allen et al., 2008
24	Salutaridinol 7-*O*-acetyltransferase (SalAT)	*Papaver somniferum*	Morphine	Kempe et al., 2009
25	Tryptophan decarboxylase	*Catharanthus roseus* plant culture	Tryptamine	Runguphan et al., 2009

compound in a plant, the appropriate heterologous genes are introduced in the plant system. Similarly, the competitive pathway of the required phytoconstituents can be reduced to eliminate the various by-products of undesirable compounds. The meticulous steps in the pathway leading to such undesirable compounds can be blocked by antisense, cosuppression, or RNA interference (RNAi) technology. Similar results can also be achieved by suppressing the genes that upregulate the pathway or by increasing catabolism (Gómez-Galera et al., 2007). Antisense technology has been widely used for blocking activity of an enzyme, but it sometimes fails when the single enzymes are coded by multigenes. In order to block the activity of such enzymes that are coded by a multigene family, RNAi technology offered an alternative that not only blocked the activity of such enzymes but also were expressed across a number of tissues and developmental stages, for example, codeinone reductase involved in the biosynthesis of morphinan alkaloid (Larkin et al., 2007). With the help of DNA-directed RNAi in transgenic opium poppy, Allen et al. (2004) have knockdown the activity of codeinone reductase, as a result of which the precursor (*S*)-reticuline was accumulated at the expense of morphine, codeine, oripavine, and thebaine. In California poppy, Fujii et al. (2007) used RNAi technology to block the activity of berberine bridge enzyme (BBE) resulting in the accumulation of (*S*)-reticuline, an important intermediate of metabolic pathways of isoquinoline alkaloid biosynthesis. Recently, RNAi technology has been used widely to accelerate the breeding of medicinal plants, where a conventional mutation breeding approach failed (Allen et al., 2004). In 2008, Allen and coworkers have shown that both overexpression and suppression of salutaridinol 7-*O*-acetyltransferase (SalAT) enzyme in opium poppy effect the accumulation of alkaloid products (Allen et al., 2008). Later on in 2009, Kempe et al. have reduced the transcript levels of the gene-encoding morphine biosynthetic enzyme, salutaridinol 7-*O*-acetyltransferase (SalAT) in *P. somniferum* by RNAi technology. The downregulation of salAT gene has resulted in the accumulation of the intermediate compounds, salutaridine and salutaridinol, in a ratio of 2:1 to 56:1 (Kempe et al., 2009). In *Artemisia annua*, the artemisinin content was increased by the suppression of enzymes involved in competitive metabolic pathways. By the help of RNAi, the key gene of sterol biosynthetic pathway, squalene synthase (SQS), was downregulated that ultimately diverted the pathway toward artemisinin biosynthesis in transgenic plants, with the highest values reaching 31.4 mg/g DW, which is about 3.14-fold higher than the control plants (Zhang et al., 2009). In addition, four competitive metabolic pathway genes like *β*-caryophyllene synthase gene (CPS), *β*-farnesene synthase gene (BFS), germacrene A synthase gene (GAS), and SQS in *A. annua* were downregulated independently by the antisense method. In anti-CPS transgenic plants, the contents of artemisinin and DHAA were increased by 77% and 132%, respectively. In anti-BFS transgenic plants, the contents of artemisinin and DHAA were increased by 77% and 54%, respectively. In anti-GAS transgenic plants, the contents of artemisinin and DHAA were enhanced by 103% and 130%, respectively. In anti-SQS transgenic plants, the contents of artemisinin and DHAA were enhanced by 71% and 223%, respectively (Lv et al., 2016). Thus, RNAi is an important technique for the production of important biomedical products by medicinal plants, which in turn can offer novel and rapid applications. In order to increase production of compounds in medicinal plants, RNAi seems to be a better technique. But the need of the hour is to understand the extent of complete effects in medicinal plant cells and to develop an inducible RNAi system with a combination controlled promoter and a convenient inducer applicable to a wide range of medicinal plants. Lastly, RNAi protocols should be used for genome-wide screening that will assist in the rapid identification of genes involved in novel compound production. Hence, RNAi is a potentially powerful tool for a wide variety of gene-silencing applications.

5.3.3 Molecular Pharming for High Production of Bioactive Compounds

The production of various recombinant proteins like industrial proteins or pharmaceuticals and other secondary metabolites in plants is known as plant molecular pharming. In this process, the active protein is synthesis within the plants and involves growing, harvesting, transport, storage, and downstream extraction and purification (De Wilde et al., 2002). The first plant-derived recombinant proteins that were produced in the transgenic plants were the human growth hormone and antibody (Hiatt et al., 1989; Barta et al., 1986). However, in 1997, the recombinant protein, avidin (an egg protein), was first time expressed in transgenic maize for commercial purpose (Hood et al., 1997). Thus, the large-scale production of these recombinant proteins within plants reveals that plants could be really turned into biofactories. It has been validated that plants have the ability to produce the complex mammalian proteins with the therapeutic activity, for example, human serum proteins, antibodies, vaccines, hormones, growth regulators, cytokines, antibodies, and enzymes (Lienard et al., 2007). Within the plants, the mammalian proteins were produced with the proper posttranslational modifications that enable these proteins to fold accurately and to maintain functional and structural integrity. As a result of these rapid-growing recombinant biopharmaceuticals, the pharmaceutical market that accounted about 10% in 2007 has been expanded to about US$ 100 billion in 2010 (Knäblein, 2005; Lowe and Jones, 2007).

Plant-Derived Recombinant Proteins

1. Plant-derived vaccine antigens
The conventional methods for the production of subunit vaccines in bacteria, yeast, and mammalian cells have some peculiar problems. As a result of that, plants were used as a platform for production of vaccines that were having potential advantages over the conventional ones, like large-scale production, reduced cost, and oral delivery option (Streatfield and Howard, 2003; Pujol et al., 2005; Buonaguro and Butler-Ransohoff, 2010) and being able to produce glycosylated subunit vaccines (Bosch and Schots, 2010). First plant-derived vaccines were developed around 20 years ago, and to date, a large number of vaccines are derived from plants (Rybicki, 2009; Tiwari et al., 2009). The hepatitis B surface antigen was expressed in transgenic potatoes (Richter et al., 2000), in tomato (He et al., 2008), in banana (Kumar et al., 2005), and in tobacco cell suspension culture (Sojikul et al., 2003). Similarly, heat-labile enterotoxin B subunit (LTB) of *Escherichia coli* was expressed in potato tubers (Lauterslager et al., 2001), in maize seed (Chikwamba et al., 2002), in tobacco (Rosales-Mendoza et al., 2009), and in soybean (Moravec et al., 2007). The cholera toxin B subunit (CTB) of *Vibrio cholerae* was expressed in several crops, including tobacco, tomato, and rice (Daniell et al., 2001; Mishra et al., 2006; Nochi et al., 2007). Besides, a large number of vaccines for veterinary purposes had also been developed in plants like avian influenza, Newcastle disease, and foot-and-mouth disease (Lentz et al., 2010; Ling et al., 2010), and hemagglutinin protein of rinderpest has been expressed in pigeon pea and peanut (Satyavathi et al., 2003). Other vaccines like human papillomavirus (Biemelt et al., 2003; Maclean et al., 2007; Giorgi et al., 2010), the Norwalk virus capsid protein (Mason et al., 1996), hemagglutinin protein from measle virus (Marquet-Blouin et al., 2003), and H5N1 pandemic vaccine candidate (D'Aoust et al., 2010) have been expressed in one or two of the following plants: tobacco, potato, and carrots. Thus, the plant-based vaccines compete with conventional methods, breaking the limits of current standard production technologies and reaching new frontiers for the plant-based production of pharmaceutical-grade proteins (Buonaguro and Butler-Ransohoff, 2010).

2. Plant-derived antibodies
Antibodies or immunoglobulins (IgGs) are serum proteins and are extensively used for prevention, detection, and treatment of diseases. The existing high cost of their production, in spite of roaring market demand, is preventing their successful introduction into the health market as therapy against infectious diseases. Thus, there is the need for the cost-effective alternatives for the large-scale production (Stoger et al., 2005). The posttranslational modification and proteolysis required for the mass production of antibodies

within the mammalian cell cultures pose the difficult challenge with respect to the complexity of the product. Plant-derived antibodies not only provide cheaper production rates as compared with the cost of mammalian cultures (0.1%–1%) and microbial culture (2%–10%) but also can assemble complex multimeric proteins with posttranslational modifications (Chen et al., 2005; Conrad and Fiedler, 1994). In 1989, the first recombinant antibodies were expressed in plants (Hiatt et al., 1989). The various antibodies that were expressed in plants were single-chain Fv fragments, Fab fragments, small immune proteins (SIP), IgGs, and chimeric secretory IgA (Ismaili et al., 2007; Xu et al., 2007). In order to produce the biologically active whole antibodies in plants, two major methods have been employed. The first is the cross-pollination of independently transformed plants expressing light or heavy chains, ensuing high yield that reaches 1%–5% of total plant protein (Hiatt et al., 1989; Ma et al., 1994). The second method is based on the cotransformation of the heavy and light chain genes on a single, two, or more expression cassettes (Düring et al., 1990; Villani et al., 2008). A secretory antibody against a surface antigen of *Streptococcus mutans* is a plant-derived antibody and is used for the protection against *S. mutans* colonization on the teeth (Ma et al., 1998). Recently, an HIV-specific monoclonal antibody was produced in maize seeds and was active at par with Chinese hamster ovary-derived counterpart, in terms of antigen-binding activity (Ramessar et al., 2008a,b). The first plant-made scFv monoclonal antibody, used in the production of a recombinant hepatitis B virus vaccine, has been commercialized in Cuba (Pujol et al., 2005).

3. Therapeutic and nutraceutical proteins
The human growth hormone was the first therapeutic human protein to be expressed in plants (Barta et al., 1986). In 1990, transgenic tobacco and potato were generated in which human serum albumin, which is normally isolated from blood, was produced for the first time (Sijmons et al., 1990). Since then, a large number of human proteins have been expressed in plants. These include erythropoietin, which promote red blood cell production (Weise et al., 2007; Musa et al., 2009); epidermal growth factor (Wirth et al., 2004; Bai et al., 2007); α-, β-, and γ-interferons, which are used in treating hepatitis B and C (Leelavathi and Reddy, 2003; Zhu et al., 1994; Arlen et al., 2007); insulin, which is used for treating diabetes (Nykiforuk et al., 2006); interleukin, which is used in treating Crohn's disease (Elías-López et al., 2008; Fujiwara et al., 2010); and human glucocerebrosidase, which is used for the treatment of Gaucher disease in genetically engineered carrot cells (Shaaltiel et al., 2007). Antimicrobial nutraceutics, for example, human lactoferrin and lysozymes, have also been successfully produced in various crops (Huang et al., 2008; Stefanova et al., 2008) and are nowadays commercially available, only as fine chemicals.

4. Nonpharmaceutical plant-derived proteins

The industrial proteins or nonpharmaceutical plant-derived proteins, most of which are enzymes, are aprotinin, avidin, peroxidase, trypsin, laccase, β-glucuronidase, cellulase, etc. The production of cellulosic ethanol that grips countless promise for the biofuel industry has been produced by the molecular pharming of cell-wall-deconstructing enzymes such as hemicellulases, cellulases, ligninases, and xylanases (Sticklen, 2008; Mei et al., 2009; Chatterjee et al., 2010). The biofuel derived from cellulosic ethanol has a potential of decreasing greenhouse gas emissions by 100% as compared with gasoline (Fulton et al., 2004). The other nonhydrolytic proteins like expansins that have also been produced as a result of molecular pharming are also involved for the production of biofuels for industrial use (Obembe, 2010). Recently, various studies have shown the production of nonpharmaceutical plant-derived proteins that involved in the production of biodegradable plastic-like compounds such as poly (3-hydroxybutyrate) (PHB), polyhydroxyalkanoate (PHA) copolymers, and cyanophycin (Conrad, 2005; Matsumoto et al., 2009). But still, only few plant-derived pharmaceuticals have been approved, and fewer still are commercially available, because of biosafety concerns and stringent governmental regulations with respect to field trials, good manufacturing practice (GMP) standards, and preclinical toxicity testing.

5.4 TRADITIONAL BREEDING PRINCIPLES AS APPLIED TO MEDICINAL PLANTS

In order to improve the yield, consistency, effectiveness, or toxicity of the medicinal or aromatic crops at the genetic level, traditional and biotechnological plant-breeding techniques play the key role. The population adapted to the particular growing conditions can be screened by selecting the viable seed-producing plants that will generate vigorous and fertile genotypes. The modern biotechnological approaches can be applied to the artificial selection process that is time-consuming, and reliable methods are available in which desired traits can be detected at the early stage of reproductive cycle; for instance, laser speckle technology has been utilized for the differentiation of viable seeds from nonviable seed (Braga et al., 2003). Chemoprofiling has been recently used by breeders for the selection of plants that are enriched with stable concentrations of desirable biologically active compounds and reduced concentrations of intrinsic toxins. For example, in the case of *G. biloba*, it has been standardized that extract should contain 24%–27% flavonoid glycosides, 6%–7% terpene lactones, and toxic component ginkgolic acid below 5 ppm. Thus, the chemoprofiling of plants at the early stage can help in the selection of genotypes that yield extracts close to the desired chemical profiles. Traditional crop breeding has been augmented by using molecular markers for selection process that has extensively influenced and fasten the crop improvement. Thus, the desirable genotypes can be recognized at the early stage by directly detecting the specific gene of the particular trait or the sequence closely linked to that trait. Recently, comparative genetics has also been applied for the improvement of medicinal plants. The complete genome sequences of *Arabidopsis* and *Oryza sativa* and several other model species, like *Medicago*, *Lycopersicon* (tomato), and *Populus* (poplar), have been used for the selection of related genus among medicinal plants. The comparative genetics has been applied to model legume *M. truncatula* (Bell et al., 2001) that is closely related to *M. sativa* and recommended in herbals for a range of conditions. Similarly, *Populus* (Strauss and Martin, 2004) and *Salix* belong to the same family, and *Salix* is the source of aspirin and, potentially, of other medicinally useful compounds. In *Cannabis*, various types of molecular markers like amplified fragment length polymorphisms and microsatellites have been developed for breeding the species as a fiber crop (Gilmore and Peakall, 2003) and for forensic use (Coyle, 2003). The specific codominant markers responsible for cannabinoid biosynthesis have also been sequenced (Mandolino et al., 2003), precisely to use them in breeding pharmaceutically useful lines.

5.5 PATHWAY ENGINEERING IN MAPs

The application of genetic engineering for the improvement of medicinal and aromatic plants is to increase the production or synthesis of active phytoconstituents but has some constrains. The metabolic pathways involved in the biosynthesis of these active compounds are poorly revealed, and the key genes of the enzymatic and regulatory steps are also not isolated. Recent studies have shown the ninefold increase in the production of scopolamine, the sedative compound from hairy root cultures of *H. niger* (black henbane), by the overexpression of its two key rate-limiting biosynthetic enzymes (Zhang et al., 2004). Similarly, in *A. belladonna*, scopolamine production was increased from its natural chemical precursor hyoscyamine by transforming and overexpressing the enzyme hyoscyamine 6-b-hydroxylase from *Hyoscyamus* (Yun et al., 1992). Similarly, much progress has been achieved in engineering synthesis of alkaloids in *P. somniferum* (Facchini et al., 2000). In *A. annua*, there has been three times increase in the production of artemisinin, a potent antimalarial and anticancer drug, by overexpressing the key biosynthetic enzyme farnesyl diphosphate synthase of artemisinin biosynthetic pathway in transgenic plants (Chen et al., 1999, 2000). By targeting transcription factors that regulate whole secondary pathways by switching them on or off as compared with targeting an individual rate-limiting enzyme reactions, it demonstrated a great promise to metabolic engineering strategy (Robbins et al., 2003). Thus, the new genomic approaches and effective gene isolation procedures applied to the plant secondary metabolism will help in the future to expand the precision of transgenic generation that will ultimately provide superior material to breeder.

5.6 ENGINEERING AGRONOMIC TRAITS IN MAPs

Medicinal plant biotechnology also aims to develop the transgenic medicinal and aromatic plants that are resistant to herbicides, pests, and diseases. The herbicide-resistant transgenic *Atropa* and *Panax* ginseng plants have been generated that are resistant to herbicides bialaphos, glufosinate, and Basta by transformation with the enzyme phosphinothricin acetyl transferase (Saito et al., 1992; Choi et al., 2003). These resistant germplasms serve as a best tool for breeding. Similarly, somatic hybridization and atrazine selection have been used to regenerate herbicide-tolerant *S. nigrum* (black nightshade) (de Donato et al., 1989). To generate the fungal-resistant plants, the fungal-resistant genes like chitinase or a thaumatin have been transformed in *P. quinquefolius* (American ginseng) (Chen and Punja, 2002; Punja and Chen, 2003). Thus, the conventional breeding incorporates variation in growth and development as compared with the modern transgenic intervention that depends on the knowledge of their genetic basis (Sharma, 1996).

5.7 PUBLIC PERCEPTION OF BIOTECHNOLOGY: IMPLICATIONS FOR MAPs

The intervention of biotechnology in science and industry is directed to bring forth the wide range of products that have the high impact for the production of food, fiber, feed fuels, and pharmaceuticals. However, there is a divided opinion about biotechnology, and the debate over the appeal of this technology is far from over. The future of food biotechnology depends upon the sole acceptance of bioengineered foods by consumers. So, the need of the hour is the broad public acceptance of genetically modified foods along wide the quality science. One of the main advantages of herbal medicine is their natural status and potentially safe and good to consume; in contrast, the transgenic plants have the high social stigma of being unnatural. One of the major concerns regarding the transgenic crops is the horizontal transfer of resistant genes from transgenic to nontransgenic plants via cross-pollination. It is worth mentioning that a field study of pollen flow from herbicide-resistant transgenic *P. somniferum* to nontransgenic varieties and weed relatives showed no gene transfer beyond 2.5 m and 20 cm, respectively (Chitty et al., 2003). Thus, we cannot use medicinal plants on a large scale without modifying the characteristics of the plant populations available to us, be they wild or domesticated. If medicinal herbs are brought into cultivation, then we can at least attempt to do this in a controlled fashion and at the same time attempt to conserve wild populations.

References

Abdin, M.Z., Israr, M., Rehman, R.U., Jain, S.K., 2003. Artemisinin, a novel antimalarial drug: biochemical and molecular approaches for enhanced production. Planta Med. 69 (4), 289–299.

Adelberg, J.W., Zhang, X.P., Rhodes, B.B., 1997. Micropropagation of *Citrullus lanatus* (Thunb.) Matsum. and Nakai (Watermelon). In: High-Tech and Micropropagation V. Springer, Berlin, Heidelberg, pp. 60–76.

Agnihotri, S., Singh, S.K., Jain, M., Sharma, M., Sharma, A.K., Chaturvedi, H.C., 2004. In vitro cloning of female and male *Carica papaya* through tips of shoots and inflorescences. Indian J. Biotechnol. 3, 235–240.

Ahuja, A., Verma, M., Grewal, S., 1982. Clonal propagation of Ocimum species by tissue culture [drug plant]. Indian J. Exp. Biol.

Ahuja, M.R., 1987. In vitro propagation of poplar and aspen. In: Cell and Tissue Culture in Forestry. Springer, The Netherlands, pp. 207–223.

Allen, R.S., Millgate, A.G., Chitty, J.A., Thisleton, J., Miller, J.A., Fist, A.J., Gerlach, W.L., Larkin, P.J., 2004. RNAi-mediated replacement of morphine with the nonnarcotic alkaloid reticuline in opium poppy. Nat. Biotechnol. 22 (12), 1559–1566.

Allen, R.S., Miller, J.A., Chitty, J.A., Fist, A.J., Gerlach, W.L., Larkin, P.J., 2008. Metabolic engineering of morphinan alkaloids by over-expression and RNAi suppression of salutaridinol 7-O-acetyltransferase in opium poppy. Plant Biotechnol. J. 6 (1), 22–30.

Ali, G., Purohit, M., Mughal, M.H., Iqbal, M., Srivastava, P.S., 1996. A rapid protocol for micropropagation of *Bacopa monniera* (L.) Wettst-an important medicinal plant. Plant Tissue Cult. Biotechnol. 2, 208–211.

Anonymous, 1969. The Wealth of India. Raw Materials. vol. III Publication and Information Directorate, CSIR, New Delhi, India.

Anonymous, 1985. The Wealth of India: A Dictionary of Indian Raw Materials and Industrial Products, revised ed. vol. 1. CSIR, New Delhi, pp. 37–47.

Anonymous, 1997. CAMP Workshop Proceedings. WWF India, Z00/CBSG, UP Forest Department.

Arlen, P.A., Falconer, R., Cherukumilli, S., Cole, A., Cole, A.M., Oishi, K.K., Daniell, H., 2007. Field production and functional evaluation of chloroplast-derived interferon-α2b. Plant Biotechnol. J. 5 (4), 511–525.

Arora, R., Bhojwani, S.S., 1989. In vitro propagation and low temperature storage of *Saussurea lappa* CB Clarke—an endangered, medicinal plant. Plant Cell Rep. 8 (1), 44–47.

Arumugam, N., Bhojwani, S.S., 1990. Somatic embryogenesis in tissue cultures of *Podophyllum hexandrum*. Can. J. Bot. 68 (3), 487–491.

Autade, R.H., Fargade, S.A., Savant, A.R., Gangurde, S.S., Choudhary, R.S., Dighe, S.S., 2016. Micropropagation of Ashwagandha (*Withania somnifera*). Biosci. Biotechnol. Res. Commun. 9 (1), 88–93.

Ayabe, M., Sumi, S., 1998. Establishment of a novel tissue culture method, stem-disc culture, and its practical application to micropropagation of garlic (*Allium sativum* L.). Plant Cell Rep. 17 (10), 773–779.

Badhwarr, L., Sharmab, K., 1963. A note on the germination of *Podophyllum* seeds. Indian For. 89, 44547.

Bagniol, S., Engelmann, F., 1992. Effect of thawing and recovery conditions on the regrowth of meristems of date palm (*Phoenix dactylifera* L.) after cryopreservation in liquid nitrogen. CryoLetters 13.

Bai, J.Y., Zeng, L., Hu, Y.L., Li, Y.F., Lin, Z.P., Shang, S.C., Shi, Y.S., 2007. Expression and characteristic of synthetic human epidermal growth factor (hEGF) in transgenic tobacco plants. Biotechnol. Lett. 29 (12).

Bajaj, Y.P.S., Furmanowa, M., Olszowska, O., 1988. Biotechnology of the micropropagation of medicinal and aromatic plants. In: Medicinal and Aromatic Plants I. Springer, Berlin, Heidelberg, pp. 60–103.

Balachandran, S.M., Bhat, S.R., Chandel, K.P.S., 1990. In vitro clonal multiplication of turmeric (*Curcuma* spp.) and ginger (*Zingiber officinale* Rosc.). Plant Cell Rep. 8 (9), 521–524.

Balunas, M.J., Kinghorn, A.D., 2005. Drug discovery from medicinal plants. Life Sci. 78 (5), 431–441.

Balz, J.P., Courtois, D., Drieu, J., Drieu, K., Reynoird, J.P., Sohier, C., Teng, B.P., Touché, A., Petiard, V., 1999. Production of ginkgolides and bilobalide by Ginkgo biloba plants and tissue cultures. Planta Med. 65 (7), 620–626.

Barta, A., Sommergruber, K., Thompson, D., Hartmuth, K., Matzke, M.A., Matzke, A.J., 1986. The expression of a nopaline synthase—human growth hormone chimaeric gene in transformed tobacco and sunflower callus tissue. Plant Mol. Biol. 6 (5), 347–357.

Barve, D.M., Mehta, A.R., 1993. Clonal propagation of mature elite trees of *Commiphora wightii*. Plant Cell Tissue Organ Cult. 35 (3), 237–244.

Begum, F., Amin, M.N., Azad, M.A.K., 2002. In vitro rapid clonal propagation of *Ocimum basilicum* L.. Plant Tissue Cult. 12, 27–35.

Bekheet, S.A., Taha, H.S., Saker, M.M., Solliman, M.E., 2007. Application of cryopreservation technique for in vitro grown date palm (*Phoenix dactylifera* L.) cultures. J. Appl. Sci. Res. 3 (9), 859–866.

Bell, C.J., Dixon, R.A., Farmer, A.D., Flores, R., Inman, J., Gonzales, R.A., Harrison, M.J., Paiva, N.L., Scott, A.D., Weller, J.W., May, G.D., 2001. The Medicago Genome Initiative: a model legume database. Nucleic Acids Res. 29 (1), 114–117.

Benson, E.E., 2004. Cryoconserving algal and plant diversity: historical perspectives and future challenges. In: Life in the Frozen State. CRC, London, pp. 299–328.

Benson, E.E., Reed, B.M., Brennan, R.M., Clacher, K.A., Ross, D.A., 1996. Use of thermal analysis in the evaluation of cryopreservation protocols for *Ribes nigrum* L. germplasm. CryoLetters 17, 347–362.

Bentley, G.R., 2000. Environmental pollutants and fertility. In: Infertility in the Modern World: Present and Future Prospects, 12. Cambridge University Press, UK, pp. 85–152.

Bernard, F., Shaker-Bazarnov, H., Kaviani, B., 2002. Effects of salicylic acid on cold preservation and cryopreservation of encapsulated embryonic axes of Persian lilac (*Melia azedarach* L.). Euphytica 123 (1), 85–88.

Bhagyalakshmi, B., Singh, N.S., 1988. Meristem culture and micropropagation of a variety of ginger (*Zingiber officinale* Rosc.) with a high yield of oleoresin. J. Horticult. Sci. 63 (2), 321–327.

Biemelt, S., Sonnewald, U., Galmbacher, P., Willmitzer, L., Müller, M., 2003. Production of human papillomavirus type 16 virus-like particles in transgenic plants. J. Virol. 77 (17), 9211–9220.

Bisht, A.K., Bhatt, A., Dhar, U., Bhatt, A., Dhar, U., 2015. Note on somatic embryogenesis and synthetic seed production in *Angelica glauca*: a valuable medicinal plant of Himalaya. J. Med. Plant Res. 9 (12), 419–425.

Bisht, S.S., Bisht, A.S., Chauhan, R.S., 2017. In-vitro mutagenesis induction to improve abiotic stress in tissue cultured plantlet of *Picrohiza kurroa* Royle ex. Benth: an endangered plant of Western Himalayas, India. Med. Aromat. Plants (Los Angel) 6 (287), https://doi.org/10.4172/2167-0412.1000287.

Bosch, D., Schots, A., 2010. Plant glycans: friend or foe in vaccine development? Expert Rev. Vaccines 9 (8), 835–842.

Braga, R.A., Dal Fabbro, I.M., Borem, F.M., Rabelo, G., Arizaga, R., Rabal, H.J., Trivi, M., 2003. Assessment of seed viability by laser speckle techniques. Biosyst. Eng. 86 (3), 287–294.

Brain, K.R., Lockwood, G.B., 1976. Hormonal control of steroid levels in tissue cultures from *Trigonella foenumgraecum*. Phytochemistry 15 (11), 1651–1654.

Breuling, M., Alfermann, A.W., Reinhard, E., 1985. Cultivation of cell cultures of *Berberis wilsonae* in 20-L airlift bioreactors. Plant Cell Rep. 4 (4), 220–223.

Briskin, D.P., 2000. Medicinal plants and phytomedicines. Linking plant biochemistry and physiology to human health. Plant Physiol. 124 (2), 507–514.

Buonaguro, F.M., Butler-Ransohoff, J.E., 2010. PharmaPlant: the new frontier in vaccines. Expert Rev. Vaccines 9 (8), 805–807.

Cavallini, A., Natali, L., Sanchez, I.C., 1991. Aloe barbadensis Mill. (= *A. vera* L.). In: Medicinal and Aromatic Plants III. Springer, Berlin, Heidelberg, pp. 95–106.

Chandra, B., Palni, L.M.S., Nandi, S.K., 2006. Propagation and conservation of *Picrorhiza kurrooa* Royle ex Benth.: an endangered Himalayan medicinal herb of high commercial value. Biodivers. Conserv. 15 (7), 23–25.

Chang, Y., Reed, B.M., 2000. Extended alternating-temperature cold acclimation and culture duration improve pear shoot cryopreservation. Cryobiology 40 (4), 311–322.

Chang, Y., Barker, R.E., Reed, B.M., 2000. Cold acclimation improves recovery of cryopreserved grass (*Zoysia* and *Lolium* sp.). CryoLetters 21, 107–116.

Charoensub, R., Phansiri, S., Sakai, A., Yongmanitchai, W., 1999. Cryopreservation of cassava in vitro-grown shoot tips cooled to-196c by vitrification. CryoLetters 20 (2), 89–94.

Chatterjee, A., Das, N.C., Raha, S., Babbit, R., Huang, Q., Zaitlin, D., Maiti, I.B., 2010. Production of xylanase in transgenic tobacco for industrial use in bioenergy and biofuel applications. In Vitro Cell Dev. Biol. Plant 46 (2), 198–209.

Chaturvedi, H.C., 1968. In vitro growth and controlled morphogenesis in callus tissue of Rauvolfia serpentina Benth (Doctoral dissertation, Ph.D. thesis). Agra University, Agra.

Chaturvedi, H.C., 1975. Propagation of *Dioscorea floribunda* from in vitro culture of single-node stem segments. Curr. Sci. 839–841.

Chaturvedi, H.C., Sinha, M., 1979. Mass propagation of *Dioscorea floribunda* by tissue culture. Extension Bulletin. National Botanical Research Institute, Lucknow, pp. 12.

Chaturvedi, H.C., Sharma, A.K., Sharma, M., Prasad, R.N., 1982. Morphogenesis, micropropagation, and germplasm preservation of some economic plants by tissue culture. In: Fujiwara, A. (Ed.), Plant Tissue Culture. Maruzen, Tokyo, pp. 687–688.

Chaturvedi, H.C., Misra, P., Sharma, M., 1984. In vitro multiplication of *Rosmarinus officinalis* L. Z. Pflanzenphysiol. 113 (4), 301–304.

Chaturvedi, H.C., Sharma, M., Sharma, A.K., Jain, M., Agha, B.Q., Gupta, P., 2004. In vitro germplasm preservation through regenerative excised root culture for conservation of phytodiversity. Indian J. Biotechnol. 3, 305–315.

Chaturvedi, H.C., Jain, M., Kidwai, N.R., 2007. Cloning of medicinal plants through tissue culture—a review. Indian J. Exp. Biol. 45 (11), 937–948.

Chaturvedi, H.C., Sinha, M., Sharma, A.K., 1977. Clonal propagation of Dioscorea deltoidea Wall. through in vitro culture of shoot apices and single node leaf cuttings. In: Atal, C.K., Kapur, B.M. (Eds.), Cultivation & Utilisation of Medicinal and Aromatic Plants. Regional Research Laboratory, India, pp. 500–505.

Chen, D.H., Liu, C.J., Ye, H.C., Li, G.F., Liu, B.Y., Meng, Y.L., Chen, X.Y., 1999. Ri-mediated transformation of *Artemisia annua* with a recombinant farnesyl diphosphate synthase gene for artemisinin production. Plant Cell Tissue Organ Cult. 57 (3), 157–162.

Chen, D.H., Ye, H.C., Li, G.F., 2000. Expression of a chimeric farnesyl diphosphate synthase gene in *Artemisia annua* L. transgenic plants via *Agrobacterium tumefaciens*-mediated transformation. Plant Sci. 155 (2), 179–185.

Chen, M., Liu, X., Wang, Z., Song, J., Qi, Q., Wang, P.G., 2005. Modification of plant *N*-glycans processing: the future of producing therapeutic protein by transgenic plants. Med. Res. Rev. 25 (3), 343–360.

Chen, W., Punja, Z., 2002. Agrobacterium-mediated transformation of American ginseng with a rice chitinase gene. Plant Cell Rep. 20 (11), 1039–1045.

Cheng, J., Raghavan, V., 1985. Somatic embryogenesis and plant regeneration in *Hyoscyamus niger*. Am. J. Bot. 580–587.

Chikwamba, R., McMurray, J., Shou, H., Frame, B., Pegg, S.E., Scott, P., et al., 2002. Expression of a synthetic *E. coli* heat-labile enterotoxin B sub-unit (LT-B) in maize. Mol. Breed. 10 (4), 253–265.

Chintapakorn, Y., Hamill, J.D., 2003. Antisense-mediated down-regulation of putrescine N-methyltransferase activity in transgenic *Nicotiana tabacum* L. can lead to elevated levels of anatabine at the expense of nicotine. Plant Mol. Biol. 53 (1–2), 87–105.

Chitty, J.A., Allen, R.S., Fist, A.J., Larkin, P.J., 2003. Genetic transformation in commercial Tasmanian cultivars of opium poppy, *Papaver somniferum*, and movement of transgenicpollen in the field. Funct. Plant Biol. 30 (10), 1045–1058.

Choi, Y., Jeong, J., In, J., Yang, D., 2003. Production of herbicide-resistant transgenic *Panax ginseng* through the introduction of the phosphinothricin acetyl transferase gene and successful soil transfer. Plant Cell Rep. 21 (6), 563–568.

Choi, Y.E., Yang, D.C., Yoon, E.S., Choi, K.T., 1998. Plant regeneration via adventitious bud formation from cotyledon explants of *Panax ginseng* CA Meyer. Plant Cell Rep. 17 (9), 731–736.

Ciddi, V., Srinivasan, V., Shuler, M.L., 1995. Elicitation of *Taxus* sp. cell cultures for production of taxol. Biotechnol. Lett. 17 (12), 1343–1346.

Conrad, U., 2005. Polymers from plants to develop biodegradable plastics. Trends Plant Sci. 10 (11), 511–512.

Conrad, U., Fiedler, U., 1994. Expression of engineered antibodies in plant cells. Plant Mol. Biol. 26 (4), 1023–1030.

Coyle, H.M., 2003. Forensic AFLP markers in marijuana. Croat. Med. J. 44, 315.

Cragg, G.M., Schepartz, S.A., Suffness, M., Grever, M.R., 1993. The taxol supply crisis. New NCI policies for handling the large-scale production of novel natural product anticancer and anti-HIV agents. J. Nat. Prod. 56 (10), 1657–1668.

Cuenca, S., Amo-Marco, J.B., Parra, R., 1999. Micropropagation from inflorescence stems of the Spanish endemic plant *Centaurea paui* Loscos ex Willk. (Compositae). Plant Cell Rep. 18 (7), 674–679.

Daniell, H., Lee, S.B., Panchal, T., Wiebe, P.O., 2001. Expression of the native cholera toxin B subunit gene and assembly as functional oligomers in transgenic tobacco chloroplasts. J. Mol. Biol. 311 (5), 1001–1009.

D'Aoust, M.A., Couture, M.M.J., Charland, N., Trepanier, S., Landry, N., Ors, F., Vézina, L.P., 2010. The production of hemagglutinin-based virus-like particles in plants: a rapid, efficient and safe response to pandemic influenza. Plant Biotechnol. J. 8 (5), 607–619.

Datta, S.K., Datta, K., Datta, P.C., 1982. In: Rao, A.N. (Ed.), Propagation of yam—*Dioscorea composita*—through tissue culture. Tissue Culture of Economically Important Plants: Proceedings of the International Symposium Held at the Botany Department, National University of Singapore, Singapore, 28–30 April, 1981. COSTED and ANBS (Asian Network for Biological Sciences), Bangalore.

Dave, A., Bilochi, G., Purohit, S.D., 2003. Scaling-up production and field performance of micropropagated medicinal herb "Safed Musli" (*Chlorophytum borivilianum*). In Vitro Cell Dev. Biol. Plant 39 (4), 419–424.

Davioud, E., Kan, C., Hamon, J., Tempé, J., Husson, H.P., 1989. Production of indole alkaloids by in vitro root cultures from *Catharanthus trichophyllus*. Phytochemistry 28 (10), 2675–2680.

Davuluri, G.R., Van Tuinen, A., Fraser, P.D., Manfredonia, A., Newman, R., Burgess, D., Bramley, P.M., 2005. Fruit-specific RNAi-mediated suppression of DET1 enhances carotenoid and flavonoid content in tomatoes. Nat. Biotechnol. 23 (7).

Daxenbichler, M.E., VanEtten, C.H., Hallinan, E.A., Earle, F.R., Barclay, A.S., 1971. Seeds as sources of L-DOPA. J. Med. Chem. 14 (5), 463–465.

De Carlo, A., Lambardi, M., 2005. Cryopreservation of citrus germplasm. In: The Role of Biotechnology. Villa Gualino, 5–7 March 2005, Turin, Italy. pp. 169–170.

de Donato, M., Mozzetti, C., Perucco, E., 1989. In: Evaluation of atrazine resistance in protoplast regenerants of *Solanum nigrum* L. International Symposium on In Vitro Culture and Horticultural Breeding. 280. pp. 263–270.

De Wilde, C., Peeters, K., Jacobs, A., Peck, I., Depicker, A., 2002. Expression of antibodies and Fab fragments in transgenic potato plants: a case study for bulk production in crop plants. Mol. Breed. 9 (4), 271–282.

Dereuddre, J., Blandin, S., Hassen, N., 1991. Resistance of alginate-coated somatic embryos of carrot (*Daucus carota* L.) to desiccation and freezing in liquid nitrogen. 1. Effects of preculture. CryoLetters.

Dexter, R., Qualley, A., Kish, C.M., Ma, C.J., Koeduka, T., Nagegowda, D.A., Clark, D., 2007. Characterization of a petunia acetyltransferase involved in the biosynthesis of the floral volatile isoeugenol. Plant J. 49 (2), 265–275.

Dhawan, S., Shasany, A.K., Naqvi, A.A., Kumar, S., Khanuja, S.P., 2003. Menthol tolerant clones of *Mentha arvensis*: approach for in vitro selection of menthol rich genotypes. Plant Cell Tissue Organ Cult. 75 (1), 87–94.

Draper, J., Mur, L.A., Jenkins, G., Ghosh-Biswas, G.C., Bablak, P., Hasterok, R., Routledge, A.P., 2001. *Brachypodium distachyon*. A new model system for functional genomics in grasses. Plant Physiol. 127 (4), 1539–1555.

Dubey, T., Guerra, D.J., 2002. Use of biotechnology for growing medicinal plants. Recent Prog. Med. Plants 5, 47–61.

Dumet, D., Benson, E.E., 2000. In: The use of physical and biochemical studies to elucidate and reduce cryopreservation-induced damage in hydrated/desiccated plant germplasm. Cryopreservation of Tropical Plant Germplasm: Current Research Progress and Application. Proceedings of an International Workshop, Tsukuba, Japan, October, 1998. International Plant Genetic Resources Institute (IPGRI), pp. 43–56.

Dumet, D., Engelmann, F., Chabrillange, N., Duval, Y., Dereuddre, J., 1993. Importance of sucrose b for the acquisition of tolerance to desiccation and cryopreservation of oil palm somatic embryos. CryoLetters 14, 243–250.

Duran-Vila, N., 1995. Cryoconservation of germplasm of Citrus. In: Cryopreservation of Plant Germplasm I. Springer, Berlin, Heidelberg, pp. 70–86.

Düring, K., Hippe, S., Kreuzaler, F., Schell, J., 1990. Synthesis and self-assembly of a functional monoclonal antibody in transgenic *Nicotiana tabacum*. Plant Mol. Biol. 15 (2), 281–293.

Elías-López, A.L., Marquina, B., Gutiérrez-Ortega, A., Aguilar, D., Gomez-Lim, M., Hernández-Pando, R., 2008. Transgenic tomato expressing interleukin-12 has a therapeutic effect in a murine model of progressive pulmonary tuberculosis. Clin. Exp. Immunol. 154 (1), 123–133.

Engelmann, F., 1997. In vitro conservation methods. In: Ford-Lloyd, B.V., Newbury, H.J., Callow, J.A. (Eds.), Biotechnology and Plant Genetic Resources: Conservation and Use. CABI, UK, pp. 119–162.

Engelmann, F., 2004. Plant cryopreservation: progress and prospects. In Vitro Cell Dev. Biol. Plant 40 (5), 427–433.

Erdei, I., Kiss, Z., Maliga, P., 1981. Rapid clonal multiplication of *Digitalis lanata* in tissue culture. Plant Cell Rep. 1 (1), 34–35.

Fabre, J., Dereuddre, J., 1990. Encapsulation-dehydration: a new approach to cryopreservation of Solanum shoot-tips. CryoLetters 11, 413–426.

Facchini, P.J., Park, S., Bird, D.A., Samanani, N., 2000. Toward the metabolic engineering of benzylisoquinoline alkaloid biosynthesis in opium poppy and related species. Recent Res. Dev. Phytochem. 4, 31–47.

Faccioli, G., 2001. Control of potato viruses using meristem and stem-cutting cultures, thermotherapy and chemotherapy. In: Virus and Virus-Like Diseases of Potatoes and Production of Seed-Potatoes. Springer, Netherlands, pp. 365–390.

Fahy, G.M., MacFarlane, D.R., Angell, C.A., Meryman, H.T., 1984. Vitrification as an approach to cryopreservation. Cryobiology 21 (4), 407–426.

Fett-Neto, A.G., Melanson, S.J., Nicholson, S.A., Pennington, J.J., DiCosmo, F., 1994. Improved taxol yield by aromatic carboxylic acid and amino acid feeding to cell cultures of *Taxus cuspidata*. Biotechnol. Bioeng. 44 (8), 967–971.

Frick, S., Chitty, J.A., Kramell, R., Schmidt, J., Allen, R.S., Larkin, P.J., Kutchan, T.M., 2004. Transformation of opium poppy (*Papaver somniferum* L.) with antisense berberine bridge enzyme gene (anti-bbe) via somatic embryogenesis results in an altered ratio of alkaloids in latex but not in roots. Transgenic Res. 13 (6), 607–613.

Fujii, N., Inui, T., Iwasa, K., Morishige, T., Sato, F., 2007. Knockdown of berberine bridge enzyme by RNAi accumulates (S)-reticuline and activates a silent pathway in cultured California poppy cells. Transgenic Res. 16 (3), 363–375.

Fujiwara, Y., Aiki, Y., Yang, L., Takaiwa, F., Kosaka, A., Tsuji, N.M., Sekikawa, K., 2010. Extraction and purification of human interleukin-10 from transgenic rice seeds. Protein Expr. Purif. 72 (1), 125–130.

Fulton, L., Howes, T., Hardy, J., 2004. Biofuels for Transport: an International Perspective. International Energy Agency, Paris, France.

Funk, C., Gügler, K., Brodelius, P., 1987. Increased secondary product formation in plant cell suspension cultures after treatment with a yeast carbohydrate preparation (elicitor). Phytochemistry 26 (2), 401–405.

Furuya, T., Ikuta, A., Syōno, K., 1972. Alkaloids from callus tissue of *Papaver somniferum*. Phytochemistry 11 (10), 3041–3044.

Gaudin, V., Vrain, T., Jouanin, L., 1994. Bacterial genes modifying hormonal balances in plants. Plant Physiol. Biochem. 32 (1), 11–29.

Gayathri, D., Ramagopal, G., 2007. In vitro propagation of *Amaryllis* and *Zephyranthes bulbous* species using household tissue culture medium. Plant Cell Biotechnol. Mol. Biol. 8 (3&4), 131–136.

Geetha, R., Gopal, G.V., 2007. Direct regeneration in *Glossocardia bosvallea*: a medicinal plant. Plant Cell Biotechnol. Mol. Biol. 8, 119–124.

Gelvin, S.B., 1998. The introduction and expression of transgenes in plants. Curr. Opin. Biotechnol. 9 (2), 227–232.

Gharyal, P.K., Maheshwari, S.C., 1990. Differentiation in explants from mature leguminous trees. Plant Cell Rep. 8 (9), 550–553.

Gilmore, S., Peakall, R., 2003. Isolation of microsatellite markers in *Cannabis sativa* L. (marijuana). Mol. Ecol. Resour. 3 (1), 105–107.

Giorgi, C., Franconi, R., Rybicki, E.P., 2010. Human papillomavirus vaccines in plant. Expert. Rev. Vaccines 9 (8), 913–924.

Giri, A., Ahuja, P.S., Kumar, P.A., 1993. Somatic embryogenesis and plant regeneration from callus cultures of *Aconitum heterophyllum* Wall. Plant Cell Tissue Organ Cult. 32 (2), 213–218.

Gómez-Galera, S., Pelacho, A.M., Gené, A., Capell, T., Christou, P., 2007. The genetic manipulation of medicinal and aromatic plants. Plant Cell Rep. 26 (10), 1689–1715.

Gonzalez-Arnao, M.T., Panta, A., Roca, W.M., Escobar, R.H., Engelmann, F., 2008. Development and large scale application of cryopreservation techniques for shoot and somatic embryo cultures of tropical crops. Plant Cell Tissue Organ Cult. 92 (1), 1–13.

Grout, B.W., 1999. Meristem-tip culture for propagation and virus elimination. Plant Cell Cult. Protoc., 115–125.

Guggenheim, M., 1913. Dioxyphenilalanin, eine neue Aminosäure aus Vicia faba. Trübner. Physiol. Chem. 88, 276.

Gulati, A., Bharel, S., Jain, S.K., Abdin, M.Z., Srivastava, P.S., 1996. In vitro micropropagation and flowering in *Artemisia annua*. J. Plant Biochem. Biotechnol. 5 (1), 31–35.

Hamilton, A.C., 2004. Medicinal plants, conservation and livelihoods. Biodivers. Conserv. 13 (8), 1477–1517.

Han, K.H., Fleming, P., Walker, K., Loper, M., Chilton, W.S., Mocek, U., Gordon, M.P., Floss, H.G., 1994. Genetic transformation of mature Taxus: an approach to genetically control the in vitro production of the anticancer drug, taxol. Plant Sci. 95 (2), 187–196.

Han, J.Y., Kwon, Y.S., Yang, D.C., Jung, Y.R., Choi, Y.E., 2006. Expression and RNA interference-induced silencing of the damma renediol synthase gene in Panax ginseng. Plant Cell Physiol. 47 (12), 1653–1662.

Hannah, M.A., Heyer, A.G., Hincha, D.K., 2005. A global survey of gene regulation during cold acclimation in *Arabidopsis thaliana*. PLoS Genet. 1 (2), e26.

Hara, M., Kobayashi, Y., Fukui, H., Tabata, M., 1991. Enhancement of berberine production by spermidine in *Thalictrum minus* cell suspension cultures. Plant Cell Rep. 10 (10), 494–497.

He, Z.M., Jiang, X.L., Qi, Y., Luo, D.Q., 2008. Assessment of the utility of the tomato fruit-specific E8 promoter for driving vaccine antigen expression. Genetica 133 (2), 207–214.

Herrera-Estrella, L., Simpson, J., Martínez-Trujillo, M., 2004. Transgenic plants: an historical perspective. Methods Mol Biol. 3–31.

Hiatt, A., Caffferkey, R., Bowdish, K., 1989. Production of antibodies in transgenic plants. Nature 342 (6245), 76–78.

Hincha, D.K., 1990. Differential effects of galactose containing saccharides on mechanical freeze-thaw damage to isolated thylakoid membranes. CryoLetters 11, 437–444.

Hirai, D., Sakai, A., 2003. Simplified cryopreservation of sweet potato [*Ipomoea batatas* (L.) Lam.] by optimizing conditions for osmoprotection. Plant Cell Rep. 21 (10), 961–966.

Hitmi, A., Barthomeuf, C., Sallanon, H., 1999. Cryopreservation of *Chrysanthemum cinerariaefolium* shoot tips. Effects of pretreatment conditions and retention of biosynthetic capacity. CryoLetters 20, 113–120.

Hood, E.E., Witcher, D.R., Maddock, S., Meyer, T., Baszczynski, C., Bailey, M., Kulisek, E., 1997. Commercial production of avidin from transgenic maize: characterization of transformant, production, processing, extraction and purification. Mol. Breed. 3 (4), 291–306.

Hossain, A.B.M.M., Ahmed, G., Debnath, R.P., Mamum, A.N.K., Roy, P.K., 1997. In: Micropropagation of patal, *Trichosanthes dioica* Roxb. Plant Tissue Culture Conference. 97. pp. 9.

Hu, C.Y., Wang, P.J., 1983. Meristem, shoot tip and bud cultures. In: Evans, D.A., Wang, W.R., Ammirato, P.V., Yamada, Y. (Eds.), Handbook of Plant Cell Culture (USA), vol. 1. Macmillan, New York, pp. 177–277.

Huang, C.N., Wang, J.H., Yan, Q.S., Zhang, X.Q., Yan, Q.F., 1995. Plant regeneration from rice (*Oryza sativa* L.) embryogenic suspension cells cryopreserved by vitrification. Plant Cell Rep. 14 (11), 730–734.

Huang, N., Bethell, D., Card, C., Cornish, J., Marchbank, T., Wyatt, D., et al., 2008. Bioactive recombinant human lactoferrin, derived from rice. In Vitro Cell. Dev. Biol. 44, 464–471.

Hunter, C.S., 1988. *Cinchona* spp.: micropropagation, and the in vitro production of quinine and quinidine. In: Medicinal and Aromatic Plants I. Springer, Berlin Heidelberg, pp. 367–387.

Indrayanto, G., Utami, W., Syahrani, A., 1999. *Costus speciosus* (Koenig) JE Smith: in vitro cultures, micropropagation, and the production of diosgenin and other phytosteroids. In: Medicinal and Aromatic Plants XI. Springer, Berlin, Heidelberg, pp. 57–77.

Inui, T., Tamura, K.I., Fujii, N., Morishige, T., Sato, F., 2007. Overexpression of Coptis japonica norcoclaurine 6-O-methyltransferase overcomes the rate-limiting step in benzylisoquinoline alkaloid biosynthesis in cultured *Eschscholzia californica*. Plant Cell Physiol. 48 (2), 252–262.

Ishida, B.K., 1988. Improved diosgenin production in *Dioscorea deltoidea* cell cultures by immobilization in polyurethane foam. Plant Cell Rep. 7 (4), 270–273.

Islam, R., Sarkar, K.P., Naderuzzaman, A.T.M., Joarder, O.I., 1994. In vitro regeneration of plants from cotyledons of *Momordica charantia* L.. Plant Tissue Cult. (Bangladesh) 4, 105–109. (in Bangladesh).

Ismaili, A., Jalali-Javaran, M., Rasaee, M.J., Rahbarizadeh, F., Forouzandeh-Moghadam, M., Memari, H.R., 2007. Production and characterization of anti-(mucin MUC1) single-domain antibody in tobacco (*Nicotiana tabacum* cultivar Xanthi). Biotechnol. Appl. Biochem. 47 (1), 11–19.

Jain, S.K., Sastry, A.R.K., 1984. The Indian Plant Red Data Book-I. Botanical Survey of India, Howrah.

Jain, V., Singh, D., Saraf, S., Saraf, S., 2003. In-vitro micropropagation of *Rauwolfia serpentine* through multiple shoot generation. Anc. Sci. Life 23 (1), 44.

Jesus, M., Martins, A.P., Gallardo, E., Silvestre, S., 2016. Diosgenin: recent highlights on pharmacology and analytical methodology. J. Anal. Meth. Chem. 2016, 16 pp.

Jordan, M.A., Wilson, L., 1998. The use and action of drugs in analyzing mitosis. Methods Cell Biol. 61, 267–295.

Josekutty, P.C., Shah, S., Prathapasenan, G., 1993. Direct and indirect organogenesis in *Coccinia indica*. J. Horticult. Sci. 68 (1), 31–35.

Kamil, W.M., Dewick, P.M., 1986. Biosynthetic relationship of aryltetralin lactone lignans to dibenzylbutyrolactone lignans. Phytochemistry 25 (9), 2093–2102.

Kaminaga, Y., Schnepp, J., Peel, G., Kish, C.M., Ben-Nissan, G., Weiss, D., et al., 2006. Plant phenylacetaldehyde synthase is a bifunctional homotetrameric enzyme that catalyzes phenylalanine decarboxylation and oxidation. J. Biol. Chem. 281 (33), 23357–23366.

Kartha, K.K., Engelmann, F., 1994. Cryopreservation and germplasm storage. In: Plant Cell and Tissue Culture. Springer, The Netherlands, pp. 195–230.

Katsumoto, Y., Fukuchi-Mizutani, M., Fukui, Y., Brugliera, F., Holton, T.A., Karan, M., Tao, G.Q., 2007. Engineering of the rose flavonoid biosynthetic pathway successfully generated blue-hued flowers accumulating delphinidin. Plant Cell Physiol. 48 (11), 1589–1600.

Kaviani, B., 2007. Effects of salicylic acid and encapsulation on enhancing the resistance of embryonic axes of Persian lilac (*Melia azedarach* L.) against cryopreservation. Int. J. Agric. Biol. (Pakistan) 9 (4), 625–627.

Kaviani, B., Abadi, D., Torkashvand, A., Hoor, S., 2009. Cryopreservation of seeds of lily [*Lilium Iedebourii* (Baker) Bioss]: use of sucrose and dehydration. Afr. J. Biotechnol. 8 (16).

Kempe, K., Higashi, Y., Frick, S., Sabarna, K., Kutchan, T.M., 2009. RNAi suppression of the morphine biosynthetic gene salAT and evidence of association of pathway enzymes. Phytochemistry 70 (5), 579–589.

Kiran, G., Venugopal, R.B., Jabeen, F.T.Z., Kaviraj, C.P., Srinath, R., 2014. Rapid regeneration of *Mentha piperita* L. from shoot tip and nodal explants. Indian J. Biotechnol. 3, 594–598.

Kline, N.S., 1954. Use of *Rauwolfia serpentina* Benth. in neuropsychiatric conditions. Ann. N. Y. Acad. Sci. 59 (1), 107–132.

Knäblein, J., 2005. Plant-based expression of biopharmaceuticals. In: Meyers, R.A. (Ed.), Encyclopedia of Molecular Cell Biology and Molecular Medicine. Wiley-VCH Verlag GmbH & Co. KGaA, Weinheim, Germany, pp. 385–410.

Kobayashi, T., Niino, T., Kobayashi, M., 2006. Cryopreservation of tobacco BY-2 suspension cell cultures by vitrification with encapsulation. Plant Biotechnol. 23 (3), 333–337.

Koblitz, H., Koblitz, D., Schmauder, H.P., Gröger, D., 1983. Studies on tissue cultures of the genus *Cinchona* L. alkaloid production in cell suspension cultures. Plant Cell Rep. 2 (3), 122–125.

Kukreja, A.K., Mathur, A.K., 1985. Tissue culture studies in *Duboisia myoporoides*; 1. Plant regeneration and clonal propagation by stem node cultures. Planta Med. 51 (02), 93–96.

Kulkarni, A.A., Thengane, S.R., Krishnamurthy, K.V., 2000. Direct shoot regeneration from node, internode, hypocotyl and embryo explants of *Withania somnifera*. Plant Cell Tissue Organ Cult. 62 (3), 203–209.

Kumar, G.S., Ganapathi, T.R., Revathi, C.J., Srinivas, L., Bapat, V.A., 2005. Expression of hepatitis B surface antigen in transgenic banana plants. Planta 222 (3), 484–493.

Lal, N., Ahuja, P.S., Kukreja, A.K., Pandey, B., 1988. Clonal propagation of *Picrorhiza kurroa* Royle ex Benth. by shoot tip culture. Plant Cell Rep. 7 (3), 202–205.

Lambardi, M., Benelli, C., Ozudogru, E.A., Ozden-Tokatli, Y., 2006. Synthetic seed technology in ornamental plants. In: Teixeira da Silva, J.A. (Ed.), Floriculture, Ornamental and Plant Biotechnology: Advances and Topical Issues. vol. II. Global Sci. Books, UK, pp. 347–354.

Larkin, P.J., Miller, J.A., Allen, R.S., Chitty, J.A., Gerlach, W.L., Frick, S., Kutchan, T.M., Fist, A.J., 2007. Increasing morphinan alkaloid production by over-expressing codeinone reductase in transgenic Papaver somniferum. Plant Biotechnol. J. 5 (1), 26–37.

Lauterslager, T.G.M., Florack, D.E.A., Van der Wal, T.J., Molthoff, J.W., Langeveld, J.P.M., Bosch, D., Hilgers, L.T., 2001. Oral immunisation of naive and primed animals with transgenic potato tubers expressing LT-B. Vaccine 19 (17), 2749–2755.

Lee, M.H., Yoon, E.S., Jeong, J.H., Choi, Y.E., 2004. Agrobacterium rhizogenes-mediated transformation of Taraxacum platycarpum and changes of morphological characters. Plant Cell Rep. 22 (11), 822–827.

Leelavathi, L., Reddy, V.S., 2003. Chloroplast expression of His-tagged GUS-fusions: a general strategy to overproduce and purify foreign proteins using transplastomic plants as bioreactors. Mol. Breed. 11, 49–58.

Lentz, E.M., Segretin, M.E., Morgenfeld, M.M., Wirth, S.A., Santos, M.J.D., Mozgovoj, M.V., Bravo-Almonacid, F.F., 2010. High expression level of a foot and mouth disease virus epitope in tobacco transplastomic plants. Planta 231 (2), 387.

Leon-Diaz, R., Meckes, M., Said-Fernandez, S., Molina-Salinas, G.M., Vargas-Villarreal, J., Torres, J., Luna-Herrera, J., Jimenez-Arellanes, A., 2010. Antimycobacterial neolignans isolated from Aristolochia taliscana. Mem. Inst. Oswaldo Cruz 105 (1), 45–51.

Leunufna, S., Keller, E.R.J., 2005. Cryopreservation of yams using vitrification modified by including droplet method: effects of cold acclimation and sucrose. CryoLetters 26 (2), 93–102.

Levitt, J., 1972. Responses of Plants to Environmental Stress, Volume 1: Chilling, Freezing, and High Temperature Stresses. Academic Press, New York, USA.

Lewis, W.H., Elvin-Lewis, M.P., 2003. Medical Botany: Plants Affecting Human Health. John Wiley & Sons, Hoboken, New Jersey.

Lienard, D., Sourrouille, C., Gomord, V., Faye, L., 2007. Pharming and transgenic plants. Biotechnol. Annu. Rev. 13, 115–147.

Ling, H.Y., Pelosi, A., Walmsley, A.M., 2010. Current status of plant-made vaccines for veterinary purposes. Expert Rev. Vaccines 9 (8), 971–982.

Lowe 3rd, J.A., Jones, P., 2007. Biopharmaceuticals and the future of the pharmaceutical industry. Curr. Opin. Drug Discov. Devel. 10 (5), 513.

Luo, J., DeNoma, J., Reed, B.M., 1995. Cryopreservation screening of Pyrus germplasm. Cryobiology 32, 558.

Lv, Z., Zhang, F., Pan, Q., Fu, X., Jiang, W., Shen, Q., Yan, T., Shi, P., Lu, X., Sun, X., Tang, K., 2016. Branch pathway blocking in Artemisia annua is a useful method for obtaining high yield artemisinin. Plant Cell Physiol. 57 (3), 588–602.

Ma, J.K.C., Lehner, T., Stabila, P., Fux, C.I., Hiatt, A., 1994. Assembly of monoclonal antibodies with IgG1 and IgA heavy chain domains in transgenic tobacco plants. Eur. J. Immunol. 24 (1), 131–138.

Ma, J.K.C., Hikmat, B.Y., Wycoff, K., Vine, N.D., Chargelegue, D., Yu, L., et al., 1998. Characterization of a recombinant plant monoclonal secretory antibody and preventive immunotherapy in humans. Nat. Med. 4 (5), 601–606.

Maclean, J., Koekemoer, M., Olivier, A.J., Stewart, D., Hitzeroth, I.I., Rademacher, T., Rybicki, E.P., 2007. Optimization of human papillomavirus type 16 (HPV-16) L1 expression in plants: comparison of the suitability of different HPV-16 L1 gene variants and different cell-compartment localization. J. Gen. Virol. 88 (5), 1460–1469.

Mahmoud, S.S., Croteau, R.B., 2001. Metabolic engineering of essential oil yield and composition in mint by altering expression of deoxyxylulose phosphate reductoisomerase and menthofuran synthase. Proc. Natl. Acad. Sci. 98 (15), 8915–8920.

Mahmoud, S.S., Williams, M., Croteau, R., 2004. Cosuppression of limonene-3-hydroxylase in peppermint promotes accumulation of limonene in the essential oil. Phytochemistry 65 (5), 547–554.

Mandolino, G., Carboni, A., Pacifico, D., Ranalli, P., De Meijer, E., 2003. In: The control of the chemical phenotype in *Cannabis sativa* L.: genetic analysis and molecular markers. XLVII SIGA Annual Congress, Verona, 24–27 September.

Marquet-Blouin, E., Bouche, F.B., Steinmetz, A., Muller, C.P., 2003. Neutralizing immunogenicity of transgenic carrot (*Daucus carota* L.)-derived measles virus hemagglutinin. Plant Mol. Biol. 51 (4), 459–469.

Mascarenhas, A.F., Hendre, R.R., Nadgir, A.L., Ghugale, D.D., Godbole, D.A., Prabhu, R.A., Jagannathan, V., 1976. Development of plantlets from cultured tissues of *Dioscorea deltoidea* Wall. Indian J. Exp. Biol. 14, 604.

Mason, H.S., Ball, J.M., Shi, J.J., Jiang, X., Estes, M.K., Arntzen, C.J., 1996. Expression of Norwalk virus capsid protein in transgenic tobacco and potato and its oral immunogenicity in mice. Proc. Natl. Acad. Sci. 93 (11), 5335–5340.

Mathur, S., Shekhawat, G.S., Batra, A., 2002. An efficient in vitro method for mass propagation of *Salvadora persica* via Apical Meristen. J. Plant Biochem. Biotechnol. 11 (2), 125–128.

Mathur, S., Sharma, S., Gupta, M.M., Kumar, S., 2003. Evaluation of an Indian germplasm collection of the medicinal plant *Bacopa monnieri* (L.) Pennell by use of multivariate approaches. Euphytica 133 (3), 255–265.

Matsumoto, K.I., Murata, T., Nagao, R., Nomura, C.T., Arai, S., Arai, Y., Shimada, H., 2009. Production of short-chain-length/medium-chain-length polyhydroxyalkanoate (PHA) copolymer in the plastid of *Arabidopsis thaliana* using an engineered 3-ketoacyl-acyl carrier protein synthase III. Biomacromolecules 10 (4), 686–690.

Matsumoto, T., Sakai, A., 2003. Cryopreservation of axillary shoot tips of in vitro-grown grape (Vitis) by a two-step vitrification protocol. Euphytica 131 (3), 299–304.

Meena, M.C., Patni, V., 2007. In vitro clonal propagation of *Citrullus colocynthis* (Linn.) Schrad: a threatened medicinal plant. Plant Cell Biotechnol. Mol. Biol. 8, 147–152.

Mei, C., Park, S.H., Sabzikar, R., Ransom, C., Qi, C., Sticklen, M., 2009. Green tissue-specific production of a microbial endo-cellulase in maize (*Zea mays* L.) endoplasmic-reticulum and mitochondria converts cellulose into fermentable sugars. J. Chem. Technol. Biotechnol. 84 (5), 689–695.

Mishra, S., Yadav, D.K., Tuli, R., 2006. Ubiquitin fusion enhances cholera toxin B subunit expression in transgenic plants and the plant-expressed protein binds GM1 receptors more efficiently. J. Biotechnol. 127 (1), 95–108.

Moravec, T., Schmidt, M.A., Herman, E.M., Woodford-Thomas, T., 2007. Production of *Escherichia coli* heat labile toxin (LT) B subunit in soybean seed and analysis of its immunogenicity as an oral vaccine. Vaccine 25 (9), 1647–1657.

Morimoto, T., Hara, Y., Kato, Y., Hiratsuka, J., Yoshioka, T., Fujita, Y., Yamada, Y., 1988. Berberine production by cultured *Coptis japonica* cells in a one-stage culture using medium with a high copper concentration. Agric. Biol. Chem. 52 (7), 1835–1836.

Moyano, E., Fornalé, S., Palazón, J., Cusidó, R.M., Bonfill, M., Morales, C., Piñol, M.T., 1999. Effect of Agrobacterium rhizogenes T-DNA on alkaloid production in Solanaceae plants. Phytochemistry 52 (7), 1287–1292.

Murashige, T., 1977. Plant cell and organ cultures as horticultural practices. Acta Hortic. (78)17–30.

Murch, S.J., Ragone, D., Shi, W.L., Alan, A.R., Saxena, P.K., 2007. In vitro conservation and micropropagation of breadfruit (Artocarpus altilis, Moracea). In: Protocols for Micropropagation of Woody Trees and Fruits. Springer, Netherlands, pp. 279–288.

Musa, T.A., Hung, C.Y., Darlington, D.E., Sane, D.C., Xie, J., 2009. Overexpression of human erythropoietin in tobacco does not affect plant fertility or morphology. Plant Biotechnol. Rep. 3 (2), 157–165.

Mustafa, M.D., Mallaiah, B., 1991. In vitro adventitious shoot formation from stem segments of snake gourd (*Trichosanthes anguina* L.). J. Mendal 8 (3), 147–148.

Nadgauda, R.S., Mascarenhas, A.F., Hendre, R.R., Jagannathan, V., 1978. Rapid multiplication of turmeric (*Curcuma longa* Linn.) plants by tissue culture. Indian J. Exp. Biol. 16, 120.

Nadgauda, R.S., Kulkarni, D.D., Mascarenhas, A.F., Jagannathan, V., 1980. In: Clonal propagation of ginger in vitro. Plant Tissue Culture, Genetic Manipulation and Somatic Hybridization of Plants. Bombay, India: Proc Natl Symp, BARC, pp. 358–368.

Nagarathnamma, M., Sudarshana, M.S., Niranjan, M.H., Pandurangamurthy, 2010. Rapid regeneration of *Enicostemma littorale* Blume from leaf and stem cultures. J. Plant Interact. 5 (1), 69–73.

Nagira, Y., Ikegami, K., Koshiba, T., Ozeki, Y., 2006. Effect of ABA upon anthocyanin synthesis in regenerated torenia shoots. J. Plant Res. 119 (2), 137–144.

Nakagawa, K., Konagai, A., Fukui, H., Tabata, M., 1984. Release and crystallization of berberine in the liquid medium of *Thalictrum minus* cell suspension cultures. Plant Cell Rep. 3 (6), 254–257.

Nakagawa, K., Fukui, H., Tabata, M., 1986. Hormonal regulation of berberine production in cell suspension cultures of *Thalictrum minus*. Plant Cell Rep. 5 (1), 69–71.

Narmatha Bai, V., Uma Rani, S., Tamilselvi, M., 1997. Callus induction and organogenesis from seedling explants of *Sesbania speciosa* Taubert ex Engl. Proc. Natl. Acad. Sci. India Sec. B 67, 39–42.

Nehra, N.S., Kartha, K.K., 1994. Meristem and shoot tip culture: requirements and applications. In: Plant Cell and Tissue Culture. Springer, Netherlands, pp. 37–70.

Nessler, C.L., 1982. Somatic embryogenesis in the opium poppy, *Papaver somniferum*. Physiol. Plant. 55 (4), 453–458.

Nishigaki, I., Rajendran, P., Venugopal, R., Ekambaram, G., Sakthisekaran, D., Nishigaki, Y., 2010. Cytoprotective role of astaxanthin against glycated protein/iron chelate-induced toxicity in human umbilical vein endothelial cells. Phytother. Res. 24 (1), 54–59.

Nochi, T., Takagi, H., Yuki, Y., Yang, L., Masumura, T., Mejima, M., et al., 2007. Rice-based mucosal vaccine as a global strategy for cold-chain-and needle-free vaccination. Proc. Natl. Acad. Sci. 104 (26), 10986–10991.

Nykiforuk, C.L., Boothe, J.G., Murray, E.W., Keon, R.G., Goren, H.J., Markley, N.A., Moloney, M.M., 2006. Transgenic expression and recovery of biologically active recombinant human insulin from *Arabidopsis thaliana* seeds. Plant Biotechnol. J. 4 (1), 77–85.

Obembe, O.O., 2010. The plant biotechnology flight: is Africa on board? Afr. J. Biotechnol. 9 (28), 4300–4308.

Orlova, I., Marshall-Colón, A., Schnepp, J., Wood, B., Varbanova, M., Fridman, E., et al., 2006. Reduction of benzenoid synthesis in petunia flowers reveals multiple pathways to benzoic acid and enhancement in auxin transport. Plant Cell 18 (12), 3458–3475.

Paek, K.Y., Yu, K.J., Park, S.I., Sung, N.S., Park, C.H., 1995. Micropropagation of *R. glutinosa* as medicinal plant by shoot tip and root segment culture. ISHS international symposium on medicinal and aromatic plants. Acta Hortic. 390, 120–123.

Panis, B., Totte, N., Van Nimmen, K., Withers, L.A., Swennen, R., 1996. Cryopreservation of banana (*Musa* spp.) meristem cultures after preculture on sucrose. Plant Sci. 121 (1), 95–106.

Panis, B., Swennen, R., Engelman, F., 2001. Cryopreservation of plant germplasm. Acta Hortic. 650, 79–86.

Park, S.U., Facchini, P.J., 2000. Agrobacterium-mediated transformation of opium poppy, *Papaver somniferum*, via shoot organogenesis. J. Plant Physiol. 157 (2), 207–214.

Park, S.U., Yu, M., Facchini, P.J., 2002. Antisense RNA-mediated suppression of benzophenanthridine alkaloid biosynthesis in transgenic cell cultures of California poppy. Plant Physiol. 128 (2), 696–706.

Park, S.U., Yu, M., Facchini, P.J., 2003a. Modulation of berberine bridge enzyme levels in transgenic root cultures of California poppy alters the accumulation of benzophenanthridine alkaloids. Plant Mol. Biol. 51 (2), 153–164.

Park, S.U., Chae, Y.A., Facchini, P.J., 2003b. Genetic transformation of the figwort, Scrophularia buergeriana Miq., an Oriental medicinal plant. Plant Cell Rep. 21 (12), 1194–1198.

Pati, P.K., Rath, S.P., Sharma, M., Sood, A., Ahuja, P.S., 2006. In vitro propagation of rose—a review. Biotechnol. Adv. 24 (1), 94–114.

Paulet, F., Engelmann, F., Glaszmann, J.C., 1993. Cryopreservation of apices of in vitro plantlets of sugarcane (Saccharum sp. hybrids) using encapsulation/dehydration. Plant Cell Rep. 12 (9), 525–529.

Pawar, P.K., Pawar, C.S., Narkhede, B.A., Teli, N.P., Bhalsing, S.R., Maheshwari, V.L., 2002. A technique for rapid micropropagation of Solanum surattense Burm. f. Ind. J. Biotechnol. 1, 201–204.

Pérez-Bermúdez, P., Seitz, H.U., Gavidia, I., 2002. A protocol for rapid micropropagation of endangered Isoplexis. In Vitro Cell Dev. Biol. Plant 38 (2), 178–182.

Prasad, R.N., Chaturvedi, H.C., 1978. In vitro induction of shoots and formation of plantlets from segments of leaf, stem and root of *Solanum xanthocarpum* Schrad & Wendl. Indian J. Exp. Biol. 16, 1121.

Prevalek-Kozlina, B., Kostović-Vranjes, V., Slade, D., 1997. In vitro propagation of *Fibigia triquetra* (DC.) Boiss., a rare stenoendemic species. Plant Cell Tissue Organ Cult. 51 (2), 141–143.

Pujol, M., Ramírez, N.I., Ayala, M., Gavilondo, J.V., Valdés, R., Rodríguez, M., Peral, R., 2005. An integral approach towards a practical application for a plant-made monoclonal antibody in vaccine purification. Vaccine 23 (15), 1833–1837.

Punja, Z.K., Chen, W.P., 2003. In: Tissue culture of American Gingseng and genetic engineering to express antifungal proteins. XXVI International Horticultural Congress: Biotechnology in Horticultural Crop Improvement: Achievements, Opportunities. 625. pp. 395–401.

Purohit, S.D., Dave, A., Kukda, G., 1994. Micropropagation of safed musli (*Chlorophytum borivilianum*), a rare Indian medicinal herb. Plant Cell Tissue Organ Cult. 39 (1), 93–96.

Ramessar, K., Rademacher, T., Sack, M., Stadlmann, J., Platis, D., Stiegler, G., Capell, T., 2008a. Cost-effective production of a vaginal protein microbicide to prevent HIV transmission. Proc. Natl. Acad. Sci. 105 (10), 3727–3732.

Ramessar, K., Sabalza, M., Capell, T., Christou, P., 2008b. Maize plants: an ideal production platform for effective and safe molecular pharming. Plant Sci. 174 (4), 409–419.

Rani, G., Grover, I.S., 1999. In vitro callus induction and regeneration studies in *Withania somnifera*. Plant Cell Tissue Organ Cult. 57 (1), 23–27.

Rech, E.L., Pires, M.J.P., 1986. Tissue culture propagation of *Mentha* spp. by the use of axillary buds. Plant Cell Rep. 5 (1), 17–18.

Redenbaugh, K., Fujii, J., Slade, D., Viss, P., Kossler, M., 1991. Artificial seeds—encapsulated somatic embryos. In: High-Tech and Micropropagation I. Springer, Berlin, Heidelberg, pp. 395–416.

Reed, B.M., Schumacher, L., Dumet, D., Benson, E.E., 2005. Evaluation of a modified encapsulation-dehydration procedure incorporating sucrose pretreatments for the cryopreservation of *Ribes germplasm*. In Vitro Cell Dev. Biol. Plant 41 (4), 431–436.

Reed, B.M., Kovalchuk, I., Kushnarenko, S., Meier-Dinkel, A., Schoenweiss, K., Pluta, S., Straczynska, K., Benson, E.E., 2004. Evaluation of critical points in technology transfer of cryopreservation protocols to international plant conservation laboratories. CryoLetters 25 (5), 341–352.

Richter, L.J., Thanavala, Y., Arntzen, C.J., Mason, H.S., 2000. Production of hepatitis B surface antigen in transgenic plants for oral immunization. Nat. Biotechnol. 18 (11), 1167.

Robbins, M.P., Paolocci, F., Hughes, J.W., Turchetti, V., Allison, G., Arcioni, S., Damiani, F., 2003. Sn, a maize bHLH gene, modulates anthocyanin and condensed tannin pathways in *Lotus corniculatus*. J. Exp. Bot. 54 (381), 239–248.

Rosales-Mendoza, S., Alpuche-Solís, Á.G., Soria-Guerra, R.E., Moreno-Fierros, L., Martínez-González, L., Herrera-Díaz, A., Korban, S.S., 2009. Expression of an *Escherichia coli* antigenic fusion protein comprising the heat labile toxin B subunit and the heat stable toxin, and its assembly as a functional oligomer in transplastomic tobacco plants. Plant J. 57 (1), 45–54.

Rout, G.R., Mohapatra, A., Jain, S.M., 2006. Tissue culture of ornamental pot plant: a critical review on present scenario and future prospects. Biotechnol. Adv. 24 (6), 531–560.

Roy, S.K., Hossain, M.Z., Islam, M.S., 1994. Mass propagation of *Rauvolfia serpentina* by in vitro shoot tip culture. Plant Tissue Cult. 4, 69–75. (in Bangladesh).

Runguphan, W., Maresh, J.J., O'Connor, S.E., 2009. Silencing of tryptamine biosynthesis for production of nonnatural alkaloids in plant culture. Proc. Natl. Acad. Sci. 106 (33), 13673–13678.

Rybicki, E.P., 2009. Plant-produced vaccines: promise and reality. Drug Discov. Today 14, 16–24.

Saito, K., Yamazaki, M., Anzai, H., Yoneyama, K., Murakoshi, I., 1992. Transgenic herbicide-resistant *Atropa belladonna* using an Ri binary vector and inheritance of the transgenic trait. Plant Cell Rep. 11 (5), 219–224.

Sajc, L., Grubisic, D., Vunjak-Novakovic, G., 2000. Bioreactors for plant engineering: an outlook for further research. Biochem. Eng. J. 4 (2), 89–99.

Sakai, A., 1993. Cryogenic strategies for survival of plant cultured cells and meristems cooled to-196 C. In: Cryopreservation of Plant Genetic Resources. JICA, Japan, pp. 1–25.

Sakai, A., 2000. In: Development of cryopreservation techniques. Cryopreservation of Tropical Plant Germplasm: Current Research Progress and Application. Proceedings of an International Workshop, Tsukuba, Japan, October, 1998. International Plant Genetic Resources Institute (IPGRI), pp. 1–7.

Sakai, A., 2004. Plant cryopreservation. In: Fuller, B., Lane, N., Benson, E.E. (Eds.), Life in the Frozen State. CRC Press, London, UK, pp. 329–345.

Sakai, A., Kobayashi, S., Oiyama, I., 1990. Cryopreservation of nucellar cells of navel orange (*Citrus sinensis* Osb. var. brasiliensis Tanaka) by vitrification. Plant Cell Rep. 9 (1), 30–33.

Sakato, K., Misawa, M., 1974. Effects of chemical and physical conditions on growth of *Camptotheca acuminata* cell cultures. Agric. Biol. Chem. 38 (3), 491–497.

Sales, E., Segura, J., Arrillaga, I., 2003. Agrobacterium tumefaciens-mediated genetic transformation of the cardenolide-producing plant Digitalis minor L. Planta Med. 69 (2), 143–147.

Salvi, N.D., George, L., Eapen, S., 2002. Micropropagation and field evaluation of micropropagated plants of turmeric. Plant Cell Tissue Organ Cult. 68 (2), 143–151.

Samanani, N., Park, S.U., Facchini, P.J., 2002. In vitro regeneration and genetic transformation of the berberine-producing plant, Thalictrum flavum ssp. glaucum. Physiol. Plant. 116 (1), 79–86.

Sambyal, M., Dogra, A., Koul, S., Ahuja, A., 2006. Rapid in vitro propagation of *Potentilla fulgens* Wall—a Himalayan Alpine herb of medicinal value. J. Plant Biochem. Biotechnol. 15 (2), 143–145.

Sato, F., Yamada, Y., 1984. High berberine-producing cultures of *Coptis japonica* cells. Phytochemistry 23 (2), 281–285.

Satyavathi, V., Prasad, V., Shaila, M., Sita, L.G., 2003. Expression of hemagglutinin protein of Rinderpest virus in transgenic pigeon pea [*Cajanus cajan* (L.) Millsp.] plants. Plant Cell Rep. 21 (7), 651–658.

Schöner, S., Reinhard, E., 1982. Clonal multiplication of *Digitalis lanata* by meristeme culture. Planta Med. 45 (07), 155.

Scottez, C., Chevreau, E., Godard, N., Arnaud, Y., Duron, M., Dereuddre, J., 1992. Cryopreservation of cold-acclimated shoot tips of pear in vitro cultures after encapsulation-dehydration. Cryobiology 29 (6), 691–700.

Sevón, N., Oksman-Caldentey, K.M., 2002. Agrobacterium rhizogenes-mediated transformation: root cultures as a source of alkaloids. Planta Med. 68 (10), 859–868.

Shaaltiel, Y., Bartfeld, D., Hashmueli, S., Baum, G., Brill-Almon, E., Galili, G., Futerman, A.H., 2007. Production of glucocerebrosidase with terminal mannose glycans for enzyme replacement therapy of Gaucher's disease using a plant cell system. Plant Biotechnol. J. 5 (5), 579–590.

Shah, R.R., Dalal, K.C., 1982. In: Fujiwara, A. (Ed.), Glycyrrhiza glabra (Liquorice): from test tube to field. Plant Tissue Culture 1982: Proceedings, 5th International Congress of Plant Tissue and Cell Culture held at Tokyo and Lake Yamanake, Japan, July 11–16, 1982. Japanese Association for Plant Tissue Culture, Tokyo.

Sharma, J.R., 1996. Biotechnology: the third phase of plant breeding with special reference to medicinal and aromatic plants. Beitrage zur Zuchtungsforschung 2, 270–281.

Sharma, N., Chandel, K.P.S., 1992. Effects of ascorbic acid on axillary shoot induction in *Tylophora indica* (Burm. f.) Merrill. Plant Cell Tissue Organ Cult. 29 (2), 109–113.

Sharma, N., Chandel, K.P.S., Srivastava, V.K., 1991. In vitro propagation of *Coleus forskohlii* Briq., a threatened medicinal plant. Plant Cell Rep. 10 (2), 67–70.

Sharma, T.R., Singh, B.M., 1995. In vitro microrhizome production in *Zingiber officinale* Rosc. Plant Cell Rep. 15 (3-4), 274–277.

Sharma, A.K., Sharma, M., Chaturvedi, H.C., 2002. Conservation of phytodiversity of Azadirachta indica A. Juss. through in vitro strategies. In: Nandi, S.K., Palni, L.M.S., Kumar, A. (Eds.), Role of Plant Tissue Culture in Biodiversity Conservation and Economic Development. Gyanodaya Prakashan, Nainital, India, pp. 513.

Sijmons, P.C., Dekker, B.M., Schrammeijer, B., Verwoerd, T.C., Van Den Elzen, P.J., Hoekema, A., 1990. Production of correctly processed human serum albumin in transgenic plants. Nat. Biotechnol. 8 (3), 217–221.

Singh, P., Singh, A., Shukla, A.K., Singh, L., Pande, V., Nailwal, T.K., 2009. Somatic embryogenesis and in vitro regeneration of an endangered medicinal plant sarpgandha (*Rauvolfia serpentina* L.). Life Sci. J. 6 (2), 57–62.

Sivanesan, I., Murugesan, K., 2005. In vitro adventitious shoot formation from leaf explants of *Withania somnifera* Dunal. Plant Cell Biotechnol. Mol. Biol. 6 (3), 4.

Sivasubramanian, S., Vallinayagam, S., PATRIC, R.D., Manickam, V.S., 2002. Micropropagation of *Plectranthus vetiveroides* (Jacob) Singh and Sharma: a medicinal plant. Phytomorphology 52 (1), 55–59.

Sojikul, P., Buehner, N., Mason, H.S., 2003. A plant signal peptide-hepatitis B surface antigen fusion protein with enhanced stability and immunogenicity expressed in plant cells. Proc. Natl. Acad. Sci. 100 (5), 2209–2214.

Sood, H., Chauhan, R.S., 2009. Development of a low cost micropropagation technology for an endangered medicinal herb (*Picrorhiza kurroa*) of North-Western Himalayas. J. Plant Sci. 4 (2), 21–31.

Srivastava, D.R., Andrianov, V.M., Piruzian, E.S., 1989. Tissue culture and plant regeneration of watermelon (*Citrullus vulgaris* Schrad. cv. Melitopolski). Plant Cell Rep. 8 (5), 300–302.

Srivastava, S.K., Srivastava, N., 2004. In vitro multiplication of *Paedaria foetida* L.—a rare medicinal plant. J. Plant Biochem. Biotechnol. 13 (1), 89–91.

Stefanova, G., Vlahova, M., Atanassov, A., 2008. Production of recombinant human lactoferrin from transgenic plants. Biol. Plant. 52 (3), 423.

Stevenson, R., 2004. Chemical harvest. Chemistry & Industry 6th September, pp. 16–18.

Stewart, G.R., Larher, F., 1980. Accumulation of amino acids and related compounds in relation to environmental stress. In: Accumulation of Amino Acids and Related Compounds in Relation to Environmental Stress. Academic Press, New York, USA, pp. 609–635.

Sticklen, M.B., 2008. Plant genetic engineering for biofuel production: towards affordable cellulosic ethanol. Nature reviews. Genetics 9 (6), 433.

Stoger, E., Ma, J.K., Fischer, R., Christou, P., 2005. Sowing the seeds of success: pharmaceutical proteins from plants. Curr. Opin. Biotechnol. 16 (2), 167–173.

Strauss, S.H., Martin, F.M., 2004. Poplar genomics comes of age. New Phytol. 164 (1), 1–4.

Streatfield, S.J., Howard, J.A., 2003. Plant-based vaccines. Int. J. Parasitol. 33, 479–493.

Strobel, G.A., Stierle, A., van Kuijk, F.J., 1992. Factors influencing the in vitro production of radiolabeled taxol by Pacific yew, *Taxus brevifolia*. Plant Sci. 84 (1), 65–74.

Sudo, H., Hasegawa, Y., Matsunaga, J., 1991. Jpn. Pat. 03,272,628.

Suffness, M. (Ed.), 1995. Taxol: Science and Applications. vol. 22. CRC Press, Boca Raton, Florida.

Sugawara, Y., Steponkus, P.L., 1990. Effect of cold acclimation and modification of the plasma membrane lipid composition on lamellar-to-hexagonal II phase transitions in rye protoplasts. Cryobiology 27, 667.

Suzuki, M., Nakagawa, K., Fukui, H., Tabata, M., 1988. Alkaloid production in cell suspension cultures of *Thalictrum flavum* and *T. dipterocarpum*. Plant Cell Rep. 7 (1), 26–29.

Takagi, H., 2000. Recent developments in cryopreservation of shoot apices of tropical species. In: Engelmann, F., Takagi, H. (Eds.), Cryopreservation of Tropical Plant Germplasm. Current Research Progress and Application. Japan International Research Center for Agricultural Sciences, Tsukuba/International Plant Genetic Resources Institute, Japan/Rome, Italy, pp. 178–193.

Takagi, H., Thinh, N.T., Islam, O.M., Senboku, T., Sakai, A., 1997. Cryopreservation of invitro-grown shoot tips of taro (*Colocasia esculenta* (L.) Schott) by vitrification. 1. Investigation of basic conditions of the vitrification procedure. Plant Cell Rep. 16 (9), 594–599.

Tal, B., Rokem, J.S., Goldberg, I., 1983. Factors affecting growth and product formation in plant cells grown in continuous culture. Plant Cell Rep. 2 (4), 219–222.

Tam, W.J., Constabel, F., Kurz, W.G., 1980. Codeine from cell suspension cultures of *Papaver somniferum*. Phytochemistry 19 (3), 486–487.

Tao, D., Li, P.H., 1986. Classification of plant cell cryoprotectants. J. Theor. Biol. 123 (3), 305–310.

Teramoto, S., Komamine, A., Bajaj, Y.P.S., 1988. Biotechnology in agriculture and forestry. Med. Aromat. Plants IV, 209–224.

Terryn, N., Van Montagu, M., Inzé, D., Goossens, A., 2006. Functional genomic approaches to study and engineer secondary metabolism in plant cell cultures. Frontis 17, 291–300.

Thangavel, K., 2008. Histochemistry Antibacterial Screening and Tissue Culture Studies on Threatened Endemic Medicinal Plants *Janakia arayalpathra* Joseph Chandras and *Decalepis hamiltonii* Wight Arn, shodhganga.inflibnet.ac.in.

Thangavel, K., Ebbie, M.G., Ravichandran, P., 2014. Biotechnology and in vitro conservation of medicinal plants. Ann. Plant Sci. 3 (6), 734–744.

Thengane, S.R., Kulkarni, D.K., Krishnamurthy, K.V., 1998. Micropropagation of licorice (*Glycyrrhiza glabra* L.) through shoot tip and nodal cultures. In Vitro Cell Dev. Biol. Plant 34 (4), 331–334.

Thinh, N.T., Takagi, H., Yashima, S., 1999. Cryopreservation of in vitro-grown shoot tips of banana (*Musa* spp.) by vitrification method. CryoLetters.

Tiwari, S., Verma, P.C., Singh, P.K., Tuli, R., 2009. Plants as bioreactors for the production of vaccine antigens. Biotechnol. Adv. 27 (4), 449–467.

Tiwari, V., Tiwari, K.N., Singh, B.D., 2000. Suitability of liquid cultures for in vitro multiplication of *Bacopa monniera* (L.) Wttst. Phytomorphology 50, 337.

Tiwari, V., Tiwari, K.N., Singh, B.D., 2001. Comparative studies of cytokinins on in vitro propagation of *Bacopa monniera*. Plant Cell Tissue Organ Cult. 66 (1), 9–16.

Tsukazaki, H., Mii, M., Tokuhara, K., Ishikawa, K., 2000. Cryopreservation of Doritaenopsis suspension culture by vitrification. Plant Cell Rep. 19 (12), 1160–1164.

Underwood, B.A., Tieman, D.M., Shibuya, K., Dexter, R.J., Loucas, H.M., Simkin, A.J., Clark, D.G., 2005. Ethylene-regulated floral volatile synthesis in petunia corollas. Plant Physiol. 138 (1), 255–266.

Uragami, A., Sakai, A., Nagai, M., 1990. Cryopreservation of dried axillary buds from plantlets of *Asparagus officinalis* L. grown in vitro. Plant Cell Rep. 9 (6), 328–331.
Van der Rest, B., Danoun, S., Boudet, A.M., Rochange, S.F., 2006. Down-regulation of cinnamoyl-CoA reductase in tomato (*Solanum lycopersicum* L.) induces dramatic changes in soluble phenolic pools. J. Exp. Bot. 57 (6), 1399–1411.
Vanisree, M., Lee, C.Y., Lo, S.F., Nalawade, S.M., Lin, C.Y., Tsay, H.S., 2004. Studies on the production of some important secondary metabolites from medicinal plants by plant tissue cultures. Bot. Bull. Acad. Sin. 45 (1), 1–22.
Verdonk, J.C., Haring, M.A., van Tunen, A.J., Schuurink, R.C., 2005. ODORANT1 regulates fragrance biosynthesis in petunia flowers. Plant Cell 17 (5), 1612–1624.
Veronese, P., Li, X., Niu, X., Weller, S.C., Bressan, R.A., Hasegawa, P.M., 2001. Bioengineering mint crop improvement. Plant Cell Tissue Org. Cult. 64 (2–3), 133–144.
Vidal, N., Sánchez, C., Jorquera, L., Ballester, A., Vieitez, A.M., 2005. Cryopreservation of chestnut by vitrification of in vitro-grown shoot tips. In Vitro Cell Dev. Biol. Plant 41 (1), 63–68.
Vijaya, S.N., Udayasri, P.V., Aswani, K.Y., Ravi, B.B., Phani, K.Y., Vijay, V.M., 2010. Advancements in the production of secondary metabolites. J. Nat. Prod. 3, 112–123.
Villani, M.E., Morgun, B., Brunetti, P., Marusic, C., Lombardi, R., Pisoni, I., Bacci, C., Desiderio, A., Benvenuto, E., Donini, M., 2008. Plant pharming of a full-sized, tumour-targeting antibody using different expression strategies. Plant Biotechnol. J. 7 (1), 59–72.
Wakhlu, A.K., Bajwa, P.S., 1986. Regeneration of uniform plants from somatic embryos of *Papaver somniferum* (opium poppy). Phytomorphology 36, 101–105.
Walkey, D.G.A., Webb, M.J.W., Bolland, C.J., Miller, A., 1987. Production of virus-free garlic (*Allium sativum* L.) and shallot (*A. ascalonicum* L.) by meristem-tip culture. J. Horticult. Sci. 62 (2), 211–220.
Wang, C.F., Duo, M.J., Chang, E., Yang, J., 1996. Essential and toxic trace elements in the Chinese medicine. J. Radioanal. Nucl. Chem. 211 (2), 333–347.
Wang, Z.C., Deng, X.X., 2004. Cryopreservation of shoot-tips of citrus using vitrification: effect of reduced form of glutathione. CryoLetters 25 (1), 43–50.
Wang, H.M., To, K.Y., 2004. Agrobacterium-mediated transformation in the high-value medicinal plant Echinacea purpurea. Plant Sci. 166 (4), 1087–1096.
Wang, Y.-L., Fan, M.-J., Liaw, S.-I., 2005. Cryopreservation of in vitro-grown shoot tips of papaya (Carica papaya L.) by vitrification. Bot. Bull. Acad. Sin. 46.
Weise, A., Altmann, F., Rodriguez-Franco, M., Sjoberg, E.R., Bäumer, W., Launhardt, H., Gorr, G., 2007. High-level expression of secreted complex glycosylated recombinant human erythropoietin in the Physcomitrella Δ-fuc-t Δ-xyl-t mutant. Plant Biotechnol. J. 5 (3), 389–401.
Wirth, S., Calamante, G., Mentaberry, A., Bussmann, L., Lattanzi, M., Barañao, L., Bravo-Almonacid, F., 2004. Expression of active human epidermal growth factor (hEGF) in tobacco plants by integrative and non-integrative systems. Mol. Breed. 13 (1), 23–35.
Wong, K.W., Loh, C.S., 1987. In vitro regeneration of plantlets in *Brassica alboglabra*. Plant Cell Tissue Organ Cult. 10 (2), 143–148.
Wróbel-Kwiatkowska, M., Starzycki, M., Zebrowski, J., Oszmiański, J., Szopa, J., 2007. Lignin deficiency in transgenic flax resulted in plants with improved mechanical properties. J. Biotechnol. 128 (4), 919–934.
Xiao, D., Singh, S.V., 2008. z-Guggulsterone, a constituent of Ayurvedic medicinal plant *Commiphora mukul*, inhibits angiogenesis in vitro and in vivo. Mol. Cancer Ther. 7 (1), 171–180.
Xu, B., Copolla, M., Herr, J.C., Timko, M.P., 2007. Expression of a recombinant human sperm-agglutinating mini-antibody in tobacco (*Nicotiana tabacum* L.). Soc. Reprod. Fertil. Suppl. 63, 465.
Yoshikawa, T., Furuya, T., 1985. Morphinan alkaloid production by tissues differentiated from cultured cells of *Papaver somniferum* (1). Planta Med. 51 (02), 110–113.
Yukimune, Y., Tabata, H., Higashi, Y., Hara, Y., 1996. Methyl jasmonate-induced overproduction of paclitaxel and baccatin III in Taxus cell suspension cultures. Nat. Biotechnol. 14 (9), 1129–1132.

Yun, D.J., Hashimoto, T., Yamada, Y., 1992. Metabolic engineering of medicinal plants: transgenic *Atropa belladonna* with an improved alkaloid composition. Proc. Natl. Acad. Sci. 89 (24), 11799–11803.

Zenkteler, M.A., 1971. Development of new plants from leaves and roots of *Atropa belladonna* L. in the in vitro culture. Acta Soc. Bot. Pol. 40 (2), 305–313.

Zhang, L., Ding, R., Chai, Y., Bonfill, M., Moyano, E., Oksman-Caldentey, K.M., Kai, G., 2004. Engineering tropane biosynthetic pathway in *Hyoscyamus niger* hairy root cultures. Proc. Natl. Acad. Sci. U. S. A. 101 (17), 6786–6791.

Zhang, L., Jing, F.Y., Li, F.P., Li, M., Wang, Y., Wang, G., et al., 2009. Development of transgenic *Artemisia annua* (Chinese wormwood) plants with an enhanced content of artemisinin, an effective anti-malarial drug, by hairpin-RNAmediated gene silencing. Biotechnol. Appl. Biochem. 52, 199–207.

Zhao, M., Dhital, S.P., Fang, Y., Khu, D., Song, Y., Park, E., Lim, H., 2005. Application of slow-freezing cryopreservation method for the conservation of diverse potato (*Solanum tuberosum* L.) genotypes. J. Plant Biotechnol. 7 (3), 1–4.

Zhu, Z., Hughes, K.W., Huang, L., Sun, B., Liu, C., Li, Y., 1994. Expression of human α-interferon cDNA in transgenic rice plants. Plant Cell Tissue Organ Cult. 36 (2), 197–204.

Further Reading

Chacko, S.M., Thambi, P.T., Kuttan, R., Nishigaki, I., 2010. Beneficial effects of green tea: a literature review. Chin. Med. 5 (1), 13.

Cole, I.B., Saxena, P.K., Murch, S.J., 2007. Medicinal biotechnology in the genus scutellaria. In Vitro Cell Dev. Biol. Plant 43 (4), 318–327.

Fukai, S., 1990. Cryopreservation of chrysanthemum shoot tips. Sci. Hortic. 45 (1-2), 167–174.

Fukai, S., Hasegawa, A., Goi, M., 2002. Polysomaty in cymbidium. Hortscience 37 (7), 1088–1091.

Li, C., Loh, C.S., Sun, W.Q., 1999. An improved dehydration protocol for cryopreservation of Brassica napus somatic embryos. CryoLetters 20, 263–268.

Ma, X.Y., Xie, C.X., Liu, C., Song, J.Y., Yao, H., Luo, K., et al., 2010. Species identification of medicinal pteridophytes by a DNA barcode marker, the chloroplast psbA-trnH intergenic region. Biol. Pharm. Bull. 33 (11), 1919–1924.

Nalawade, S.M., Sagare, A.P., Lee, C.Y., Kao, C.L., Tsay, H.S., 2003. Studies on tissue culture of Chinese medicinal plant resources in Taiwan and their sustainable utilization. Bot. Bull. Acad. Sin. 44, 79–98.

Paek, K.Y., Yu, K.J., Park, S.I., Sung, N.S., Park, C.H., 1994. In: Micropropagation of *Rehmannia glutinosa* as medicinal plant by shoot tip and root segment culture. International Symposium on Medicinal and Aromatic Plants. vol. 390, pp. 113–120.

Sharma, A.K., Sharma, M., Chaturvedi, H.C., 1999. In: Conservation of phytodiversity of *Azadirachta indica* A. Juss. through in vitro strategies. National Symposium on Role of Plant Tissue Culture in Biodiversity Conservation and Economic Development. GB Pant Institute of Himalayan Environment and Development, Almora, pp. 1–13.

Sivanesan, I., Murugesan, K., 2008. An efficient regeneration from Nodal Expiants of *Withania somnífera* Du nal. Asian J. Plant Sci. 7 (6), 551–556.

Yuan-Long, W.A.N.G., Ming-Jen, F.A.N., Song-Iuan, L.I.A.W., 2005. Cryopreservation of in vitro-grown shoot tips of papaya (*Carica papaya* L.) by vitrification. Bot. Bull. Acad. Sin. 46.

CHAPTER

6

Quality Assurance and Quality Control of Medicinal and Aromatic Herbs

OUTLINE

https://doi.org/10.1016/B978-0-08-102227-6.00006-1

6.1 INTRODUCTION

In most traditional systems of medicine, herbal medicines form an important part as they play an important role particularly in primary health care, nutraceuticals, and cosmetics (Alyaiev, 2007). The official and commercial interest has increased due to the significant contribution made by herbal medicines to human health. In order to treat various diseases, about 80% of world populations still rely on the use of herbal drugs. There has been an increase in reports describing adverse health effects and variable quality, efficacy, and contents of herbal products (Taylor, 2004), since the use of herbal products as food and feed supplements and as phytotherapeutic drugs. Due to this popularity, more governments are considering policies on the appropriate use of herbal medicines. A resolution was adopted in 1989 by the World Health Assembly on traditional medicine and modern health care as in some countries; herbal medicines are widely used by a significant percentage of population for their health care. The member states of this resolution are insisted to introduce measures for the regulation and control of medicinal plant products. Allergic reactions, interactions with conventional drugs, and intrinsic toxicity are some of the side effects of herbal products. Other reasons for the urgency are the preparation and manufacturing of the herbs, such as misidentification of plants, the lack of standardization, contamination, substitution and adulteration of plants, failure of good manufacturing practice, and incorrect preparations and/or dosages (Calixto, 2000). A schedule should be set up to implement the guidelines step by step as it may not be possible to implement some aspects of the guidelines immediately. These guidelines represent a set of generic principles that can be flexibly implemented by individual countries. These guidelines are believed to prove a useful technical, managerial, and administrative tool for countries planning to begin or continue promoting the appropriate use of herbal medicines in their health-care systems.

Analytic techniques are required to confirm the identity, quality, purity, potency, safety, and efficacy of the herbal drugs or it's standardize extracts or pure active compound (Zafar et al., 2005). A predefined amount of quantity, quality, and therapeutic effect of ingredients in each dose comes under the system of standardization. The herbal product needs to be drug tested in order to be scientifically authenticated and characterized to ensure reproducibility in the manufacturing of the product (Shrikumar et al., 2006). The expanding industry accompanied by calls for assurance of quality, efficacy, and safety meets the increased demand for herbal products (Cardellina, 2002; Ernst, 2006).

6.2 STANDARDIZATION OF MAPs—CONCEPTS AND SCOPE

Being safe and effective are in general the basic requirements of all medicines whether they are synthetic or of plant origin (EMEA, 2005; World Health Organization, 1988a,b, 1990, 1996a,b, 1998a,b, 2000a,b). The process of converting plants or plant parts in phytopharmaceuticals by means of simple harvesting, drying, and storing leads to the final term "herbal drugs" (EMEA, 1998). Differences in growth, geographic location, and time of harvesting cause variability among the herbal drugs. The process of developing and agreeing upon technical terms can come under standardization of herbal medicines. A set of standards or inherent characteristics; constant parameters; and definitive, qualitative, and quantitative values are prescribed that carry an assurance of quality, efficacy, safety, and reproducibility. Experimentation and observations lead to specific standards, which form the process of prescribing a set of characteristics exhibited by the particular herbal medicine depicting standardization as a tool in the quality control process.

In both developing and developed countries, there has been an increase of public interest and acceptance of natural therapies during the past decade. About 80% of the world's population, especially in the developing countries, uses herbal medicine as their source of primary health care due to poverty and limited access to modern medicine (Bodeker et al., 2005; Mukherjee, 2002; Farnsworth et al., 1985; Bisset, 1994). Traditional medical practice in these communities is often viewed as an integral part of their culture. It is believed in the west that in order to live healthier lives, like our ancestors, people perform and are attracted to herbal therapies. Herbal medicines are referred to as balanced and moderate approach to healing. Billions of dollars are spend on herbal medicines by individuals who use them as home remedies and over-the-counter drugs. This represents a substantial proportion of the global drug market (World Health

Organization, 1990, 1999a,b, 2002a,b, 2005; Blumenthal, 2000; Blumenthal, 1998; Roberts and Tyler, 1997; Farnsworth et al., 1985). A required dose of herbal preparations is taken over a period of time to achieve the desired benefit. Most herbal preparations are safe for consumption as believed, but some herbs with most biologically active compounds could be toxic with lethal side effects (Bisset, 1994). The use of herbal medicine has become more challenging than necessary due to the variability of the constituents in herbs or herbal preparations due to genetic, cultural, and environmental factors. For instance, the availability and quality of the raw materials are frequently problematic, the active principles are diverse and may be unknown, and quality of different batches of preparation may be difficult to control and ascertain. Most countries launch their herbal products into the market without proper scientific evaluation and without any mandatory safety and toxicological studies. This practice leads to consumers buying herbal products without prescription and are unaware of the potential hazards in an inferior products that lead to believe there is no effective machinery to regulate manufacturing practices and quality standards. The most important prerequisites for the production of quality drug are a well-defined and constant composition.

Given the nature of products of plant origin, which are not usually constant and are dependent on and influenced by many factors, ensuring consistent quality of products is vital for the survival and success of the industry (Bauer, 1998).

The initial vital task is to develop such measures that identify the entire ingredients as presently it's very difficult to identify the presence of all ingredients as it's known, for which various chromatographic and spectrophotometric methods and evaluation of physiochemical properties can be tried. Wherever possible, these methods can be applied for quantitative estimation of bioactive group of compounds like alkaloids, flavonoids, and polyphenolic components or estimation of particular compound (Wani et al., 2007).

6.3 THE NEED FOR STANDARDIZATION—PRODUCERS' AND CONSUMERS' PERSPECTIVE

As the dangers and the shortcoming of modern medicine are getting more apparent, there is a shift toward the use of medicine of herbal origin, in the global perspective (Folashade et al., 2012). Medication that guarantees purity, safety, potency, and efficacy to the consumers is the cardinal responsibility of the regulatory authorities. The regulatory authorities strictly follow various standards of quality prescribed for raw materials and finished products in pharmacopoeias, formularies, and manufacturing operation through statutory imposed good manufacturing practices

(Shinde et al., 2016). Whether included in modern system of medicine or one of the traditional systems, these procedures logically would apply to all types of medication. The lack of standard quality control profile is one of the impediments in its acceptance, although herbal products have become increasingly popular throughout the world. Efficacy and safety has implications in the quality profile of the constituents of the final product of herbal medicine (Rasheed et al., 2013, 2014). It is difficult to establish quality control parameter though modern analytic techniques are expected to help in circumventing this problem due to the complex nature and inherent variability of the constituents of plant-based drugs. Additionally, the constituents responsible for the claimed therapeutic effects are frequently unknown or only partly explained. This is further complicated by the use of combination of herbal ingredients, used in traditional practice. Commonly, five different herbal ingredients are used in one product. Thus, batch-to-batch variation starts from the collection of raw material itself in the absence of any reference standard for identification that multiplies during storage and further processing. Hence, from cultivation of medicinal plant to its clinical application, standardization should encompass the entire field.

In this respect, internationally recognized guidelines for their quality assessment and quality control are necessary as plant materials and herbal remedies derived from them represent substantial portion of global market (Madhav et al., 2011).

6.4 STANDARDIZATION AND QUALITY CONTROL OF HERBAL CRUDE DRUGS—PROCESSES AND PROCEDURES

According to World Health Organization (1992, 1996a,b), the process involved in the physicochemical evaluation of crude drug covers various aspects, such as selection and handling of crude material, safety, efficacy and stability assessment of finished product, documentation of safety and risk based on experience, provision of product information to consumer, and product promotion; all comes under standardization and quality control of herbals. Heed is normally paid to such quality indexes as follows (Folashade et al., 2012):

1. Macro and microscopic examination: This includes evaluation of adulterants and identification of right variety.
2. Foreign organic matter: This involves removal of matter other than source plant to get the drug in pure form.
3. Ash values: The criteria to judge the identity and purity of crude drug are total ash, sulfated ash, water-soluble ash, acid-insoluble ash, etc.

4. Moisture content: In order to estimate the actual weight of drug material, proper measurement of moisture content helps in reducing errors. Low moisture suggests better stability against degradation of product.
5. Extractive values: These are suggestive weights of the extractable chemical constituents of crude drug under different solvents' environment.
6. Crude fiber: It is a criterion for judging purity and helps to determine the woody material component.
7. Qualitative chemical evaluation: It covers identification and characterization of crude drug with respect to phytochemical constituent. It also employs different analytic techniques to detect and isolate the active constituents. Botanical identification, extraction with suitable solvents, purification, and characterization of the active constituents of pharmaceutical importance come under phytochemical screening techniques.
8. Chromatographic examination: Major chemical constituents are used as markers for identification of crude drug.
9. Quantitative chemical evaluation: The amount of the major classes of constituents is estimated.
10. Toxicological studies: This is to establish the absence or presence of potentially harmful microorganisms, toxic elements, safety studies in animals like LD50, and microbial assay. This also helps to determine the pesticide residues.

The specific aims of such investigation in assuring herbal quality are as varied as the processes employed as the processes mentioned above involve a wide array of scientific investigations, which include physical, chemical, and biological evaluation employing various analytic methods and tools.

6.4.1 Physical Evaluation

To ensure both identity and purity, detailed botanical, macroscopic, and microscopic descriptions of the physical characteristics of each plant that can be used are present in each monograph (Folashade et al., 2012). Detailed illustrations and photographic images that provide visual documentation of accurately identified material are accompanied in each description. Some important parameters used in the evaluation of drugs are mainly physical contents such as elasticity in fibers, viscosity of drugs containing gums, swelling factor for mucilage containing materials, froth number of saponin drugs, congealing point of volatile and fixed oils, melting and boiling points, and water. To determine the fluorescence of extracts of some drugs, ultraviolet light is used. Physical constants such as alkaloids, volatile oils, and fixed oils are extensively applied to the active principles of drugs. A few of them are moisture content, viscosity, melting point, solubility, optical rotation, refractive index, ash values, extractive values, volatile oil content, foreign organic matter, and swelling factor (Raman, 2006).

6.4.2 Microscopic Evaluation

A thorough physical examination is required for full and accurate characterization of plant material (Wallis, 1999). For assuring the identity of the material and as an initial screening test for impurities, microscopic analyses of plants are critical (Folashade et al., 2012).

Qualitative microscopy: Organized drugs that are known by their histological characters through transverse section (TS) or longitudinal section (LS) or radial longitudinal section (RLS) or tangential longitudinal section (TLS) are identified by this method.

Stomata: Many types of stomata are distinguished by the forms and arrangement of the surrounding cells, for example, (a) anomocytic (ranunculaceous) irregular celled, *Digitalis*; (b) anisocytic (cruciferous) unequal celled, *Datura*; (c) diacytic (caryophyllaceous) cross celled, *Mentha*; and (d) paracytic (rubiaceous) parallel celled, *Senna*.

Trichomes: Trichomes are divided and subdivided as follows:

Covering trichomes—(a) unicellular trichomes, nux vomica, *Canabis*; (b) uniseriate multicellular unbranched trichomes, *Datura*; (c) biseriate multicellular unbranched trichomes, *Calendula officinalis*; (d) multiseriate multicellular unbranched trichomes, male fern; and (e) multicellular branched trichomes, *Verbascum Thapsus*.

Glandular trichomes—(a) unicellular glandular trichomes, vasaka, and (b) multicellular glandular trichomes, *Digitalis purpurea*.

Quantitative microscopy: It involves different parameters like the following:

Palisade ratio: It is defined as average number of palisade cells beneath each epidermal cell, for example, *Atropa belladonna* (6–10) and *D. Lanata* (2.5–6.5).

Stomatal number: It is defined as average number of stomata per square millimeter area of epidermis, for example, *A. belladonna* (from 6.0 to 14–37.5 (upper surface) and from 62.5 to 93–174 (lower surface)):

(a) Stomatal index: It is the percentage that the number of stomata forms to the total number of epidermal cells. It is calculated by

$$\text{S.I.} = S^{*}100/(E+S)$$

where S.I. is the stomatal index; *S* is the number of stomata per unit area; *E* is the number of epidermal cells in the same unit area, for example, *A. belladonna*, from −2.3–3.9 to 10.5 (upper surface) and from 20.2 to 21.7–23.0 (lower surface), and *D. Purpurea*, from −1.6–2.7 to 4.0 (upper surface) to 19.2–25.2

(b) Vein islet number: The average number of vein islet per square millimeter of the leaf surface midway between the midrib and the margin—(i) *D. Lanata*, 2.0–8.0, and (ii) *D. Purpurea*, 2.0–5.5.

(c) Vein termination number: The average number of vein terminations per square millimeter of the leaf surface midway between the midrib and the margin—(iii) *A. belladonna*, 6.3–10.3, and (iv) *Atropa acuminata*, 1.4–3.5.

6.4.3 Chemical Evaluation

Screening, isolation, identification, and purification of the chemical components and determination of the active constituent in a drug by chemical tests come under chemical evaluation (Sasidharan et al., 2011). In order to assess the potency of vegetable material in terms of its active principles, chemical analysis of the drug is done. To determine the identity of the drug substance and possible adulteration, color reaction tests are done under chemical screening.

The following are various methods of chemical evaluation:

1. *Instrumental methods*: Various instruments are used for evaluation like colorimetry, fluorimetry, and spectrophotometry.
2. *Chemical constants tests*: For the identification of fixed oils and fats, tests like acid value, iodine value, and ester value are used.
3. *Individual chemical tests*: These are the tests that are used for identifying particular drugs.
4. *Microchemical tests*: These tests are carried on slides. For example, eugenol in clove oil is precipitated as potassium eugenate crystals.

6.4.4 Microbial Contaminants and Aflatoxins

A broad variety of microbial contaminants, represented by bacteria, fungi, and viruses, may be associated with medicinal plants. Necessarily, this microbiological background depends on several environmental factors and exerts an important impact on the overall quality of herbal products and preparations. In the establishment of modern hazard analysis and critical control point (HACCP) schemes, risk assessment of the microbial load of medicinal plants has therefore become an important subject (Folashade et al., 2012).

A number of bacteria and molds, often originating in the soil, are normally present in herbal drugs. Additional contamination may be caused by poor methods of harvesting, cleaning, drying, handling, and storage, as may be the case with *Escherichia coli* or *Salmonella* spp., while a large range of bacteria and fungi are from naturally occurring microflora, aerobic spore-forming bacteria that frequently predominate. Pharmacopoeias and WHO (World Health Organization, 1998a,b, 2000a,b) have guidelines for laboratory procedures investigating microbial contaminations, and limit values can also be found in the sources mentioned. Determining the total aerobic microbial count, the total fungal count, and the total Enterobacteriaceae count, together with tests for the presence of *E. coli*, *Staphylococcus aureus*, *Shigella*, and *Pseudomonas aeruginosa* and *Salmonella* spp., generally forms a complete procedure. *E. coli* and *Salmonella* spp. should be absent from herbal preparations, which is specified by the European Pharmacopoeia. The levels of microbial contamination are usually much higher in materials of vegetable origin than synthetic products, and the requirements for microbial contamination in the European Pharmacopoeia allow higher levels of microbial contamination in herbal remedies than in synthetic pharmaceuticals. The method of processing of the drug may also impact the allowed contamination level. For example, if the final herbal preparation involves boiling with water, higher contamination levels are permitted. The presence of fungi should be carefully investigated and/or monitored, since some common species produce toxins, especially aflatoxins. As per guidelines (World Health Organization, 2000a,b), aflatoxins in herbal drugs can be dangerous to health even if they are absorbed in minute amounts. Aflatoxin-producing fungi sometimes build up during storage (De Smet et al., 1992). World Health Organization (2000a, 2000b) has published procedures for the determination of aflatoxin contamination in herbal drugs. TLC is used for confirmation after a thorough cleanup procedure. As such, bacterial endotoxins and mycotoxins in herbal remedies at times may also be an issue (De Smet et al., 1997, 1992), in addition to the risk of bacterial and viral contamination.

Enzyme activity is enhanced during withering, transforming some of the constituents to other metabolites not initially found in the herb. These newly formed constituents along with the molds such as *Penicillium nigricans* and *P. jensi* may then have adverse effects (De Smet et al., 1992). Certain plant constituents are susceptible to chemical transformation by contaminating microorganisms that leads to believe that medicinal plants from some countries may be contaminated with toxigenic fungi (*Aspergillus* and *Fusarium*).

6.4.5 Biological Evaluation and Purity Determination

A mixture of bioactive compounds constitutes the biological potency of an herbal product, while the relative property of a single compound can vary from batch to batch (Sagar et al., 2005). In order to evaluate and standardize certain drugs, pharmacological activity has been applied. Strength of the drug or their preparations can be indicated by the assays on living animal and on their intact or isolated organs known as biological assays or bioassay. Each monograph includes standards for purity and other qualitative indexes. In order to deem a material worthy for clinical use, according to the guidance of global pharmacopoeial and regulatory frameworks, a thorough characterization and fulfilling of certain criteria needs to be done. For instance, the purity of pharmaceutical grade materials is rigorously defined. Basically, the structural characterization and the quantification of the impurities, frequently down to 0.1% w/w level, define the purity assessment of pharmaceutical grade materials (Pauli et al., 2014). As trace compounds can lead to false conclusions, purity assessment is most vital in case of herbal compounds to which a biological activity is ascribed.

6.5 DETECTION OF ADULTERATION IN MAPs PRODUCTS BY MODERN GENETIC AND EPI-GENETIC APPROACHES

An overview and description of the various techniques that have been used for genome-based authentication of medicinal plants are presented in Table 6.1.

6.6 GOOD AGRICULTURAL/MANUFACTURING PRACTICES

Source and quality of raw materials, good agricultural practices (GAP), and good manufacturing practices are other steps involved in quality control and the standardization of herbal medicines. These practices play a critical role in guaranteeing the quality and stability of herbal preparations (World Health Organization, 1988a,b, 1992, 2000a,b, 2003, 2004; EMEA, 2002; Blumenthal, 1998; Roberts and Tyler, 1997). Accepting GAP as an integral part and maintaining prevailing conditions during growth determine the quality of a plant product, which cover seed selection, growth conditions, fertilizer application, harvesting, drying, and storage.

The factors that greatly affect the quality and hence the therapeutic value of herbal medicines are the use of fresh plants, age and part of plant collected, period, time and method of collection, temperature of processing, exposure to light, availability of water, nutrients, drying, packing, transportation of raw material, and storage. Apart from these criteria, factors such as the method of extraction, contamination with microorganisms, heavy metals, and pesticides can alter the quality, safety, and efficacy of herbal drugs (Folashade et al., 2012).

Most of these factors can be minimized by using cultivated plants under controlled conditions instead of those collected from the wild (Eskinazi et al., 1999; Blumenthal, 1998; Bauer, 1998). Variations in composition are caused by the processes that continue for long periods from

TABLE 6.1 Detection of Adulteration in MAP Products by Modern Genetic and Epigenetic Approaches

Name	Acronym	Explanation	References
Polymerase chain reaction	PCR	PCR is generally used for quick amplification of DNA as an in vitro method. Each strand of DNA is copied with the help of heat-stable DNA polymerase, as it uses two primers, namely, forward and reverse that are complementary to the 5′ and 3′ flanking sequences of the DNA. A three-step reaction is conducted in a programmable thermocycler that involves denaturation of template DNA, annealing of primers, and synthesis of new strands. Different temperatures are maintained at every step	Saiki et al. (1985, 1988) and Dieffenbach and Dveksler (2003)
Allele-specific diagnostic PCR		A common primer for PCR reaction is used along with a primer with allele-specific 3′ ends and labeled with different fluorochromes at their 5′ end. Gel electrophoresis or capillary electrophoresis with an automated DNA sequencer is used to analyze the amplicon	Gupta et al. (2001)
Amplification refractory mutation system	ARMS	The primers designed for PCR reactions are specific to their target sequence as their 3′ ends are complementary. This variation of the PCR is based on the mismatch 3′ end residues causing no amplification	Newton et al. (1989)
Amplified fragmented length polymorphism	AFLP	This technique involves digestion of genomic DNA digested with restriction enzymes. The next step involves a ligation reaction, where specific oligonucleotide adapters are added to the ends of the fragments. PCR reaction is used to selectively amplify based on the complementary of the adapter and restriction sites	Vos et al. (1995)
Arbitrarily primed PCR	AP-PCR	It's identical to RAPD, but sets of two longer (> 18 nucleotides) primers of arbitrary sequences are used	
Direct amplification of length polymorphism	DALP	It involves the use of variable forward primers that contain a universal core sequence at their 5′ end and a constant reverse primer resulting in multiple amplicons in a PCR reaction. Amplicons are separated by gel electrophoresis, isolated and directly sequenced	Desmarais et al. (1998)
Multiplex PCR		Parallel amplification is attained in a single PCR reaction using multiple forward and reverse primers	Dieffenbach (2003)

(Continued)

TABLE 6.1 Detection of Adulteration in MAP Products by Modern Genetic and Epigenetic Approaches—cont'd

Name	Acronym	Explanation	References
PCR-selective restriction	PCR-SR	Restriction enzymes are used to cut the amplicon attained after a gene-specific PCR reaction. Gel electrophoresis is used for analysis	Desmarais et al. (1998)
Randomly amplified polymorphic DNA	RAPD	A single short (10 nucleotides) primer with arbitrary sequence is used to amplify genomic DNA (gDNA). Gel electrophoresis is used to analyze the resulting multiple amplicons of different lengths	Williams et al. (1990)
Sequence-characterized amplified region	SCAR	Different amplicons attained by RAPD are sequenced, and amplicon-specific primers are designed for use in PCR	Paran and Michelmore (1993)
Restriction length polymorphism	RFLP	Sequence-specific endonucleases are used to generate a number of small fragments of different lengths of genomic DNA. Gel electrophoresis is used to separate the fragments according to their size. DNA fingerprint of the source is obtained by the results of electrophoresis	
DNA microarray		A DNA is covalently fixed to a solid surface usually glass slide or a synthetic membrane in the form of minute spots. These spots contain DNA of specific sequence known as probe that is hybridized to fluorescently labeled target DNA. The next step involves washing and analyzing the hip in a microarray reader. This process is referred as DNA microarray or gene microarray	

DNA sequencing		A heat-stable DNA polymerase and fluorescent-labeled dideoxynucleotides are used in a thermocycler, the process collectively known as DNA sequencing. The resulting polymerase products are separated according to length using capillary electrophoresis, detected by laser-induced fluorescence, and automatically analyzed by computer software. Older methods making use of radioactively labeled nucleotides and gel electrophoresis are still in use and may be the only option, when access to automated sequencers is not available	Kretz et al. (1994)
Inter-simple sequence repeat anchored-PCR	ISSR-PCR	Primers anchored at simple sequence repeat (SSR) sequences (e.g., CACACACA) are used to amplify the DNA regions between the flanking SSR	Kretz et al. (1994)
Multiplex amplification refractory mutation system	MARMS	Multiplex PCR using a common primer and multiple mutation-specific primers as used in ARMS	Fortina et al. (1992)
Simple sequence repeat polymorphism	SSR	Simple sequence repeats (SSRs) or microsatellites are small sequence motifs consisting of two or more nucleotides (e.g., CA and ATG), which repeat in tandem (e.g., CACACA and ATGATGATG). These repeats vary in length (e.g., CACACA vs. CACACACACACA) and are ubiquitously and randomly distributed in all eukaryotic genomes. The length polymorphisms can be easily detected by gel electrophoresis of amplicons generated by PCR using unique pairs of primers flanking the repeat	Weber and May (1989)

collection to marketing, thus destroying the active principles by enzymes. Thus, proper standardization and quality control of both the raw material and the herbal preparations should be conducted.

6.7 CRITICAL FACTORS AFFECTING THE QUALITY CONTROL OF HERBAL DRUG

Herbal ingredients that have not been authenticated properly can lead to major public health issues, thus highlighting the problems associated with unregulated herbal products. Herbal ingredients must be accurately identified by macroscopic and microscopic comparison with authentic material or accurate descriptions of authentic herbs (Houghton, 1998).

6.7.1 Analytical Methods and Chromatography

Critical to compliance with any monograph standard is the need for appropriate analytic methods for determining identity, quality, and relative potency. There are a plethora of analytic methods available. However, it is often difficult to know which is the most appropriate to use, but critical among known analytic tools in monograph standardization is chromatography (Folashade et al., 2012).

The separation of molecules based on differences in their structure and/or composition is studied through chromatography. In general, chromatography involves moving a preparation of the materials to be separated, "the test preparation," over a stationary support (Stahl, 2005). The similar molecules are separated by the fact that the molecules in the test preparation will have different interactions with the stationary support. The test molecules that have weaker interactions with the support will move faster than those having tighter interactions with the support. In this way, different types of molecules can be separated from each other as they move over the support material. A variety of supports, including immobilized silica on glass plates (thin-layer chromatography—TLC), very sensitive high-performance thin-layer chromatography (HPTLC), volatile gases (gas chromatography), paper (paper chromatography), and liquids that may incorporate hydrophilic, insoluble molecules (liquid chromatography), can be used to carry out chromatographic separations (Anonymous, 1998). In order to analyze a broad number of compounds both efficiently and cost-effectively, HPTLC is used, which is a valuable quality assessment tool for the evaluation of botanical materials. Additionally, numerous samples can be run in a single analysis, thereby dramatically reducing analytic time. With the use of HPTLC, the same analysis can be viewed collectively in different wavelengths of light, thereby providing a more complete profile of the plant.

6.7.2 Quantitative Analysis

The primary goal of the analysis is to provide validated methods to be used to quantify the compounds most correlated with pharmacological activity or qualitative markers (Wani et al., 2007). The most appropriate quantitative analytic method, accompanying chromatograms, is desirable.

6.7.3 Control of Starting Material

In order to ensure reproducible quality of herbal medicinal products, control of the starting materials is essential (De Smet, 2004; Gaedcke and Steinhoff, 2003; World Health Organization, 2002a,b; Phillipson, 1995). The following points are to be considered in the control of starting materials:

6.8 AUTHENTICATION AND REPRODUCIBILITY OF HERBAL INGREDIENTS

It is imperial to refer the herbal ingredients by their binomial Latin names of genus and species; only permitted synonyms should be used. Herbal ingredients must be accurately identified by macroscopic and microscopic comparison with authentic material or accurate descriptions of authentic herbs (Houghton, 1998). Even when correctly authenticated, it is important to realize that different batches of the same herbal ingredient may differ in quality due to a number of factors such as the following (Folashade et al., 2012):

1. *Inter- or intraspecies variation*: The variation in constituents may be related to the country of origin and is mostly genetically controlled.
2. *Environmental factors*: Environmental factors like climate, altitude, and other conditions under which it was cultivated can affect the quality of herbal ingredients.
3. *Time of harvesting*: Concentrations of constituents in a plant can vary during the growing cycle or even during the course of a day; therefore, for some herbs, the optimum time of harvesting should be specified as it is known.
4. *Plant part used*: There occurs variation of active constituents between plant parts, and it is a common practice to adulterate an herbal ingredient with parts of the plant not normally utilized. In addition, an "exhausted" plant material that has been previously subjected to extraction is sometimes used as adulterants to increase the weight of a batch of herbal ingredient.

5. *Postharvesting factors*: The quality of an herbal ingredient can be greatly affected by storage conditions and processing treatments. Inappropriate storage after harvesting can result in microbial contamination, and processes such as drying may result in a loss of thermolabile active constituents.

6.8.1 Adulteration/Substitution

The major drawback in the commercial use of natural products is the adulteration and substitution of herbal drugs. It is the main problem in herbal industry. One of the greatest drawbacks in promotion of herbal products is adulteration in market samples. The practice of substituting original crude drug partially or fully with other substances that is either free from or inferior in therapeutic and chemical properties or addition of low-grade or spoiled drugs or entirely different drug similar to that of original drug substituted with an intention of enhancement of profits that do not confirm with the official standards (Ansari, 2011; Kokate and Gokhale, 2004) is termed as adulteration. Adulteration may be defined as mixing or substituting the original drug material with other spurious, inferior, defective, spoiled, useless parts of the same or different plant or harmful substances or drug.

Direct or Intentional Adulteration

Direct or intentional adulteration is done purposely that usually includes practices in which an herbal drug is substituted partially or fully with other inferior products. Many different inferior commercial varieties are used as adulterants due to morphological resemblance to the authentic herb that may or may not have any chemical or therapeutic potential (Afaq, 2013). Adulteration is caused in case of substituting "exhausted" drugs of the plant material with the same plant material devoid of the active constituents. This practice is most common in the case of volatile oil-containing materials, where the dried exhausted material resembles the original drug but is free of the essential oils (Jaya Preethi et al., 2014). Foreign matters such as other parts of the same plant with no active ingredients, sand and stones, manufactured artifacts, and synthetic inferior principles are used as substitutes (Billore, 2005). The practice of intentional adulteration is mainly encouraged by traders who are reluctant to pay premium prices for herbs of superior quality and hence are inclined to purchase only the cheaper products. This encourages producers and traders to sell herbs of inferior quality.

Indirect or Unintentional Adulteration

Sometimes, an authentic drug partially or fully devoid of the active ingredients may enter the market in the absence of proper means of evaluation. Geographic sources, growing conditions, processing, and storage are all factors that influence the quality of the drug (Ansari, 2011; Kokate and Gokhale, 2004). Herbal collectors and suppliers are also responsible for some of the herbal adulteration due to the carelessness. The correct part of genuine plant should be collected. Other less valuable part of the genuine plant should not be collected. Moreover, collection should be carried out at a proper season and time when the active constituents reach maximum. *Datura strumarium* leaves should be collected during flowering stage, and wild cherry bark in autumn, etc. collection from other plants by ignorance, due to similarity in the appearance and color and the lack of knowledge, may lead to adulteration. For example, in place of *Aconitum napellus*, the other *A. deinorhizum* may be collected, or in place of *Rhamnus purshiana* (cascara bark), *R. californica* is generally collected. Confusion existing in the common vernacular name of different plants in various states of India may lead to this type of adulteration. Often, in different states, the same plant is known by different vernacular names, while quite different drugs are known by the same name. This creates confusion that is best illustrated by Punarnava and Brahmi. The Indian pharmacopoeia drugs *Trianthema portulacastrum* L. and *Boerhavia diffusa* L. are both known by the same vernacular name "Punarnava." Deterioration, especially during storage, leads to the loss of the active ingredients, the production of metabolites with no activity and, in extreme cases, the production of toxic metabolites. Physical factors such as air (oxygen), humidity, light, and temperature can bring about deterioration directly or indirectly (Poornima, 2010). These factors, alone or in combination, can lead to the development of organisms such as molds, mites, and bacteria.

6.8.2 Identity and Purity

In order to try to ensure the quality of licensed herbal medicines, it is essential not only to establish the botanical identity of an herbal ingredient but also to ensure batch-to-batch reproducibility. Thus, in addition to macroscopic and microscopic evaluation, identity tests are necessary. Such tests include simple chemical tests, for example, color or precipitation and chromatographic tests. TLC is commonly used for identification purposes, but for herbal ingredients containing volatile oils, a gas-liquid chromatographic test may be used (Springfield et al., 2005). Although the aim of such tests may be to confirm the presence of active principles, it is frequently the case that the nature of the active principle has not been established. In such instances, chemical and chromatographic tests help to provide batch-to-batch comparability, and the chromatogram may be used as a "fingerprint" for the herbal ingredient by demonstrating the profile of some common plant constituents such as flavonoids, alkaloids, and terpenes.

A lot of quality determinants are critically examined for identity and purity of the concerned herb. Determinants such as purity and chemical constituents are very important. Criteria such as type of preparation, sensory properties, physical constants, adulteration, contaminants, moisture, ash content, and solvent residues have to be checked to prove the identity. Macroscopic and microscopic examinations can be used to achieve identity. Voucher specimens are reliable reference sources. Changes in the physical appearance of the plant can be caused by disease outbreaks, which in turn lead to incorrect identification (De Smet, 1999). Sometimes, an incorrect botanical quality with respect to the labeling can be a problem. For instance, in the 1990s, a South American product labeled as "Paraguay tea" was associated with an outbreak of anticholinergic poisoning in New York. A subsequent chemical analysis revealed the presence of a class of constituents that was different from the metabolites normally found in the plant from which Paraguay tea is made.

Another method to ensure product identity and purity is assaying for those herbal ingredients with known active principles. In order to set the criterion for the minimum accepted percentage of active substances, an assay should be established. Specificity for individual chemical substances and high-pressure liquid chromatography and gas-liquid chromatography should be prioritized, wherever possible in such assays. In order to determine the total content of a group of closely related compounds, where such assays have not been established, nonspecific classical methods such as titration or colorimetric assays may be used. Purity is associated with the safe use of drugs and deals with factors such as values, contaminants (e.g., foreign matter in the form of other herbs), and heavy metals. Modern purity evaluation also includes microbial contamination, aflatoxins, radioactivity, and pesticide residues due to the application of improved analytic methods. Photometric analysis, TLC, high-performance liquid chromatography (HPLC), and gas chromatography (GC) are some analytic methods that can be employed in order to establish the constant composition of herbal preparations. Different concepts such as "normalization versus standardization" have to be applied in order to establish relevant criteria for uniformity in case where the active principles of the preparation are known or unknown.

Since in most herbal drugs the active constituents are not known, content assay is the most difficult area to perform in quality control. Sometimes, markers can be used. In pharmacopoeias, where no active constituent or marker can be defined for the herbal drug, the percentage extractable matter with a solvent may be used as a form of assay. The incorporation of the extracting solvent depends on the nature of the compounds involved and might be deduced from the traditional uses. For instance, using an herbal drug to make a tea, the hot-water extractable matter, expressed as milligrams per gram of air-dried material, may serve this purpose (World Health

Organization, 1996a,b, 1998a,b). Steam distillation is a special form of assay for the determination of essential oils. When the active constituents (e.g., sennosides in *Senna*) or markers (e.g., alky amides in *Echinacea*) are known, a vast array of modern chemical analytic methods such as ultraviolet/visible spectroscopy (UV/VIS), TLC, HPLC, GC, mass spectrometry (MS), or a combination of GC and MS (GC-MS) can be employed (Watson, 1999).

6.8.3 Contaminants of Herbal Ingredients

High-quality herbal ingredients should be free from insects, animal matter, and excreta. It is usually not possible to remove completely all contaminants; hence, specifications should be set in order to limit them (Folashade et al., 2012):

1. *Ash values*: Incineration of an herbal ingredient produces ash that constitutes inorganic matter. Treatment of the ash with hydrochloric acid results in acid-insoluble ash that consists mainly of silica and may be used to act as a measure of soil present. Limits may be set for ash and acid-insoluble ash of herbal ingredients.
2. *Foreign organic matter*: It is not possible to collect an herbal ingredient without small amounts of related parts of plant or other plants. Standards should be set in order to limit the percentage of such unwanted plant contaminants.
3. *Microbial contamination*: Faulty growing, harvesting, storage, or processing increases the chances of aerobic bacteria and fungi in plant material that are normally present. Herbal ingredients with high starch content may be prone to increased microbial growth. Herbal ingredients are also contaminated by pathogenic organisms including *Enterobacter*, *Enterococcus*, *Clostridium*, *Pseudomonas*, *Shigella*, and *Streptococcus*. European Pharmacopocia gives nonmandatory guidance on acceptable limits (Barnes et al., 2007) that makes it essential that limits be set for microbial contamination.
4. *Pesticides*: Dichlorodiphenyltrichloroethane (DDT) or other chlorinated hydrocarbons, organophosphates, carbamates, or polychlorinated biphenyls may contaminate herbal ingredients, particularly those grown as cultivated crops. The European Pharmacopoeia includes details of test methods together with mandatory limits for 34 potential pesticide residues (Barnes et al., 2007). So limit tests are vital for acceptable levels of pesticide contamination of herbal ingredients.
5. *Fumigants*: Ethylene oxide, methyl bromide, and phosphine have been used to control pests that contaminate herbal ingredients. The use of ethylene oxide as a fumigant with herbal drugs is no longer permitted in Europe (Barnes et al., 2007).

6. *Toxic metals*: Lead, cadmium, mercury, thallium, and arsenic have been shown to be contaminants of some herbal ingredients. Limit tests for such toxic metals are essential for herbal ingredients.
7. *Radioactive contamination*: A certain degree of exposure is inevitable (AOAC, 2005; World Health Organization, 2000a,b; De Smet et al., 1992) as there are many sources of ionization radiation, including radionuclides, occurring in the environment.
8. *Other contaminants*: It is imperial to limit other contaminants such as endotoxins and mycotoxins as standards increase for the quality of herbal ingredients, which will be utilized to ensure high quality for medicinal purposes (Barnes et al., 2007).

6.8.4 Standardization of Herbal Medicines

As botanical extracts made directly from crude plant materials show substantial variation in composition, quality, and therapeutic effects, standardization involves adjusting the herbal drug preparation to a defined content of a constituent or a group of substances with known therapeutic activity by adding excipients or by mixing herbal drugs or herbal drug preparations. After subjecting to rigorous quality controls during all phases of the growing, harvesting, and manufacturing processes, standardized extracts become high-quality extracts containing consistent levels of specified compounds. Dietary supplements have no regulatory definition for standardization, which results in multiple meanings of the term "standardization" (Folashade et al., 2012).

Following a recipe is not sufficient for a product to be called standardized as some manufacturers use the term standardization incorrectly to refer to uniform manufacturing practices. Therefore, the presence of the word "standardized" on a supplement label does not necessarily indicate product quality. Marker substances should be established for analytic purposes and standardization in case the active ingredients are unknown. Chemically defined constituents of an herbal drug that are vital for the quality of the finished product are known as marker substances (Madhav et al., 2011). The compounds that are responsible for the pharmacological effects in the body are ideal chemical markers. There are two categories of standardization: in the first category, "true" standardization, a definite phytochemical or group of constituents, is known to have activity, for example, *Ginkgo* with its 26% *Ginkgo* flavones and 6% terpenes. These products are considered as phytopharmaceuticals; they are highly concentrated and no longer represent the whole herb, and in many cases, they are vastly more effective than the whole herb. This process may result in the loss of efficacy, and the potential for adverse effects and herb-drug interactions may increase. The second category of standardization is based on the manufacturers' guarantee for the presence of a certain percentage of marker compounds that are not indicators of therapeutic activity or quality of the herb.

Quality of the active pharmaceutical ingredients (APIs), in case of herbal drug preparations, is directly influenced by the production and primary processing of the medicinal plant or herbal drug (AOAC, 2005). Adequate quality assurance system is needed due to the inherent complexity of naturally growing medicinal plants and the limited availability of simple analytic techniques to identify and characterize the active constituents solely by chemical or biological means. Cultivation, harvesting, primary processing, handling, storage, packaging, and distribution process also require this assurance. It is immensely important to establish good agricultural, harvesting, and manufacturing practices for herbal starting materials in order to minimize the undesirable factors such as deterioration and contamination through adulteration, especially microbial contamination (Barnes et al., 2007). While as the producers, processors, and traders of medicinal plants or herbal drugs have an obligation and a role to play, at the same time, manufacturers and suppliers of herbal products should adhere to quality control standards and good manufacturing practices as presently, only a handful of manufacturers adhere to complete quality control and good manufacturing procedures including microscopic, physical, chemical, and biological analysis. Organizations such as National Agency for Food and Drug Administration and Control (NAFDAC) help safeguard Nigerians' health, and organizations such as Health Canada help safeguard Canadians' health by carrying out premarket reviews of all drugs before they are authorized for sale. In order to ensure that products are free of unsafe ingredients and actually contain the ingredients indicated on the labels, the products available in the market are analyzed regularly.

Due to the lack of regulation, the potency and quality of an individual herbal product may be unclear. It is obvious that for a given plant product, its quality will also be determined by the prevailing conditions during the growth cycle of the plant. Therefore, for cultivated plants, the GAP system has been introduced, under which each step, including seed selection, growing conditions, use of fertilizers, optimization of harvest time, harvesting, and drying, has to adhere to a set of criteria. It is likely that GAP procedures will become an integral part of quality control in the near future (GAP, 2006).

6.8.5 Microscopic Evaluation

Traditionally, visual evaluations have been used for quality control of herbal drugs, but today, microscopic evaluation forms the basic step in the identification of herbs of varied forms and textures and detection of adulterants. Visual evaluation is required to ascertain the taxonomic identification of the plant or part used. This seldom needs more than a simple magnifying lens (Trease and Evans, 2002). For determining the correct species and/or the correct part, microscopic analysis is needed. For example, pollens can be used for identifying the species in the case of flowers, and leaf stomata can be used to identify the plant part used. This evaluation is of utmost importance when different parts of the same plant are to be used for different treatments, like stinging nettle (*Urtica urens*); aerial parts are used to treat rheumatism, and roots are applied for benign prostate hyperplasia (AOAC, 2005).

6.8.6 Foreign Matter

Herbal drug formulations should only be made from the plant parts mentioned in the literature and should be free from other parts of the same plant or other plants. They should be entirely free from foreign matters like molds or insects, excreta, visible contaminant (sand and stones), poisonous and harmful foreign matter, and chemical residues.

Insects and invisible toxin-producing microbial contaminants are the other potential contaminants of herbal medicines (WHO, 2004, 2003; EMEA, 2002). Macroscopic examination can easily be employed to determine the presence of foreign matter, although microscopy is indispensable in certain special cases (e.g., starch deliberately added to dilute the plant material). Furthermore, the detection method varies according to the nature of the foreign material present. For example, TLC is often needed to detect the contaminants (AOAC, 2005; World Health Organization, 1998a,b, 1999a,b).

6.8.7 Ash Content

Total ash is the measure of the total amount of material left after burning. To determine ash content, the plant material is burnt, and the residual ash is measured as total and acid-insoluble ash. The latter is the residue obtained after boiling the total ash with dilute hydrochloric acid and burning the remaining insoluble matter. This procedure measures the amount of silica present, especially in the form of sand and siliceous earth (AOAC, 2005).

6.8.8 Heavy Metals

Heavy metal contamination in herbal medicines occurs by different ways. The presence of heavy metals like mercury, lead, copper, cadmium, and arsenic in herbal can pose severe health hazards to the user (AOAC, 2005; World Health Organization, 1998a,b; De Smet et al., 1992). The potential intake of the toxic metal can be estimated on the basis of the level of its presence in the product and the recommended or estimated dosage of the product. This potential exposure can then be put into a toxicological perspective by comparison with the so-called provisional tolerable weekly intake (PTWI) values for toxic metals, which have been established by the Food and Agriculture Organization of the World Health Organization (FAO-WHO) (De Smet, 1999; WHO, 1981, 1979). Many pharmacopoeias contain heavy metal detection methods that are usually based on color reactions with special reagents such as thioacetamide or diethyldithiocarbamate, and the amount present is estimated by comparison with a standard (WHO, 1988a). Instrumental analyses have to be employed when the metals are present in trace quantities and in admixture or when the analyses have to be quantitative. Generally, the main methods commonly used are atomic absorption spectrophotometry (AAS), inductively coupled plasma (ICP), and neutron activation analysis (NAA) (Watson, 1999).

6.8.9 Pesticide Residues

It is important that herbs and herbal products are free of these chemicals or at least are controlled for the absence of unsafe levels even though there are no serious reports of toxicity due to the presence of pesticides and fumigants (De Smet et al., 1992). Agricultural practices such as spraying, treatment of soils during cultivation, and administering of fumigants during storage lead to accumulation of pesticide residues in herbal drugs. However, it may be desirable to test herbal drugs for broad groups in general, rather than for individual pesticides.

Analysis of total organic chlorine can be used to measure chlorine as many pesticides contain chlorine. In an analogous way, insecticides containing phosphate can be detected by measuring total organic phosphorus.

Samples of herbal material are extracted by a standard procedure, impurities are removed by partition and/or adsorption, and individual pesticides are measured by GC, MS, or GC-MS. Some simple procedures have been published by the WHO, and the European Pharmacopoeia has laid down general limits for pesticide residues in medicine (World Health Organization, 1996a,b, 1998a,b, 2000a,b; De Smet, 1999; AOAC, 2005).

6.8.10 Radioactive Contamination

As a consequence of a major nuclear accident, caused by radio nuclides, dangerous contamination in the form of radiations can be widespread and lethal, to which WHO, in close cooperation with several other international organizations, has developed guidelines. These publications emphasize that the health risk, in general, due to radioactive contamination from naturally occurring radio nuclides is not a real concern but those arising from major nuclear accidents such as the nuclear accident in Chernobyl and Fukushima may be serious and depend on the specific radionuclide, the level of contamination, and the quantity of the contaminant consumed. Taking into account the quantity of herbal medicine normally consumed by an individual is unlikely to be a health risk. Therefore, at present, no limits are proposed for radioactive contamination (AOAC, 2005; World Health Organization, 2000a,b; De Smet et al., 1992).

6.8.11 Validation and Release of Herbal Products

Fakers selling adulterated herbal products are common in both developed and developing countries, which causes the validation of herbal products a major public health concern. Despite the existence of certain guidelines in some individual countries and those outlined by the WHO, there is no control by the government agencies. Drug control administrators should periodically monitor the quality and efficacy and ensure scientific validation as the herbal products are marketed as therapeutic agents and irrespective of whether the products really have any positive effects to cure and reduce the severity of the disease (Ferenczi-Fodor et al., 2001). The production of impure or adulterated herbal products would be controlled by the introduction of scientific validation and would eventually ensure their rational use. This could also lead to the regulation of the industry so that only qualified physicians and health providers are allowed to prescribe the medication.

Standards for herbal drugs are outlined in several monographs of the principal pharmacopoeias. Standards are defined and available, and the analytic procedures used are fully validated in an official monograph published in a pharmacopoeia. This is of major importance, since validation can be a rather time-consuming process (Pharmacopeia, 1990).

As per the definition, validation is the process of proving that an analytic method is acceptable for its intended purpose. Guidelines from the US Pharmacopeia (USPC, 1994–2001), the International Conference on Harmonization (ICH), and the US Food and Drug Administration (FDA) provide a framework for performing such validations. Studies on specificity, linearity, accuracy, precision, range, detection, and quantitative limits, depending on whether the analytic method used is qualitative or quantitative, are generally included in validation investigations (De Smet, 1997). Also, of utmost importance is the availability of standards. Reliable reference samples of the plant must be available for macroscopic and microscopic procedures in general. This problem will normally be solved by a defined botanical source (e.g., voucher specimens). Standards for chromatographic procedures are less easy to obtain, while characteristic plant constituents, either active or markers, are seldom available commercially.

Sometimes, an LC-MS approach can be referred to as a mode of characterization. Going one step further, elucidations to prove its definite structure will not be easy, after isolation of such a compound. The method often employed is to use readily available compounds that behave similarly in the chosen chromatographic systems and to calculate retention values and/or times toward these compounds as a standard. Qualitative chemical examination is designed to detect and isolate the active ingredients. TLC and HPLC are the main analytic techniques commonly used. In cases when active ingredients are not known or too complex, the quality of plant extracts can be assessed by a "fingerprint" chromatogram (De Smet, 1997).

Labelling of Herbal Products

The finished herbal product is as vital as the quality of consumer information about the product. The risk of inappropriate uses and adverse reactions are reduced by warnings on the packet or label (De Smet, 1997). The product label is considered as the primary source of information on herbal products. Presently, there is no organization or government body that certifies herb or a supplement as being labeled correctly due to which herbal remedy labels often cannot be trusted to reveal what is in the container. Published analyses of herbal supplements have found significant differences between what is listed on the label and what is in the bottle and studies of herbal products have shown that consumers have less than a 50% chance of actually getting what is listed on the label. Since there is no legal definition of the word "standardized," on a product label, it is no guarantee of higher product quality. The lack of consistent labeling on herbal products can be a source of consumer frustration, as consumers are often left on their own to decide what is safe and effective for them. Certain information such as "the product has been manufactured according to pharmacopoeia standards," listing of active ingredients and amounts, and directions such as serving quantity (dosage) and frequency of intake of the drug must be in the label (Folashade et al., 2012).

It is recommended that the following is printed on the product label in the official language(s) used by the countries or areas:

(a) Name of product
(b) Name and quantity (in dry weight when relevant) of active ingredient(s)
(c) Dosage form

(d) Directions for use including indications, dosage, mode of administration, duration of use, age group limitations, and use during pregnancy and lactation
(e) Warning statements and relevant contraindications; adverse effects, if any; and overdose information when relevant
(f) Batch number
(g) Expiry date
(h) Storage conditions
(i) Name and address of manufacturers and/or importers
(j) Registration or notification (listing) number

The scientific name of active ingredient(s), in addition to the common name in the language of preference of the national regulatory authority, should be used. The label and package insert should be "user-friendly." Easy and understandable information should be provided. The drug regulatory authority may provide to industry directions on labeling and on allowable indications and claims.

References

Afaq, S. H., 2013. A comparative introduction of the Unani and Tibetan medical traditions, AyurVijnana 6 (accessed 25 January).

Alyaiev, R.V.D., 2007. Pharmacognosy and Phytochemistry, first ed. Vol. I. Carrier Publication, New Delhi, pp. 80–102.

Anonymous, 1998. Quality Control Methods for Herbal Materials (Updated Edition of Quality Control Methods for Medicinal Plant Materials). World Health Organization, Geneva, p. 2011.

Ansari, S.H., 2011. Essentials of Pharmacognosy. Birla Publications Pvt Ltd, New Delhi, pp. 10–16. Horwitz, W. (1975).

AOAC (Eds.), 2005. Official Methods of Analysis of AOAC International. 18th ed. AOAC International, Gaithersburg, MD.

Barnes, J., Anderson, L.A., Phillipson, J.D., Newall, C.A., 2007. Herbal Medicines. Vol. 459. Pharmaceutical Press, London, pp. 1–23.

Bauer, R., 1998. Quality criteria and standardization of phytopharmaceuticals: can acceptable drug standards be achieved? Drug Inform. J. 32 (1), 101–110.

Billore, K.V., 2005. Database on medicinal plants used in Ayurveda. CCRAS.

Bisset, H. (Ed.), 1994. Herbal Drugs and Phytopharmaceuticals. Medpharm Scientific Publishers, UK.

Blumenthal, M., 1998. Therapeutic Guide to Herbal Medicines. American Botanical Council, USA.

Blumenthal, M., 2000. Ephedra update: industry coalition asks FDA to adopt national labeling guidelines on Ephedra; offers cooperative research with NIH. Herbal Gram 50, 64–65. En Ephedra_sinica, Ephedraceae, medicines, legislation (EBBD, 190205549).

Bodeker, C., Bodeker, G., Ong, C.K., Grundy, C.K., Burford, G., Shein, K., 2005. World Health Organization. (2005). WHO Global Atlas of Traditional, Complementary and Alternative Medicine (Vol. 2). World Health Organization.

Calixto, J.B., 2000. Efficacy, safety, quality control, marketing and regulatory guidelines for herbal medicines (phytotherapeutic agents). Braz. J. Med. Biol. Res. 33 (2), 179–189.

Cardellina, J.H., 2002. Challenges and opportunities confronting the botanical dietary supplement industry. J. Nat. Prod. 65 (7), 1073–1084.

De Smet, P.A., 1997. Adverse effects of herbal remedies. Adv. Drug React. Bull. 183 (1), 695–698.

De Smet, P.A., 1999. Overview of herbal quality control. Drug Inform. J. 33 (3), 717–724.
De Smet, P.A., 2004. Health risks of herbal remedies: an update. Clin. Pharmacol. Ther. 76 (1), 1–17.
De Smet, P.A.G.M., Keller, K., Hansel, R., Chandler, R.F., 1992. Aristolochia species. Adverse Effects of Herbal Drugs. Springer-Verlag, Heidelberg.
De Smet, P.A.G.M., Keller, K., Hansel, R., Chandler, R.F., 1997. Adverse Effects of Herbal Drugs 129 (3), 137–145. Springer-Verlag, Heidelberg.
Desmarais, E., Lanneluc, I., Lagnel, J., 1998. Direct amplification of length polymorphisms (DALP), or how to get and characterize new genetic markers in many species. Nucleic Acids Res. 26 (6), 1458–1465.
Dieffenbach, C.W., Dveksler, G.S., 2003. PCR Primer: A Laboratory Manual, second ed. Cold Spring Harbor Laboratory Press, USA.
EMEA, 1998. Quality of Herbal Medicinal Products. Guidelines. European Agency for the Evaluation of Medicinal Products (EMEA), London. In: Substances, Excipients, and Dosage Forms, third ed. World Health Organization, Geneva, p. 3.
EMEA, 2002. Points to Consider on Good Agricultural and Collection Practice for Starting Materials of Herbal Origin. EMEA/HMPWP/31/99 Review, European Agency for the Evaluation of Medicinal Products (EMEA), London.
EMEA, 2005. Guidelines on Quality of Herbal Medicinal Products/Traditional Medicinal Products. EMEA/CVMP/814OO Review, European Agency for the Evaluation of Medicinal Products (EMEA), London.
Ernst, E., 2006. Herbal medicines—they are popular, but are they also safe? Eur. J. Clin. Pharmacol. 62 (1), 1–2.
Eskinazi, D., Blumenthal, M., Farnsworth, N., Riggins, C.W., 1999. Botanical Medicine: Efficacy. Quality Assurance, and Regulation. Mary Ann Liebert, Inc., New York.
Farnsworth, N.R., Akerele, O., Bingel, A.S., Soejarto, D.D., Guo, Z., 1985. Medicinal plants in therapy. Bull. World Health Organ. 63 (6), 965.
Ferenczi-Fodor, K., Végh, Z., Nagy-Turák, A., Renger, B., Zeller, M., 2001. Validation and quality assurance of planar chromatographic procedures in pharmaceutical analysis. J. AOAC Int. 84 (4), 1265–1276.
Folashade, O., Omoregie, H., Ochogu, P., 2012. Standardization of herbal medicines: a review. Int. J. Biodivers. Conserv. 4 (3), 101–112.
Fortina, P., Dotti, G., Conant, R., Monokian, G., Parrella, T., Hitchcock, W., Surrey, S., 1992. Detection of the most common mutations causing beta-thalassemia in Mediterraneans using a multiplex amplification refractory mutation system (MARMS). Genome Res. 2 (2), 163–166.
Gaedcke, F., Steinhoff, B., 2003. Quality assurance of herbal medicinal products. In: Herbal Medicinal Products. Medpharm GmbH Scientific Publishers, Stuttgart, pp. 37–66. 81–88.
GAP, 2006. Good Agricultural and Collection Practice for Herbal Raw Materials. http://www.herbal-ahp.org/06_1208_AHPA-AHP_GACP.pdf.
Gupta, P.K., Roy, J.K., Prasad, M., 2001. Single nucleotide polymorphisms: a new paradigm for molecular marker technology and DNA polymorphism detection with emphasis on their use in plants. Curr. Sci. 80 (4), 524–535.
Houghton, P.J., 1998. Establishing identification criteria for botanicals. Drug Inform. J. 32 (2), 461–469.
Jaya Preethi, P., Padmini, K., Lohita, M., Swetha, K., Priyanka, B., Vengal Rao, P., 2014. Adulterants and substitutes of foods and herbs: a review, IJMCA, **4** (4), 213–217.
Kokate, C.K., Gokhale, S.B., 2004. Pharmacognosy. Nirali Prakashan, Delhi.
Kretz, K., Callen, W., Hedden, V., 1994. Cycle sequencing. Genome Res. 3 (5), S107–S112.
Madhav, N.S., Upadhyaya, K., Bisht, A., 2011. Phytochemical screening and standardization of polyherbal formulation for dyslipidemia. Int. J. Pharm. Pharm. Sci. 3 (3), 235–238.
Mukherjee, P.K., 2002. Quality Control of Herbal Drugs: An Approach to Evaluation of Botanicals. Business Horizons Publication, New Delhi.

Newton, C.R., Graham, A., Heptinstall, L.E., Powell, S.J., Summers, C., Kalsheker, N., Markham, A.F., 1989. Analysis of any point mutation in DNA. The amplification refractory mutation system (ARMS). Nucleic Acids Res. 17 (7), 2503–2516.

Paran, I., Michelmore, R.W., 1993. Development of reliable PCR-based markers linked to downy mildew resistance genes in lettuce. Tag Theor. Appl. Genet. 85 (8), 985–993.

Pauli, G.F., Chen, S.N., Simmler, C., Lankin, D.C., Gödecke, T., Jaki, B.U., Napolitano, J.G., 2014. Importance of purity evaluation and the potential of quantitative 1H NMR as a purity assay: miniperspective. J. Med. Chem. 57 (22), 9220–9231.

Pharmacopeia US, 1990. The United States Pharmacopeia, XXII, NF XVII. United States Pharmacopeia, Rockville, MD, p. 1788.

Phillipson, J.D., 1995. Global trend and market size of herbal medicine in primary health care. In: Chan, K.L., Hussin, A.H., Sadikun, A., Yuen, K.H., Asmawi, M.Z., Ismail, Z. (Eds.), Trends in Traditional Medicine Research. Universiti Sains Malaysia, Penang.

Poornima, B., 2010. Adulteration and substitution in herbal drugs a critical analysis. IJRAP 1 (1), 8–12.

Raman, N., 2006. Phytochemical Techniques, first ed. New India Publications, New Delhi.

Rasheed, A., Satyanarayana, K.V., Gulabi, P.S., Rao, M.S., 2013. Chemical and pharmacological standardization of Ashwagandhadi lehyam: an ayurvedic formulation. J. Complement. Integrat. Med. 10 (1), 1–9.

Rasheed, A., Naik, M., Haneefa, M., Pillanayil, K., Kumar, A., Pillai, R., Azeem, A.K., 2014. Formulation, characterization and comparative evaluation of Trivanga bhasma: a herbo-mineral Indian traditional medicine. Pak. J. Pharm. Sci. 27 (4).

Roberts, J.E., Tyler, V.E., 1997. Tyler's Herbs of Choice. The Therapeutic Use of Phytomedicinals. The Haworth Press, New York.

Sagar, B.P., Zafar, R., Panwar, R., Kumar, V., Mangla, A., 2005. Herbal drugs standardization. Indian Pharm. 4 (35), 19–22.

Saiki, R.K., Scharf, S., Faloona, F., Mullis, K.B., Horn, G.T., Erlich, H.A., Arnheim, N., 1985. Enzymatic amplification of b-globin genomic sequences and restriction site analysis for diagnosis of sickle cell anemia. Science 230 (4732), 1350–1354.

Saiki, R.K., Chang, C.A., Levenson, C.H., Warren, T.C., Boehm, C.D., Kazazian Jr, H.H., Erlich, H.A., 1988. Diagnosis of sickle cell anemia and β-thalassemia with enzymatically amplified DNA and nonradioactive allele-specific oligonucleotide probes. N. Engl. J. Med. 319 (9), 537–541.

Sasidharan, S., Chen, Y., Saravanan, D., Sundram, K.M., Latha, L.Y., 2011. Extraction, isolation and characterization of bioactive compounds from plants' extracts. Afr. J. Tradit. Complement. Alternat. Med. 8 (1).

Shinde, J.S., Khurde, S.S., Suchita, S.L., Chavan, S.S., Hulmajge, S.B., 2016. Need of polyherbal formulations and its standardization: a review. World J. Pharm. Pharm. Sci. **11** (5), 526–533.

Shrikumar, S., Maheshwari, U., Sughanti, A., Ravi, T.K., 2006. WHO guidelines for standardization of herbal drugs. Pharminfo Net 2, 78–81.

Springfield, E.P., Eagles, P.K.F., Scott, G., 2005. Quality assessment of South African herbal medicines by means of HPLC fingerprinting. J. Ethnopharmacol. 101 (1), 75–83.

Stahl, E., 2005. Thin-Layer Chromatography—A Laboratory Handbook, second ed. Springer-Verlag, Berlin Heidelberg.

Taylor, D.A., 2004. Botanical supplements: weeding out the health risks. Environ. Health Perspect. 112 (13), A750.

Trease, G.E., Evans, W.C., 2002. Pharmacognosy, 15th ed. W.B. Saunders, Edinburgh, London, New York, Philadelphia, St. Louis, Sydney, Toronto.

Vos, P., Hogers, R., Bleeker, M., Reijans, M., Lee, T.V.D., Hornes, M., Zabeau, M., 1995. AFLP: a new technique for DNA fingerprinting. Nucleic Acids Res. 23 (21), 4407–4414.

Wallis, T.E., 1999. Textbook of Pharmacognosy, fifth ed. CBS Publishers & Distributors, New Delhi.
Wani, M.S., Parakh, S.R., Dehghan, M.H., 2007. Herbal medicine and its standardization. Pharm. Rev. 5 (6).
Watson, D.G., 1999. Pharmaceutical Analysis. Churchill Livingstone, London.
Weber, J.L., May, P.E., 1989. Abundant class of human DNA polymorphisms which can be typed using the polymerase chain reaction. Am. J. Hum. Genet. 44 (3), 388.
Williams, J.G., Kubelik, A.R., Livak, K.J., Rafalski, J.A., Tingey, S.V., 1990. DNA polymorphisms amplified by arbitrary primers are useful as genetic markers. Nucleic Acids Res. 18 (22), 6531–6535.
World Health Organization (Eds.), 1979. The International Pharmacopeia, General Methods of Analysis. third ed. World Health Organization.
World Health Organization (Eds.), 1981. The International Pharmacopeia, Quality Specifications. third ed. World Health Organization.
World Health Organization, 1988a. Quality Control Methods for Medicinal Plant Materials. World Health Organization, Geneva.
World Health Organization, 1988b. The International Pharmacopeia, Quality Specifications for Pharmaceutical.
World Health Organization, 1990. The Use of Essential Drugs. Eighth report of the WHO Expert committee. World Health Organization, Geneva.
World Health Organization, 1992. Quality Control Methods for Medicinal Plant Materials. World Health Organization, Geneva.
World Health Organization, 1996a. Quality Assurance of Pharmaceuticals: A Compendium of Guidelines and Related Materials, Good Manufacturing Practices and Inspection. World Health Organization, Geneva.
World Health Organization, 1996b. Guidelines for the Assessment of Herbal Medicines. WHO Technical Report Series, World Health Organization, Geneva.
World Health Organization, 1998a. Quality Control Methods for Medicinal Plant Materials. World Health Organization, Geneva.
World Health Organization, 1998b. Basic Tests for Drugs, Pharmaceutical Substances, Medicinal Plant Materials and Dosage Forms. World Health Organization.
World Health Organization, 1999a. Quality Control Methods for Medicinal Plant Materials. World Health Organization.
World Health Organization, 1999b. WHO Monographs on Selected Medicinal Plants. World Health Organization.
World Health Organization, 2000a. General Guidelines for Methodologies on Research and Evaluation of Traditional Medicine. World Health Organization.
World Health Organization, 2000b. The WHO Recommended Classification of Pesticides by Hazard and Guidelines to Classification 2000–2002 (WHO/PCS/01.5). International Programme on Chemical Safety, World Health Organization.
World Health Organization, 2002a. WHO Monographs on Selected Medicinal Plants. World Health Organization.
World Health Organization, 2002b. General Guidelines for Methodologies on Research and Evaluation of Traditional Medicine. World Health Organization.
World Health Organization, 2003. WHO Guidelines on Good Agricultural and Collection Practices (GACP). World Health Organization.
World Health Organization, 2004. Guidelines on Good Agricultural and Collection Practices (GACP) for Medicinal Plants. World Health Organization.
World Health Organization, 2005. WHO Global Atlas of Traditional, Complementary and Alternative Medicine. World Health Organization.
Zafar, R., et al., 2005. Herbal standardization: the Indian. Pharmacist 4 (36), 21–25.

Further Reading

David Phillipson, J., 1992. In: Quality assurance of medicinal plants. WOCMAP I-Medicinal and Aromatic Plants Conference: Part 1 of 4 333, pp. 117–122.
Kirtikar, K.R., 1918. Indian Medicinal Plants. Vol. 4. Bishen Singh Mahendra Pal Singh, Dehradun.
Kirtikar, K.R., Basu, B.D., 2000. Indian Medicinal Plants. Vol. I, pp.503–507.
Krishnamurthy, K., 2004. An Advanced Textbook on Biodiversity: Principles and Practice.
Schena, M., Heller, R.A., Theriault, T.P., Konrad, K., Lachenmeier, E., Davis, R.W., 1998. Microarrays: biotechnology's discovery platform for functional genomics. Trends Biotechnol. 16 (7), 301–306.
Thirumalai, D., Paridhavi, M., Gowtham, M., 2011. An overview of standardization of herbal drugs. Int. J. Rev. Life Sci. 1 (3), 167–170.
World Health Organization, 1991. Guidelines for the Assessment of Herbal Medicines.
Zietkiewicz, E., Rafalski, A., Labuda, D., 1994. Genome fingerprinting by simple sequence repeat (SSR)-anchored polymerase chain reaction amplification. Genomics 20 (2), 176–183.

CHAPTER

7

Marketing Indigenous Herbs: Connecting Link Between International Trade and Himalayan Livelihood

OUTLINE

Himalayan Phytochemicals
https://doi.org/10.1016/B978-0-08-102227-6.00007-3

7.1 INTRODUCTION

There are huge diversity and immense wealth of MAPs (medicinal and aromatic plants) along with their folklore practice in Asia since primeval times. As per reports of World Health Organization (WHO), target of "health for all" is not possible without the utilization of herbal medicine (Maheswari, 2011; Gupta et al., 2014; Kala, 2015). Due to overwhelming demand of herbal-based drugs in developing countries, there are hints that consumers in developed countries are becoming disillusioned with modern medicine and hence are hunting new alternatives (Ramawat and Goyal, 2008; Mazid et al., 2012). Developed countries exhibit higher inclination toward herbal-based drugs; for example, the consumption of the herb *Hypericum perforatum* is 61 million dollars double the allopathic medicine Prozac 30 mg used as antidepressant in Germany (Schmidt and Butterweck, 2015; Hritcu and Cioanca, 2016). Enhanced proclivity toward herbal-based drugs by consumers in both developed and developing countries has instilled awareness by multinational pharmaceutical industry in bioprospecting (Yadav, 2013; Subramoniam, 2014). Due to improper legislation or ineffectual international agreements on conservation and sustainable utilization of biodiversity, there is consequent indiscriminate harvesting of Himalayan MAPs leading to enormous diminution of natural resources (Arora and Gupta, 2014; Purkayastha, 2016). India has about 8% of world biodiversity covering maximum MAPs that can act protagonists in the development of herbal-based drugs (Shukla, 2014; Chen et al., 2016; Vijayan et al., 2017). For already specified biodiversity of Indian Himalaya, we now require authenticated identification and validated development of herbal-based drugs by all stakeholders including policy makers for extensive export (Rao et al., 2015; Phondani et al., 2016). In order to address the requirements of each stakeholder, all of the stakeholders need to integrate their holistic approach for inclusive expansion of herbal sector. So far, there are no effective national policies that could assist druggists and health professionals comprising both conventional and modern practitioners to legalize market and certify consumer safety with conservation, intellectual property protection, and sustainable utilization of MAPs (Shahidullah, 2016). Collection and trade of MAPs and their value products are source of employment for major proportion of Indian Himalaya. Microstudies confirm that maximum employees in herbal sector are women (Singh, 2014). MAPs constitute major fraction of nonwood forest products that contribute 80% to India's net export income yearly (reference). Herbal products contribute about US$ 62 in international market that is known to grow to US$5 trillion by 2050, and India contributes only 0.5% to world herbal trade (Bhattacharya et al., 2014). This demonstrates escalating growth of herbal sector in international market in the coming years. In order to maximize the contribution of India in herbal sector, India needs to capitalize its MAP cultivation and improve postharvest and processing technologies and proper framing legislation policies for marketing and patenting of MAPs (Maikhuri et al., 2017a,b). The local and marginal farmers from remote places of Indian Himalaya are implicated in growing herbal enterprise in India. The state and the national government need to advance herbal enterprise with high enthusiasm (Bawa et al., 2007; Torri, 2010). Despite an immense economic importance of MAPs, herbal sector is recognized as subsector in India. There has been no executive or regulatory authority that regulates constitution of MAPs till the establishment of National Medicinal Plants Board in 2000 (Sati, 2005; Torri, 2008). No nodal agency is there to monitor and assess medicinal plant sector, and diverse organizations manipulate MAPs in varied characteristics sans any objective and coordination, thus resulting to inconsistency of instant continuation of underutilization and overexploitation (Gautam et al., 2004; Bera, 2010). Moreover, the lack of coordination between farmers and scientists has led to significant research gap. There is almost no research group working on socioeconomic and policy characteristics of medicinal plants, for instance, that that exists in respect to agrotechnology, biotechnology, etc. Actually, scientists working on basic sciences have carried socioeconomic surveys of MAPs leading to generalized viewpoint consequently advancing toward messy evaluation lacking any deeper understanding of MAP sector (Krishnan et al., 2011). This chapter will describe current status of medicinal plant sector in terms of an assessment of opportunities and constraints in medicinal plant sector particularly under the changing global scenario. We will highlight issues concerning medicinal plant management and marketing herbal products and policy initiatives at national and international levels. Finally, the chapter summarizes the overall evaluation and recommendations for appropriate measures and instrument for speeding up expansion of the herbal sector.

7.2 MEDICINAL PLANT MANAGEMENT AND MARKETING—CONSERVATION AND LIVELIHOOD CHALLENGES

Western Ghats are main hub for distribution of enormous supply of MAPs across India (Kala, 2005; Kumar et al., 2011a,b,c). Nearly 20% of MAPs are procured from Indian subcontinent, approximately 4500 plant species out of which 3500 are of medicinal significance (Verma et al., 2012). Domestication of MAPs in their native habitats is an imperative intervention for meeting market demands (Singh, 2006). However, it is a very intricate process to cultivate MAPs at farm level due to the lack of specific agroclimatic characteristics (Van de Kop et al., 2006; Kala, 2009). Moreover, the phytochemical composition and the overall metabolic profile are modulated by varied agroclimatic characteristics, thereby affecting their therapeutic potential in Ayurveda (Skaria, 2007; Taneja, 2009; Rawat and Tewari, 2015). Further, phenological and ontogenetic variation alters composition of herbs procured for the formulation of drugs (Kumar et al., 2001; Bruni and Sacchetti, 2009; Verma and Shukla, 2015). These are some of the constraints that hinder the cultivation of MAPs at mass level (reference). Additionally, technological innovations and institutional management further limit the choice of MAP cultivation (Rao et al., 2004; Kala et al., 2006; Bajaj, 2012). So, there is an immediate requirement for prioritization of assessing which MAPs should be planned for mass cultivation (Verma et al., 2010; Sujatha et al., 2011). Hence, there is a need to prioritize MAPs on the basis of commercial and therapeutic potential. But there cannot be single criteria for selecting the most beneficial MAP for advancement of socioeconomic aspects of diverse stakeholders (Debnath et al., 2006a,b; Phondani et al., 2011; Mazid et al., 2012). In this context, Indian Scientific Advisory Committee on herbal products has recommended that the government should focus on cultivation and marketing of about 45 MAPs for over 25 years (Singh et al., 2006). This committee has prioritized about 7–11 MAPs for intensive farming for the next 5 years keeping in view their market demands and their endemic nature (Farooquee et al., 2004; Bawa et al., 2007). In order to implement cultivation program, we need to develop effective cultivation and postharvest protocols followed by clinical trials and integration of national level associations for each MAP (Bajaj, 2012; Kumar et al., 2014; Singh, 2017a,b).

7.2.1 Development of the Chain: Factors Limiting Medicinal-Plant Cultivation

Different governmental agencies from individual states of India have developed initiatives for conservation and mass cultivation of MAPs; for example, Uttaranchal has developed several policies to promote mass cultivation of MAPs (Kuniyal et al., 2015a,b; Verma et al., 2016). These policies are executed by different governmental agencies such as Horticulture Department, Forest Department, and Department of Rural Development together with numerous research institutes (Wani et al., 2016; Deb et al., 2013). In order to popularize mass cultivation of MAPs, we need to create awareness and promote technological interventions that could accelerate mass propagation of MAPs (Chhetri et al., 2015). We can also facilitate cultivation through supply of quality planting material to farmers and dispense the funds via loans and subsidies on agricultural tools for establishment of herbal parks and nurseries (Chauhan et al., 2014; Upadhyay et al., 2016; Maikhuri et al., 2017a). Farmers are still reluctant to adopt these policies due to high risk and transaction costs, improper framing of stakeholders, and requirement for establishment of parks and institutional infrastructure (Kuniyal et al., 2014; Pangriya, 2015).

7.2.2 Long Gestation Period and High Risk

Majority of MAPs have long gestation periods that can be germinated only following 3 years especially high-altitude alpine herbs (Kumar et al., 2014). Marginal farmers having limited landholdings and inadequate credit are more hesitant to convert major section of their land to medicinal plant production (Singh and Vidyasagar, 2015). Cultivation of MAPs is highly risky due to relatively novel initiatives and new technological innovations, and other plans are yet to be fully developed (Thillaivanan and Samraj, 2014; Sen and Chakraborty, 2017). Numerous communities involved in the cultivation of MAPs are conventional traders. Farming is relatively a novel initiative, and risk of collapse is specifically high. With risk of crop failure, farmers experience severe market-associated risks and problems (Uniyal et al., 2002; Gupta, 2010). Mostly, farmers have no secured markets and best price for quality material. Moreover, farmers lack data pertaining to demand and pricing, resulting to local traders extending all expenses to farmers (Kala et al., 2004; Van de Kop et al., 2006; Shahidullah and Haque, 2010).

7.2.3 Transaction Costs

Majority of Indian states exhibiting high diversity are remote, having no or relatively lower level of communication and restricted access to new agricultural innovations and agroprocessing facilities (Kala et al., 2004; Shrivastava et al., 2008). In Uttaranchal, the transaction costs for rural production and local entrepreneurs are high even though some of these costs compensate by conducive agroclimatic conditions for mass cultivation of MAPs and high value-to-weight ratio (Debnath et al., 2006a,b; Schippmann et al., 2006).

7.2.4 Social Capital and Values

Herbal trade is a new initiative that relies totally on mass production of quality herbs, and Uttaranchal is still lagging in this sector due to the lack of well-developed market (reference). Currently, the possibility of economic coordination opportunism is quite high, as local traders put all cultivation expenses to farmers or producers (Olsen and Bhattarai, 2005; Booker et al., 2012; Joshi and Joshi, 2014). The trade and farmers are not able to coordinate, hence retreating from the communal agreements (Maiti and Geetha, 2013; Choudhary et al., 2014). So, our prime efforts are required to reinforce the effective networking between all stakeholders involved in medicinal plant chain (Bera, 2010; Gularia, 2012). Effective social networking can build trust and support collaboration in reducing risks and transaction costs.

7.2.5 Institutional Infrastructure

Another major constraint delimiting mass cultivation of MAPs is due to improper institutional infrastructure prevalent in maximum Indian countries (Dhar et al., 2000; Kala and Sajwan, 2007). This is true for institutions that contribute to technical facilitation and assisting market expansion (Singh, 2006; Mittal and Singh, 2007). In order to grow, MAPs require proper interaction of abiotic and biotic factors along with association of related species (Samal et al., 2003; Shameem and Kangroo, 2011). There is urgent need to develop new technological interventions associated with cultivation, harvesting, and quality control (Canter et al., 2005; Kala, 2009). There is enormous dearth of institutional infrastructure that could facilitate marketing links to farmers for growing medicinal plant sector (Gulati et al., 2007). There is either no or meager cooperation among the research institutes working on MAPs leading to duplication of efforts and incompetent utilization of limited resources (Bhattacharya et al., 2008; Kala, 2009). Likewise, there is scarce association between these institutions, agricultural extension institutions, and farmers. This restricts both suitability of technological interventions and its dissemination (Olsen and Helles, 2009; Verma et al., 2012).

7.2.6 Overcoming Constraints by Public-Private Partnerships

Effective public and private partnerships are imperative in eliminating numerous bottlenecks. There are few public-private relations that have been already initiated in Uttaranchal that intend to surmount some of the current blockages to advancement of medicinal plant sector (Sridharan and Viswanathan, 2008; Spielman et al., 2010). We need to involve contract farming including both national and international market (Kumar et al., 2011a,b,c; Mazid et al., 2012). In Uttaranchal, collaboration between farmers, industry, and research institute was initiated in the small village Gheshe. This partnership was started by High Altitude Plant Physiology Research Centre (HAPPRC), an imperative research center for medicinal herbs in Uttaranchal. Initially, HAPPRC focused on mass cultivation of vegetables and provided farmers with different modalities so as to assess current technologies (reference). Farmers cooperated with researchers from HAPPRC and started cultivation of MAPs like *Picrorhiza kurroa* (kutki) and *Saussurea lappa* (kuth) (Sharma et al., 2010; Bisht et al., 2011; Kuniyal et al., 2015a,b). HAPPRC provided quality material like viable seeds and seedlings to farmers and assisted in marketing by providing local company that led to tripartite arrangement between the farmers' groups, HAPPRC and Dhawan International, a Delhi-based firm. Farmers are agreed to provide quality plant material organically grown following natural conditions (Kala, 2009; Amujoyegbe et al., 2012). This agreement further aids in commercialization of cultivation technology as developed by public sector research. In order to link village farmers with international market, government institutions link farmers with international company by publicizing their products through different networking sites (Butola and Badola, 2008; Mishra et al., 2009; Bhatt et al., 2015; Chen et al., 2016).

7.3 INTERNATIONAL HERBAL MARKET SYSTEM AND TRIBAL KNOWLEDGE

India is still lagging behind in international herbal market exploiting about only 7000 where as traditional Chinese medicine (TCM) exploits over ~5000 plant species (Pan et al., 2014a,b). The reason behind their laggard is their improper marketing and commercialization in random fashion (Pandey et al., 2013; Gupta et al., 2014). Indian herbal sector fetches around ~US$ 1 billion worldwide, whereas a Chinese herbal medical market earns ~US$ 19 billion across the global market for herbal of worth US$ 62 billion (Vyuptakesh, 2013; Rajpurohit and Jhang, 2015). There is an immense prospective for the advancing sector of Indian pharmaceutical through technological innovations, patents, and trademarks (Greenhalgh, 2016; Abrol et al., 2017). India harbors an immense diversity of MAPs and huge traditional knowledge of Ayurveda that entails treatment for various diseases that can be exploited completely by India (Awate et al., 2015). If Indian herbal sector develops completely, it could emerge as economically viable market particularly in the health-care, pharmaceutical, and cosmetic sector (Rentala et al., 2014; Scalera et al., 2015). India has an immense prospective in global market, and it can emerge as global leader in export of quality MAPs. TCM is more popular due to wide research market and diverse bioefficacy studies carried out by the United States on these herbs (Pandey et al., 2013). Indian herbal sector is not validated authentically due to lack of double blind trails and inadequate scientific analysis on their utilization and significance (reference). On the contrary, Chinese medicine is completely validated, and they are ready to export their medicinal products through proper government agencies. Chinese have not only ready-to-export medicinal herbs but also manual and book containing practice of using these medicinal herbs (Acharya, 2015; Ahmad et al., 2016). However, Indian government has not yet advanced their export and import market of crude drugs. Indian government has not yet exchanged their traditional knowledge of Ayurveda globally; however, some of the Indian authors have documented some scriptures in American audience (Singh, 2017a,b). TCM is based on theory of yin and yang and five elements and unique tradition of acupuncture; therapeutic massage has evolved and has been used across the world (Krishnakumar et al., 2015; Pradeep, 2016). Ayurveda relies on three doshas, wherein significance is given to dietary limitations and significance of oil massages and various purification procedures has limited popularity (Bhattacharya et al., 2014; Wal et al., 2016). Among these various practices of Ayurveda, Panchkarma is most widely used in that patients are subjected to strict dietary regime. In the United States, these programs are quite costly and need immense time to accommodate it in daily program. On the contrary, quality of Chinese herbal products is much better than Indian drugs, and practice of acupuncture takes only half an hour, which can fit into any schedule (Khan, 2014; Jaiswal et al., 2016). Quality standardization and effective production and authenticated evaluation are entirely lacking in India, while China has upgraded its drug development sector and successfully surmounted several hindrances via advancement of its traditional medicine using government sponsored Good Agriculture Practice (GAPs) and Good Manufacturing Practices (GMPs) (Aneesh et al., 2009; Patwardhan et al., 2015). GAPs focus on screening of elite germplasm that exhibits high concentration of active components (Patra et al., 2010). In addition to this, GAPs ascertain standards for deleterious contaminants, for example, heavy metals, pesticide residues, and microbes in plants (Zhang et al., 2012; Prakash et al., 2013). Though China has wide market in herbal medicine, the popularity of yoga, that is, an Ayurvedic practice, is too immense in international market (Satyanarayana, 2010; Pandey et al., 2013). Proper directive on Ayurveda like Chinese medicine together with laissez-faire strategy that appeared to have marked this field up to now could mean that bits and bobs of Ayurveda will simply be included into the wide-ranging ground of complementary and alternative medicine (CAM) (Saks, 2015). Indian and Chinese medicine have some common grounds; for example, Chinese herbs are imported to India, for example, green tea, cinnamon, and essential oils such as bergamot, citronella, geranium, spearmint, vetiver, anise, cinnamon, eucalyptus, and ginger oil (Singh et al., 2004; Patwardhan et al., 2005). The herbal companies that have acquired patent rights for herbal products in India include Himalaya Drug Company (HDC), Emami, Aswini, Ayur, Dabur, and Cholayil Pharmaceuticals (Kamboj, 2000). The Government of India has also initiated programs that advance traditional Indian medicine, for example, AYUSH (Ayurveda, Yoga, Unani, Siddha, and Homeopathy) that was established in March 1995 to advance indigenous systems (Dubey et al., 2004; Verma and Singh, 2008). The main concern includes education, standardization of drugs, improvement in availability of raw materials, research and development, information, communication, and extensive implementation in the national system for delivering health care. There are several central research institutes as well such as Central Council of Indian Medicine that executes and coordinates teaching and training programs and Central Council for Research in Ayurveda and Siddha that associates with interdisciplinary research (Sahoo et al., 2010). During the recent years, India has developed superlative capability and infrastructure for synthesis, isolation and structure analysis, bioefficacy, and toxicological testing and pharmacokinetics along with advanced agrotechnological innovations for cultivation of medicinal herbs (Umashanker and Shruti, 2011; Saini, 2011). India is scaling toward upliftment of technological interventions. India has some leading institutes like Central Drug Research Institute (CDRI), Central Institute of Medicinal and Aromatic Plants (CIMAP), and National Botanical Research Institute (NBRI) at Lucknow; Indian Institute of Integrative Medicine (IIIM) at Jammu and Srinagar; and North East Institute of Science and Technology (NEIST), Jorhat, National Chemical Laboratory, at Pune that are actively involved in research on MAPs. Majority of these institutes are involved in standardization and characterization of active metabolite, quality control of farmers, and conservation of endangered species (Mani, 2006; Verma et al., 2007). Ayurvedic medicines are plant-based that are evaluated through thorough testing on human (43%) and animals (62%), for example, guggul, brahmi, ashwagandha, amlaki, guduchi, kutki, shatavari, and shunthi (Panda, 2004). Monographs of some medicinally important plants include clove, guggul, opium, mentha, senna, ashwagandha, and 258 Ayurvedic drugs in the pharmacopoeia of India (1996). Currently, Ayurveda exploits about 1100 MAPs collected from wild habitat, and among them, 60 species have high market demand (Pan et al., 2014a,b; Sengar et al., 2017). Tribal communities of India are dependent on herbal wealth for their livelihood. However, local traders amend quality raw material with adulteration and contamination in the process. Adulteration of herbal drugs is a major problem that hampers their supply and demand in international market (Singh et al., 2015; Keshari et al., 2015). India like China should adopt GAPs for the utilization of authenticated raw material and include standardization covering harvesting, processing, transportation, and storage (Rawat and Tewari, 2015). Chemoprofiling and molecular characterization are main objective for authenticated identification of germplasm via DNA fingerprinting and molecular markers (reference). Indian herbal is scaling up export from 100 to 450crores (Dhar et al., 2013). Though Chinese medicine has crossed 2000 crores, the prospective of India is advancing on large scale (Farooq, 2013; Kakarla et al., 2015). However, Indian herbal drugs are still inconsistent in developing market in developed countries like Europe and the United States. In addition to incompetent standardization of herbal drugs, there is lack of promotion of herbal drugs in international market particularly in regulated markets

(Yadav and Dixit, 2008; Ram and Abdin, 2008; Sahil, 2016). International pharmacopoeia has included several drugs following standardization of ethnopharmacology and preservation of traditional medicine. Traditional medical practices can provide holistic strategy to drug design and innumerable probable objectives for scientific evaluation (Shankar, 2010; Patwardhan and Vaidya, 2010). Recent technological interventions such as computerized separation techniques, high-throughput screening, and metabolomic approach are transforming drug discovery, and this will facilitate to rediscovery of traditional drug (Patwardhan, 2000; Mohanta et al., 2003; Katiyar et al., 2012; Rather et al., 2013).

7.4 EMERGING MARKETS FOR MAPs

Economic potential of MAPs in diverse climatic regimes and possible profit procured by farmers through cultivation of MAPs are quite high (Mukherjee et al., 2007; Ali et al., 2008; Kumar et al., 2011a,b,c). Numerous research studies evaluated that cultivation of high-altitude MAPs could produce high yields having profit worth Rs. 7150–55,000 per hectare (Mishra et al., 2016). This profit scales up to annual income Rs. 120,000 per hectare via mixed cropping of high-altitude MAPs. Economic profit can be procured from low-altitude MAPs as observed from Amarkantak region of Madhya Pradesh that exhibited extensive returns for four economically important species, *Curcuma angustifolia* (Rs. 48,000), *Rauvolfia serpentina* (Rs. 54,000), *Acorus calamus* (Rs. 27,000), and *Chlorophytum tuberosum* (Rs. 13,000) (Ahmad et al., 2006; Kumar et al., 2011a,b,c). Though there are previous research reports that document different aspects of cultivating MAPs, there is a need to elucidate all aspects that regulate proper planning for effective conservation and utilization of MAPs in lieu of their environmental and aesthetic values (reference). Research studies demonstrate that subjects concerned with suitability and possibility of cultivation, marketing, trade, and bioprospecting subjects should be discussed in all inclusive manner (Singh et al., 2005). According to WHO reports 2010, international herbal trade is estimated to be US$ 250 billion at a rate of 7% per annum and expected to grow further by 2020 (WHO, 2011). Western Himalaya is the backbone for 50% of drugs enlisted in British Pharmacopoeia that are procured from raw material of high-altitude MAPs (Khare, 2008). This Indian Himalaya contributes not only 80% to Ayurveda but also 46% to Unani and 33% to allopathic system of medicines and provides a main share to the rural and tribal market (Sheikh, 2014). In order to advance Indian herbal trade and increase annual turnover of Himalayan herbs, chain of export markets should be extended beyond prime markets of Mumbai, Delhi, Chennai, and Tuticorin (Larsen and Olsen, 2007; Pan et al., 2014a,b). There is the lack of proper and regulated marketing opportunities working exclusively in their supply chain (Kala et al., 2006; Ved and Goraya, 2007; Sahoo et al., 2013). However, at present, Indian herbal trade is quite intricate, unorganized, traditional, and lacking proper regulations. Further, there is no well-organized local, regional, or national level data pertaining to market modalities. Majority of data are unorganized, inadequate, and unrivaled. Major constraints in expansion of herbal trade include the following:

- The lack of cataloging of Himalayan MAPs.
- Ambiguous trade names are another constraint that hampers marketing of MAPs; for example, Ashok is a trade name for two botanical species, *Saraca indica* and *Polyalthia longifolia*. Similarly, Chirayata is a trade name that is given to two botanical species, *Andrographis paniculata* and *Swertia chirata*.
- Misidentification of raw material for drug development. Since MAPs are harvested and marketed in dried form, it is cumbersome to identify herbs in dried form.

7.5 DEMANDS FOR MAPs

International herbal trade is expansively channeled through India. Maximum MAPs of about 2000 tons worth 2 million dollars are harvested from Karnali Zone of Nepal. Among them, "jatamansi" (spikenard) or *Nardostachys jatamansi* is the most significant evaluated against both volume and value (CECI, 1999). Karnali zone is a major economic zone in terms of herbal trade in NTFPs, but this market is limited due to inadequacy due to meticulous and time-consuming harvesting modes by poorest people belonging to lower cast and hence more vulnerable to exploitation by local traders (Karki, 2000). The demand of Indian herbs is further limited by delimited marketing opportunities in remote areas where local markets depend on traditional medicines and only local traders can formulate those herbal drugs (Singh, 2006; Regmi and Joshi, 2008). The medicinal herbs with high market demands include *P. kurroa, Delphinium himalayas, Rheum australe, Selinum tenuifolium,* and *Eulophia* (Sher et al., 2017; Dar et al., 2017; Sharma et al., 2017). Most of the medicinally important herbs are sold to airport traders so as to get maximum profit from foreigners (Evans, 2009; Booker et al., 2012). Mostly, the airport traders commonly known as "*pansari*" arrange transport, fix price, and then sell the raw material to Ayurvedic drug manufacturing companies like Dabur, Baidyanath, Himalaya, and Zandu in India (Gouri et al., 2004). Majority of drug manufacturing companies formulate herbal drugs following traditional knowledge, develop new pharmacological innovations, and market them at high price. The profit margins for varied herbs are represented in Table 7.1. The tables demonstrate that properly

TABLE 7.1 Description of Medicinal and Aromatic Plants With Estimated Consumption Per Annum

Plant name	Estimated consumption per annum (tons)
Aconitum ferox	78
Aconitum heterophyllum	94
Berberis aristata	378
Nardostachys jatamansi	456
Picrorhiza kurroa	320
Saussurea costus	220
Glycyrrhiza glabra	4000
Acorus calamus	150

legalized drugs fetch handsome profits of about 19% and 43% to local and village traders, respectively, while banned drugs like jatamansi roots fetch profits only to regional traders like Nepalganj traders (Acharya, 2005; Kunwar et al., 2009). Current demand for MAP-based raw material across the world is approximately US$ 14 billion per year as per records of WHO and is likely to rise 15%–25% and reach to worth US$ 5 trillion in 2050 (Poudel, 2007; Chauhan et al., 2014). This overescalating demand on herbal drugs has led to indiscriminate harvesting of MAPs that eventually leads to loss of diversity. The western belt of Himalaya has exhibited more vigorous depletion in diversity of major medicinal herbs like Himalayan yew (*Taxus baccata*) that is profusely harvested for procurement of Taxol used against treatment of ovarian cancer (Padalia et al., 2015). Similarly, numerous other critically endangered MAPs include *Aconitum heterophyllum, Nardostachys grandiflora, Dactylorhiza hatagirea, Polygonatum verticillatum, Gloriosa superba, Arnebia benthamii,* and *Megacarpoea polyandra* that are overexploited for therapeutic potential (Kala et al., 2006; Singh et al., 2014a,b; Rawat et al., 2016). There is dearth of organized research reports on supply, demand, and overall marketing of MAPs in herbal sector.

7.6 GLOBAL MARKETING OPPORTUNITIES

Indian Himalaya possesses 8% of the world's total biodiversity exhibiting major areas rich in MAPs having high metabolite concentration. According to WHO, international market is known to grow further from US$ 62 billion to US$ 5 trillion by the year 2050 (Singh et al., 2014a,b; Srinivasan and Srinivasan, 2016), but India contributes 0.5% share in herbal sector (Kuniyal et al., 2015a,b; Bhushan et al., 2016). Keeping in view the extent of biodiversity, the main task for expansion of international marketing will include validated identification of MAPs and directive for development of products exhibiting high export potential (reference). India has already set new policy strategies for the expansion of herbal sector; however, there is further need to strategize research on farming the policy for advancement of MAPs (Jain and Sharma, 2015; Kala, 2017). For expansion of herbal sector, we need to integrate innovative economic position, pragmatic strategy, and effectual development policy (Sharma and Garg, 2013). However, existing evidences are insufficient to completely include intricate themes of this sector. There is an immense need to evaluate socioeconomic and policy research analysis to fill the gaps in comprehending the dynamic of MAPs (Sen et al., 2011; Doke et al., 2013; Phondani et al., 2016). The most promising herbal markets worldwide like EU, the United States, Canada, Australia, Singapore, and Japan (Karki and Nagpal, 2004; Aneesh et al. 2009), and emerging new markets include Brazil, Argentina, Mexico, China, and Indonesia (reference). The marketing of MAPs as nutraceutical has huge prospective due to their utilization in dietary supplements, for example, European Union, Japan, and the United States (reference). Currently, India's share in global herbal market is a miniscule 0.5% at $ 358.60 million, and estimated global market is about $ 70 billion (Alam and Belt, 2009; Mishra et al., 2010a,b). As per records, India's export of AYUSH and value-added products of medicinal plants during 2015–16 was $ 358.60 million as reported by the leading newspaper Indian Express (Indian Express, 2017). India has worked out possibilities for expansion of market further by creating awareness related to international opportunities to export AYUSH products in varied forms like dietary supplements and nutraceutical. In order to globalize Indian herbal sector, India has signed MoU with WHO to develop global protocol for advancing traditional and complimentary medicines that would facilitate marketing of MAPs internationally (Sahoo et al., 2010; Sahoo and Manchikanti, 2013). During the recent years, there is increased inclination toward utilization of traditional and complementary medicine around the globe. Both developing and underdeveloped countries depends on MAPs for their health care; nearly 80% of African population rely on herbal wealth, whereas in China, traditional medicine system contributes 30%–50% in health-care sector (Nikam et al., 2012; Katiyar et al., 2012). Similarly, developed countries like Australia, Canada, Germany, the United States, Belgium, and France utilize herbal medicine 48%, 70%, 80%, 42%, 39%, and 76%, respectively (Sen et al., 2011; Chan et al., 2012). Indian herbal sector is popularized on global level such as Ayurveda that is well recognized in Asian countries such as Nepal, Sri Lanka, and Bangladesh (Sahoo et al., 2010; Ghosh and General, 2013). In Japan, the Society of Ayurveda was founded in 1969 by Osaka Medical School, and practice of Ayurveda is popular in Japan from the last 40 years. Ayurveda is practiced in Thailand and Myanmar, and many international countries have started education and training courses in Ayurveda such as Argentina, Brazil, Venezuela, Chile, Nicaragua, Costa Rica, Guatemala, Germany, Austria, Switzerland, France, Czech Republic, Greece, and Israel (Basu et al., 2007; Bhattacharya et al., 2014). Countries like South Africa, UAE, Russia, Sweden, Indonesia, Netherlands, Italy, Spain, Australia, New Zealand, and Hungary have acknowledged Ayurveda (Sheth, 2005; Meier zu Biesen, 2017). Expansive globalization of herbal sector has led to huge pressure on species richness like *R. serpentina, Podophyllum hexandrum, Aconitum ferox, N. grandiflora, S. chirata,* and *Saussurea costus* (OFDC, 2006). The Convention on International Trade in Threatened Species (CITES) treaty documented 11 plant species restricted in India for foreigners, for example, *R. serpentina, N. grandiflora, Dioscorea deltoidea,* and *P. kurroa* (EXIM Bank of India 2003, op cit, p. 80). Further, the Biodiversity Diversity Act, 2002 has several provisions so as to ensure that the countries' bioresources are not exposed to the risk of undue commercial exploitation. These species can be exported only if they are not collected from wild habitats unless Forest Department has permitted so as to avert any danger for extinction of plant species (reference). Himalayan herbs that need conservation for meeting demand and supply of international market are presented in Table 7.2.

TABLE 7.2 Description of Supply and Demand Requirements of Different Economically Important MAPs

Equal demand and supply	High demand/low supply	Low demand/low supply Equal demand supply relation	Low demand/high supply Demand lower than supply
Cedrus deodara	*Aconitum heterophyllum*	*Hippophae rhamnoides*	*Berberis aristata*
Betula utilis	*Dactylorhiza hatagirea*	*Pinus roxburghii*	*Urtica dioica*
Valeriana jatamansi	*Angelica glauca*		
	Nardostachys grandiflora		
	Saussurea costus		
	Taxus baccata subsp. *wallichiana*		
	Podophyllum hexandrum		
	Swertia chirayita		

7.7 EXPORTS AND IMPORTS

Indian marketing intelligence has documented some agencies dealing with the trading of traditional Ayurvedic medicines like *Dasamularishta* and *Ashokarishta* procured from local pharmacies of Odisha that is then exported to different countries by simply swapping the local trade name from bottles (Rath, 2005; Sadhu et al., 2015). The roots of *R. serpentina* and *nagarmutha* oil are also exported without any direct linkage between importer and primary supplier; Bhuin-kadamb seeds and senna leaves are procured from local farmers of Odisha, sold to exporters from Tuticorin, and then marketed at higher price in international market (Rath, 2005). There are meager examples of exports wherein there is direct link between exporter and primary supplier such as *G. superba* that was directly exported from Odisha by S. K. Das of Sabuja Viplab (Shukla and Pandey, 2015). Besides having high medicinal value, this medicinal plant is valued for its ornamental value with maximum cost of production Rs. 5 in Odisha as against US$ 8–10 per tuber in Europe (Anilkumar et al., 2012; Sharma, 2013). Major constraints in expansion of herbal trade at international level are scarcity of data pertaining to source of collection and primary supplier for export product (reference). Moreover, there is no validated data pertaining to evaluation of export linkages in Indian herbal production. According to local trading agencies of India, there is little or no impact of globalization on supply and utilization of MAPs (Chhetri, 2014). As per records of forest departments from states of Odisha, annual production of *indrajava* was about 71 quintals during 1988–89 and dropped drastically to 0.87 tons during 1994–95, evaluated for 10 years of cultivation (Rath, 2005; Bera, 2010). Similarly, total production of *sunari* bark was reported 76.2 tons during 1999; however, harvesting of tree bark has been banned since 2000, but marketing of *sunari* is still going on, and local traders observed higher demands of this herb as compared with supply (Rath, 2005; Nanyunja and Baguma, 2005). The list of Himalayan MAPs banned is represented in Table 7.3. Similarly, the utilization of "*Ashok bark*" exceeds its cultivation not due to globalization impact but its excessive use as an alternative and traditional medicine. For example, Ayurvedic medicine *Ashokarishta*, the accepted Ayurvedic preparation that summoned its contribution in women's health, is based on this bark (Biswas et al., 2010, Pandey, 2015). This bark is an ingredient of majority of commercially important products that even some consumers make adulterations to overexploit it. Though production of raw material was declined drastically, there remained more or less constant over the past 10 years due to the lack of compensatory rise in demand and market price of the bark (Sen et al., 2011; Kunwar et al., 2013). *R. serpentina* roots are recognized as a main crude drug for export long before globalization commenced, and the indefensible supply led to depletion of the resource as in the Saharpada area of Keonjhar District

TABLE 7.3 List of Himalayan Herbs Banned for Marketing

Plant name
Saussurea costus
Rauvolfia serpentine
Podophyllum hexandrum
Nardostachys jatamansi
Dioscorea deltoidea
Taxus wallichiana
Aconitum
Gentiana kurroo
Nardostachys grandiflora
Swertia chirayita
Picrorhiza kurroa

(Goenka, 2007; Ramawat and Goyal, 2008; Bindu et al., 2014). During the last decade, its export has been banned by the Government of India, and even the Odisha Government declared it as a lease-barred plant in March 2000. Still, the commercial exploitation and supply of its roots have not been blocked basically because the domestic demand is still promising and ban on export has rather resulted in a decrease in its market/procurement price (Ved and Goraya, 2007; Bhardwaj et al., 2011; Booker et al., 2012). Local trading without proper linkage between exporter and primary supplier has led to huge financial risk in marketing at both national and international level. The trade and commercialization of Himalayan herbs from the western belt are still underway. At present, there is no concrete report on demand and supply of Himalayan herbs procured from western belt albeit it is species-rich and has enormous diversity in MAPs.

So far, the value of export is concerned, and both India and China have the same problem that most of their export is in the form of low-value-added products that lower the price. However, speaking ethically, the current trend is against our age-old tradition of glorification of medicinal plants. Market has become the ultimate goal now, and self-reliance in local health care through the conservation and propagation of medicinal plants is of secondary importance. Hence, many people cultivate medicinal plants not for their consumption but for some use elsewhere. Of course, even during the British period, people used to cultivate poppy, etc. for the end use elsewhere, but at that time, such cases were rather isolated; but now, thanks to the market-oriented policy of the government, this is going to become a national phenomenon, for example,

Safed Musli. And since it is the market that decides which plant is to be cultivated or propagated, the role of local health-care traditions becomes marginal in the conservation of medicinal plants, and hence, many traditionally important plants are actually ignored. Globalization is essentially market centric, and this market-oriented approach is ultimately derogatory to India's rich and powerful tradition of local health care. And the sufferers of its impact would be the lower income group in particular, though the middle class would also be affected. China emerged as world's no. 1 exporter of MAPs with export worth US$ 200 million in 2003 (EXIM Bank of India 2003), while as in terms of export-import, Hong Kong along with China had the largest share (21%) in import market followed by the United States (14%) and Japan (10%) in 2001 (EXIM Bank of India 2003). The most leading market of herbal products in Europe is Germany followed by France, the United Kingdom, and Italy, while the United States is the largest marketplace for essential oil and herbal tea (ICMHHP3, 2004).

7.8 CHALLENGES

During the recent years, Indian herbal sector has established it as major hub of biodiversity having varied climate and with validated system of medicine, Indian Systems of Medicine (ISM), and developing pharmaceutical industry (Kala et al., 2006; Patwardhan and Vaidya, 2010). Hence, India can be enumerated as best seat for law, environment, and Development journal carrying basic and application-oriented research in this field (Sahoo et al., 2011; Subramoniam, 2014). Actually, besides information technology and biotechnology, research in medicinal plants along with potentials of both these technologies seems the most sustainable growth sector in the future (Kala, 2009, Yadav and Goyal, 2015). In spite of these abundant resources, there are various constraints that hamper the growth of herbal sector. These constraints include changeability in price, struggle against synthetics, exporters' noncooperation with rules and regulations of importing countries and resultant repudiation, quality limitations, scattered data with suppliers across world trade in MAPs, specified number of herbal suppliers and traders with potential settlement influence in comparison with the growers, irregular supply, and irrelevant collection system and prolonged storage resulting in below-average levels of active constituents and resulting in inflation in the price of their derivatives, which are some of the main limitation experienced by this sector (Sahoo and Manchikanti, 2013; Pan et al., 2014a,b; Mishra et al., 2016). Value-chain addition at different stages in medicinal herbs is carried out by various governmental institutions and universities working under the Government of India including the Indian Council of Agricultural Research (ICAR), Department of Science and Technology (CSIR), Ministry of Health and Family Welfare (CCRAS), Ministry of Environment and Forests (through ICFR) with state forest departments and corporate sector, and Defense Research and Development Organization (DRDO). According to the Scientific Advisory Committee to the Cabinet, research and development (R&D) in medicinal sector is inundated with the following problems.

Indian Himalayan is yet not fully explored; there are few MAPs without any preliminary reports on taxonomic identification and chemical and molecular characterization.

- Some of medicinal herbs are privileged to cater more attention like *S. lappa*, *P. kurroa*, *Ocimum sanctum*, *A. ferox*, and *Berberis aristata*, while herbs like *A. benthamii* and *Thymus serpyllum* are not having even validated taxonomic, value addition, and conservation studies.
- Inadequate intrainstitutional linkages resulting in limited data on previous research projects or there is duplicity in the generation of research project reports.
- The mandates of government institutions are too discreet and scattered without any single focus.
- The lack of cooperation exists between government institutions, local growers, and industry. Hence, research carried in laboratories does not reach farmers and hence distant from demands of industry. Consequently, the need of the hour is an integrated strategy that concentrates on diverse themes in the supply chain right from farm to firm and consumers.
- Improper agricultural practices, uneven accessibility, and incoherent quality of raw material.
- Shortage of R&D of high-yielding varieties, product, and process development.
- Incompetent processing techniques and substandard quality control procedures.
- The lack of regulatory mechanism in the trade of medicinal plants and the inflated growth of synthetic drugs.
- Export market is hampered by the scarcity of validated documentation of the therapeutic values of medicines, and their formulation is a main limitation in export predominantly to countries of EU and the United States.
- Dearth of supposed accessibility for exports in terms of quality, quantity, and price.
- The lack of data pertaining to the total world trade, demand dynamics in the overseas market, and marketing techniques.
- The lack of awareness on international and explicit regulations of importing countries prevailing imports of such products, including quality and intellectual property right (IPR) issues.

7.9 MARKETING STRATEGIES

Majority of Himalayan resources are collected from the wild so as to ensure livelihood of tribal communities. Indiscriminate harvesting and higher market demand drag several plant species to the verge of extinction leading to negative impact on the incomes of the underprivileged communities of the Himalayan region (Sridharan and Viswanathan, 2008; Athreye et al., 2009). For specified collection, effective communities or authentic government measures are necessary to be implemented (Chaturvedi et al., 2007; Khare et al., 2013). Indiscriminate collection and strict regulations seriously affected species richness thereby declining market supply (Kuniyal et al., 2015a,b). Mass cultivation of MAPs offers an immense enterprise for economic development (reference). Possibilities of public-private collaboration are explored via action research program and address the issues that affect the livelihood of the poor, including farmers with small landholdings, income opportunities for women, and food security of underprivileged section of rural society (Singh and Singh, 2009; Krishnakumar et al., 2015). However, mass cultivation of MAPs has not been implemented completely due to numerous hindrances like small landholdings, the lack of agricultural practices, and low-quality plant material (Jain and Rao, 2015; Srivastava, 2016). Moreover, natural constraints in mass cultivation of MAPs include long gestation period, high mortality, and inadequate institutional infrastructure to offer technical and marketing support, high transaction costs, and insufficient social capital (Kumari and Gupta, 2014; Gupta, 2016). Public-private collaboration is frequently recommended as a method of lowering these costs to satisfactory levels, decreasing risk, and securing market access to small producers (Singh, 2006; Maheswari, 2011). Such collaborations can overcome most of the constraints and provide a potential mechanism for ascertaining the requirements indispensable for the expansion of chains that are in the preliminary stages of development (Ghosh and Bhattacharya, 2009; Torri, 2011; Jain and Rao, 2016). Public-private alliance is still underway in India, and there is need to promote collaborations supporting mass cultivation of MAPs. Several initiatives need to be established prior to implementation of programs that could fascinate the public for cultivation of MAPs. These initiatives include the following:

1. Inclination of public to invest in capacity building prerequisite for technical and marketing support.
2. Establishment of collaborations between farmers with public institutions and private companies in order to gather data pertaining to mass cultivation of MAPs and build entrepreneurship.
3. Construction of social capital so as to advance efficiency of chain network between primary suppliers and traders and other

protagonists involved in chain development of MAPs vertically (producer-industry) plus horizontally (advancing the producer organizations) so as to augment public reliance and capacity to cooperate and establish entry ticket to market.

4. Generating higher inclination toward cultivated plant material. Generally, big-scale exporters are inclined toward hiring farmers to get primary supply, while the private sector has meager inclination toward joint ventures as it is extensively content with supply of plant produce irrespective of legal scrutiny. Exporters incline toward cultivated material due to the lack of data pertaining to origin of raw material because of scattered or missing documentation of MAPs. The functional significance of exports is too meager due to lower importance of export, and it confines impact on the chain.

5. Enforcement of legal policies for collection of MAPs from wild habitat can increase inclination of private sector toward mass cultivation of MAPs so as to maintain supply of raw material at constant rate. Collaborations will generate data that could be reproduced in other mountainous areas, and this would provide risky impulsion to agricultural diversification furnishing higher income generation for farmers (reference).

7.10 CONCLUSIONS

The major challenge in making herbal trade a positive factor for advancement and expansion of developing countries is exploiting the potential of MAPs. We need to accelerate the advancement in MAP cultivation through ascertainment of positive impact on socioeconomic development of herbal sector. The main factor leading to laggard growth of India is its poverty in spite of huge diversity. There is an urgent need to make people not only aware of their natural resources but also focus on prospects of herbal sector. Local growers should be acquainted with emerging profits of modern technology along with credit supplies. National policies should be implemented for effective provisions of ensuring equitable benefit sharing for all stakeholders. Global marketing of herbs can be ensured by maintaining inflow of raw material supply through mass cultivation of MAPs. Mass cultivation is primary first toward the development of herbal sector, which needs to be followed by value addition and promotion of MAPs. Majority of MAPs are exported outside the country prior to any legalized scrutinization. Moreover, indiscriminate harvesting and the lack of coordination between farmers and scientists create huge gap in the expansion of herbal sector. There is still misalignment between primary supplier and exporter. Mass cultivation of MAPs is associated with cultural significance linked with folklore plants, and the following factors required for success of the sector:

1. Selection and screening of MAPs with enormous market value.
2. Identification of suitable area exhibiting feasible ecological condition for cultivation of screened MAPs and comparatively economic development.
3. Optimization of favorable ecological factors that could enhance productivity and production of herbal and medicinal plants via alliance between scientists and farmers.
4. Identification of suitable market/potential buyer that can offer better returns and increase inclination of farmers toward MAPs.
5. Cultivation of MAPs in barren areas and extension of government institutes that investigate MAPs.
6. Integration of traditional knowledge with modern techniques for the development of modern drug discovery.
7. Facilitation of agronomic and agricultural strategies for mass multiplication of MAPs.
8. Enhancement of production and productivity for ascertaining the cluster strategy and low-cost processing.
9. Facilitation of communication at rural and tribal sections of Indian Himalaya.
10. Upgradation of forest policies that aid conservation and sustainable utilization of MAPs.
11. Establishment of technological interventions and scientific innovations to advance the mass production of active metabolites in MAPs and to investigate avenues for production of herbal-based drugs.

References

Abrol, D., Prajapati, P., Singh, N., 2017. Globalization of the Indian pharmaceutical industry: implications for innovation. J. Inst. Econ. 3 (2), 327–365.

Acharya, C.K., 2015. Ethnic uses and conservation practices of some plants with major medicinal potential by the tribal community in India. Int. J. Life Sci. 9 (6), 1–7.

Acharya, S.P., 2005. Developing Medicinal and Aromatic Plant based Livelihood Options for Traditional Gathering Communities. Western Uplands Poverty Alleviation Project, Nepalgunj.

Ahmad, I., Aqil, F., Ahmad, F., Owais, M., 2006. Herbal medicines: prospects and constraints. In: Modern Phytomedicine: Turning Medicinal Plants Into Drugs. John Wiley & Sons, UK.

Ahmad, S., Gangoo, S.A., Sultan, S.M., Dar, M.D., 2016. Ethnoveterinary practices and use of herbal medicine by pastoralists of himalaya: a review. Ann. Biol. 32 (2), 260–268.

Alam, G., Belt, J., 2009. Developing a Medicinal Plant Value Chain: Lessons From an Initiative to Cultivate Kutki (Picrorhiza Kurrooa) in Northern India. KIT Working Papers Series (WPS. C5). KIT Working Papers Series C5, Amsterdam.

Ali, S.S., Kasoju, N., Luthra, A., Singh, A., Sharanabasava, H., Sahu, A., Bora, U., 2008. Indian medicinal herbs as sources of antioxidants. Food Res. Int. 41 (1), 1–15.

Amujoyegbe, B.J., Agbedahunsi, J.M., Amujoyegbe, O.O., 2012. Cultivation of medicinal plants in developing nations: means of conservation and poverty alleviation. Int. J. Med. Arom. Plants 2, 345–353.

Aneesh, T., Hisham, M., Sekhar, M., Madhu, M., Deepa, T., 2009. International market scenario of traditional Indian herbal drugs-India declining. Int. J. Green Pharm. 3 (3), 184.

Anilkumar, D., Joshi, H., Nishteswar, K., 2012. Review of *Glycyrrhiza glabra* (*yastimadhu*)—a broad spectrum herbal drug. Pharma Sci. Monit. 3 (4), 3171.

Arora, S., Gupta, R.D., 2014. Soil related constraints in crop production and their management options in North Western Himalayas. J. Soil Water Conserv. 13 (4), 305–312.

Athreye, S., Kale, D., Ramani, S.V., 2009. Experimentation with strategy and the evolution of dynamic capability in the Indian pharmaceutical sector. Ind. Corp. Chang. 18 (4), 729–759.

Awate, S., Larsen, M.M., Mudambi, R., 2015. Accessing vs sourcing knowledge: a comparative study of R&D internationalization between emerging and advanced economy firms. J. Int. Bus. Stud. 46 (1), 63–86.

Bajaj, Y.P.S., 2012. Medicinal and Aromatic Plants I (vol. 4). Springer Science & Business Media, Hedielberg, UK.

Basu, S.K., Thomas, J.E., Acharya, S.N., 2007. Prospects for growth in global nutraceutical and functional food markets: a Canadian perspective. Aust. J. Basic Appl. Sci. 1 (4), 637–649.

Bawa, K.S., Joseph, G., Setty, S., 2007. Poverty, biodiversity and institutions in forest-agriculture ecotones in the Western Ghats and Eastern Himalaya ranges of India. Agric. Ecosyst. Environ. 121 (3), 287–295.

Bera, S.K., 2010. Globalisation and Sustainable Exports of Indian Medicinal and Aromatic Plants: A Protection Study. MPRA is a Munich Personal RePEc Archive hosted by the Munich University Library in Germany.

Bhardwaj, R.K., Sikka, B.K., Singh, A., Sharma, M.L., Singh, N.K., Arya, R., 2011. In: Challenges and constraints of marketing and export of Indian spices in India. Proc. International Conference on Technology and Business Management. pp. 28–30.

Bhatt, R.K., Raghuvanshi, M.S., Kalia, R.K., 2015. Achieving sustainable livelihood in cold arid regions of India through multienterprise options. Ann. Arid Zone 54 (3&4), 1–12.

Bhattacharya, P., Prasad, R., Bhattacharyya, R., Asokan, A., 2008. Towards certification of wild medicinal and aromatic plants in four Indian states. Unasylva 230 (59), 35–44.

Bhattacharya, R., Reddy, K.R.C., Mishra, A.K., 2014. Export strategy of Ayurvedic products from India. Int. J. Ayurvedic Med. 5 (1), 125–128.

Bhushan, I., Kumar, A., Patel, J.S., Yadav, R.P., Singh, S., Meena, R., Meena, V.S., 2016. The Indian Himalayan ecosystem as source for survival. In: Conservation Agriculture. Springer, Singapore, pp. 367–380.

Bindu, S., Rameshkumar, K.B., Kumar, B., Singh, A., Anilkumar, C., 2014. Distribution of reserpine in Rauvolfia species from India—HPTLC and LC-MS studies. Ind. Crop. Prod. 62, 430–436.

Bisht, V.K., Kathait, A.S., Negi, J.S., Bhandari, A.K., Rana, C.S., 2011. Development of market supply chain for *Picrorhiza kurroa* and *Saussurea costus* in the district Chamoli of the Uttarakhand state, India: a case study. Rep. Opin. 3, 14–17.

Biswas, A., Bari, M.A., Roy, M., Bhadra, S.K., 2010. Inherited folk pharmaceutical knowledge of tribal people in the Chittagong hill tracts. Bangladesh. Ind. J. Tradit. Knowl. 9, 77–89.

Booker, A., Johnston, D., Heinrich, M., 2012. Value chains of herbal medicines—research needs and key challenges in the context of ethnopharmacology. J. Ethnopharmacol. 140 (3), 624–633.

Bruni, R., Sacchetti, G., 2009. Factors affecting polyphenol biosynthesis in wild and field grown St. John's Wort (*Hypericum perforatum* L. *Hypericaceae/Guttiferae*). Molecules 14 (2), 682–725.

Butola, J.S., Badola, H.K., 2008. Propagation conditions for mass multiplication of three threatened Himalayan high value medicinal herbs. Int. Plant Genet. Resour. Newsl. 153, 43–47.

Canter, P.H., Thomas, H., Ernst, E., 2005. Bringing medicinal plants into cultivation: opportunities and challenges for biotechnology. Trends Biotechnol. 23 (4), 180–185.

CECI, 1999. Subsector analysis of high-altitude NTFPs in the Karnali zone. Canadian Centre for International Studies and Co-operation (CECI), Kathmandu, 26 pp.

Chan, K., Shaw, D., Simmonds, M.S., Leon, C.J., Xu, Q., Lu, A., Williamson, E.M., 2012. Good practice in reviewing and publishing studies on herbal medicine, with special emphasis on traditional Chinese medicine and Chinese materia medica. J. Ethnopharmacol. 140 (3), 469–475.

Chaturvedi, K., Chataway, J., Wield, D., 2007. Policy, markets and knowledge: strategic synergies in Indian pharmaceutical firms. Tech. Anal. Strat. Manag. 19 (5), 565–588.

Chauhan, R.S., Tiwari, D., Bisht, A.S., Shukla, A., 2014. Ex situ conservation of medicinal and aromatic plants in Bharsar, Uttarakhand, India. Med. Plant Int. J. Phytomed. Relat. Ind. 6 (4), 282–292.

Chen, S.L., Yu, H., Luo, H.M., Wu, Q., Li, C.F., Steinmetz, A., 2016. Conservation and sustainable use of medicinal plants: problems, progress, and prospects. Chin. Med. 11 (1), 37.

Chhetri, D.R., 2014. Medicinal Plants of the Himalaya: Production Technology and Utilization. Agrobios, Jodhpur, India.

Chhetri, B.K., Ali, N.A.A., Setzer, W.N., 2015. A survey of chemical compositions and biological activities of Yemeni Aromatic Medicinal Plants. Medicines 2 (2), 67–92.

Choudhary, D., Kala, S.P., Todaria, N.P., Dasgupta, S., Kollmair, M., 2014. Drivers of exploitation and inequity in non-timber forest products (NTFP) value chains: the case of Indian Bay Leaf in Nepal and India. Dev. Policy Rev. 32 (1), 71–87.

Dar, A.K., ul Hassan, W., Lone, A.H., Haji, A., Manzoor, N., Mir, A.I., 2017. Study to assess high demand and high commercial value medicinal plants of Jammu and Kashmir India—with special focus on routes of procurement and identification. Int. J. 6 (2), 2576–2585.

Deb, S., Lynrah, M.M., Tiwari, B.K., 2013. Technological innovations in shifting agricultural practices by three tribal farming communities of Meghalaya, Northeast India. Trop. Ecol. 54, 133–148.

Debnath, M., Malik, C.P., Bisen, P.S., 2006a. Micropropagation: a tool for the production of high quality plant-based medicines. Curr. Pharm. Biotechnol. 7 (1), 33–49.

Debnath, M., Malik, C.P., Bisen, P.S., 2006b. Micropropagation: a tool for the production of high quality plant-based medicines. Curr. Pharm. Biotechnol. 7 (1), 33–49.

Dhar, N., Rana, S., Bhat, W.W., Razdan, S., Pandith, S.A., Khan, S., Lattoo, S.K., 2013. Dynamics of withanolide biosynthesis in relation to temporal expression pattern of metabolic genes in *Withania somnifera* (L.) Dunal: a comparative study in two morpho-chemovariants. Mol. Biol. Rep. 40 (12), 7007–7016.

Dhar, U., Rawal, R.S., Upreti, J., 2000. Setting priorities for conservation of medicinal plants—a case study in the Indian Himalaya. Biol. Conserv. 95 (1), 57–65.

Doke, R.M., Mahale, A.M., Karanjkar, A.R., Amge, S.A., 2013. Recent trends for neutraceuticals in the Indian market. IAJPR 3 (5), 4001–4020.

Dubey, N.K., Kumar, R., Tripathi, P., 2004. Global promotion of herbal medicine: India's opportunity. Curr. Sci. 86 (1), 37–41.

Evans, W.C., 2009. Trease and Evans' Pharmacognosy E-Book. Elsevier Health Sciences, UK.

Farooq, S., 2013. Genetic Diversity of *Podophyllum hexandrum Royle.* and *Hypericum Perforatum* L. Using Molecular and Morphological Markers. Doctoral Dissertation, SKUAST.

Farooquee, N.A., Majila, B.S., Kala, C.P., 2004. Indigenous knowledge systems and sustainable management of natural resources in a high altitude society in Kumaun Himalaya, India. J. Hum. Ecol. 16 (1), 33–42.

Gautam, P.L., Singh, B.B., Saxena, S., Sharma, R.K., 2004. Collection, conservation and utilization of plant genetic resources. In: Plant Breeding. Springer, Netherlands, pp. 657–690.

Ghosh, A., Bhattacharya, S., 2009. Retailing in traditional and herbal medicine: facts and future. Pharmacogn. Rev. 3 (5), 135.

Ghosh, S.P., General, F.D.D., 2013. In: Promotion of medicinal and aromatic plants in the Asia-Pacific region. Expert Consultation on Promotion of Medicinal and Aromatic Plants in the Asia-Pacific Region: Proceedings, 33.

Goenka, A.H., 2007. Rustom Jal Vakil and the saga of Rauwolfia serpentina. J. Med. Biogr. 15 (4), 195–200.

Gouri, M.S., Morrison, E., Mayers, J., 2004. Policy Influences on Forest-Based Livelihoods in Himachal Pradesh, India. International Institute for Environment and Development, London.

Greenhalgh, C., 2016. Science, technology, innovation and IP in India: new directions and prospects. Econ. Chang. Restruct. 49 (2-3), 113–138.

Gularia, C., 2012. Economics of Production and Marketing of Important Medicinal and Aromatic Plants in Mid Hills of Himachal Pradesh. Doctoral Dissertation.

Gulati, A., Minot, N., Delgado, C., Bora, S., 2007. Growth in high-value agriculture in Asia and the emergence of vertical links with farmers. In: Global Supply Chains: Standards and the Poor: How the Globalization of Food Systems and Standards Affects Rural Development and Poverty. CABI Head Office, Oxon, UK, pp. 98–108.

Gupta, P.D., Daswani, P.G., Birdi, T.J., 2014. Approaches in fostering quality parameters for medicinal botanicals in the Indian context. Indian J. Pharm. 46 (4), 363.

Gupta, R., 2016. Disruptive marketing and brand building: a case study of Patanjali, Ayurveda Limited. AEBM 3 (2), 225–231.

Gupta, R.K., 2010. Medicinal and Aromatic Plants. 234. CBS Publishers and Distributors, New Delhi, 499.

Hritcu, L., Cioanca, O., 2016. Prevalence of use of herbal medicines and complementary and alternative medicine in Europe. In: Herbal Medicine in Depression. Springer International Publishing, Switzerland, pp. 135–181.

ICMHHP3, 2004. Globalisation, Global Trend in Herbal Market, and The Impact Thereof on Medicinal Plants in Orissa.

Jain, A.K., Sharma, B.K., 2015. Developments in the field of Ayurveda—past to present. Ayushdhara 1 (2), 51–64.

Jain, R., Rao, B., 2015. Critical analysis of India's National Mission on Medicinal Plants (NMMP) in providing access to quality botanical drugs to improve public health. J. Ayurveda Integr. Med. 6 (3), 198.

Jain, R., Rao, B., 2016. In: Causal-loop and SWOT analysis of Indian Herbal Industry. International Conference on Inter Disciplinary Research in Engineering and Technology. Vol. 26. p. 30.

Jaiswal, Y., Liang, Z., Zhao, Z., 2016. Botanical drugs in Ayurveda and traditional Chinese medicine. J. Ethnopharmacol. *194*, 245–259.

Joshi, B.C., Joshi, R.K., 2014. The role of medicinal plants in livelihood improvement in Uttarakhand. Int. J. Herb. Med. 1 (6), 55–58.

Kakarla, L., Katragadda, S.B., Botlagunta, M., 2015. Morphological and chemoprofile (liquid chromatography-mass spectroscopy and gas chromatography-mass spectroscopy) comparisons of *Cyperus scariosus* R. Br and *Cyperus rotundus* L. Pharmacogn. Mag. *11* (Suppl 3), S439.

Kala, C.P., 2005. Indigenous uses, population density, and conservation of threatened medicinal plants in protected areas of the Indian Himalayas. Conserv. Biol. 19 (2), 368–378.

Kala, C.P., 2009. Medicinal plants conservation and enterprise development. Med. Plants 1 (2), 79–95.

Kala, C.P., 2015. Medicinal and aromatic plants: boon for enterprise development. JARMAP 2 (4), 134–139.

Kala, C.P., 2017. Traditional health care systems and herbal medicines. Eur. J. Environ. Public Health 1 (1), 03.

Kala, C.P., Sajwan, B.S., 2007. Revitalizing Indian systems of herbal medicine by the National Medicinal Plants Board through institutional networking and capacity building. Curr. Sci. 93 (6), 797–806.

Kala, C.P., Dhyani, P.P., Sajwan, B.S., 2006. Developing the medicinal plants sector in northern India: challenges and opportunities. J. Ethnobiol. Ethnomed. 2 (1), 32.

Kala, C.P., Farooquee, N.A., Dhar, U., 2004. Prioritization of medicinal plants on the basis of available knowledge, existing practices and use value status in Uttaranchal, India. Biodivers. Conserv. 13 (2), 453–469.

Kala, C.P., Dhyani, P.P., Sajwan, B.S., 2006. Developing the medicinal plants sector in northern India: challenges and opportunities. J. Ethnobiol. Ethnomed. 2 (1), 32.

Kamboj, V.P., 2000. Herbal medicine. Curr. Sci. 78 (1), 35–39.

Karki, M.B., 2000. In: Commercialisation of natural resources for sustainable livelihoods: the case of forest products. Growth, Poverty Alleviation and Sustainable Resource Management in the Mountain Areas of South Asia, Feldafing, Deutsche Stiftung fur internationale Entwicklung, pp. 293–320.

Karki, M., Nagpal, A., 2004. Study on Marketing Opportunities for Medicinal, Aromatic, and Dye Plants in South Asia. IDRC-Medicinal and Aromatic Plants Program in Asia (MAPPA), 114 p.

Katiyar, C., Gupta, A., Kanjilal, S., Katiyar, S., 2012. Drug discovery from plant sources: an integrated approach. Ayu 33 (1), 10.

Keshari, D.S., Bishnupriya, M., Panda, P.K., Arun, J.B., 2015. A study of various market samples of spikenard (*Nardostachys jatamansi* dc.) with special reference to its pharmacognostic & phytochemical aspects. GJRMI 4 (3), 46.

Khan, H., 2014. Medicinal plants in light of history: recognized therapeutic modality. J. Evid. Based Complement. Alternat. Med. 19 (3), 216–219.

Khare, A., Mukerjee, S., Goyal, T., 2013. Social influence and green marketing: an exploratory study on Indian consumers. J. Cust. Behav. 12 (4), 361–381.

Khare, C.P., 2008. Indian Medicinal Plants: An Illustrated Dictionary. Springer Science & Business Media, Heidelberg, Germany.

Krishnakumar, J., Yanagida, J.F., Anitha, V., Balakrishnan, R., Radovich, T.J., 2015. Non-timber forest products certification and management: a socioeconomic study among the Kadars in Kerala, India. Environ. Dev. Sustain. 17 (4), 837–858.

Krishnan, P.N., Decruse, S.W., Radha, R.K., 2011. Conservation of medicinal plants of Western Ghats, India and its sustainable utilization through in vitro technology. In Vitro Cell Dev. Biol. Plant 47 (1), 110–122.

Kumar, A., Kaul, B.L., Verma, H.K., 2001. Phenological observations on root yield and chemical composition in different morphotypes of *Withania somnifera*. J. Med. Aro. Plant. Sci. 23, 21–23.

Kumar, G.P., Kumar, R., Chaurasia, O.P., Singh, S.B., 2011a. Current status and potential prospects of medicinal plant sector in trans-Himalayan Ladakh. J. Med. Plant Res. 5 (14), 2929–2940.

Kumar, S.N., Aggarwal, P.K., Rani, S., Jain, S., Saxena, R., Chauhan, N., 2011b. Impact of climate change on crop productivity in Western Ghats, coastal and northeastern regions of India. Curr. Sci. 101 (3), 332–341.

Kumar, S., Kumar, R., Khan, A., 2011c. Medicinal plant resources: manifestation and prospects of life-sustaining healthcare system. Cont. J. Biol. Sci. 4 (1), 19–29.

Kumar, V., Babu, S., Revale, A.K., Meena, R.K., Ranjan, M.K., Desai, B.S., 2014. Cultivation of medicinal plants in natural ecosystem in Gujarat (India): constraints and conservation need. JPDS 6 (3), 425–435.

Kumari, P., Gupta, N., 2014. A case study on: the critical inhibitions in successful rural marketing in India. IJERMT 3 (2), 17–21.

Kuniyal, C.P., Bhatt, V.P., Bhatt, V.P., Butola, J.S., Sundriyal, R.C., 2014. Promoting nursery enterprise in high altitude villages: a participatory approach for conservation and commercialization of Himalayan threatened medicinal plants. J. Med. Plant Res. 8 (48), 1399–1407.

Kuniyal, C.P., Bisht, V.K., Negi, J.S., Bhatt, V.P., Bisht, D.S., Butola, J.S., Singh, S.K., 2015a. Progress and prospect in the integrated development of medicinal and aromatic plants (MAPs) sector in Uttarakhand, Western Himalaya. Environ. Dev. Sustain. 17 (5), 1141–1162.

Kuniyal, C.P., Bisht, V.K., Negi, J.S., Bhatt, V.P., Bisht, D.S., Butola, J.S., Singh, S.K., 2015b. Progress and prospect in the integrated development of medicinal and aromatic plants (MAPs) sector in Uttarakhand, Western Himalaya. Environ. Dev. Sustain. 17 (5), 1141–1162.

Kunwar, R.M., Mahat, L., Acharya, R.P., Bussmann, R.W., 2013. Medicinal plants, traditional medicine, markets and management in far-west Nepal. J. Ethnobiol. Ethnomed. 9 (1), 24.

Kunwar, S. C., Ansari, A. S., & Luintel, H., 2009. Non-Timber Forest Products Enterprise Development: Regulatory Challenges Experienced in the Koshi Hills of Nepal. Forest Action Discussion Paper, 4.

Larsen, H.O., Olsen, C.S., 2007. Unsustainable collection and unfair trade? Uncovering and assessing assumptions regarding Central Himalayan medicinal plant conservation. Biodivers. Conserv. 16 (6), 1679–1697.

Maheswari, J., 2011. Patenting Indian medicinal plants and products. Indian J. Sci. Technol. 4 (3), 298–301.

Maikhuri, R.K., Negi, V.S., Rawat, L.S., Pharswan, D.S., 2017b. Bioprospecting of medicinal plants in Nanda Devi biosphere reserve: linking conservation with livelihood. Curr. Sci. 113 (4), 571–577.

Maikhuri, R.K., Phondani, P.C., Rawat, L.S., Jha, N.K., Maletha, A., Bahuguna, Y.M., Kandari, L.S., 2017a. Conservation and management strategies of medicinal plant resources through action research approaches in Indian Himalaya. IJST 41 (3), 771–777.

Maiti, S., Geetha, K.A., 2013. In: Country status report on medicinal and aromatic plants in India. Expert Consultation on Promotion of Medicinal and Aromatic Plants in the Asia-Pacific Region: Proceedings. pp. 101.

Mani, S., 2006. The sectoral system of innovation of Indian pharmaceutical industry.

Mazid, M., Khan, T.A., Mohammad, F., 2012. Medicinal plants of rural India: a review of use by Indian folks. Indo Global J. Pharm. 2 (3), 286–304.

Meier zu Biesen, C., 2017. From coastal to global: the transnational flow of Ayurveda and its relevance for Indo-African linkages. Glob. Public Health 1–16.

Mishra, M., Kotwal, P.C., Prasad, C., 2009. Harvesting of medicinal plants in the forest of central India and its impact on quality of raw materials: a case of Nagpur District, India. Int. J. Ecol. 16, 35–42.

Mishra, P., Kumar, A., Nagireddy, A., Mani, D.N., Shukla, A.K., Tiwari, R., Sundaresan, V., 2016. DNA barcoding: an efficient tool to overcome authentication challenges in the herbal market. Plant Biotechnol. J. 14 (1), 8–21.

Mishra, S.B., Rao, C.V., Ojha, S.K., Vijayakumar, M., Verma, A., 2010a. An analytical review of plants for anti diabetic activity with their phytoconstituent and mechanism of action. Int. J. Pharm. Sci. Res. 1 (1), 1647–1652.

Mishra, G., Srivastava, S., Nagori, B.P., 2010b. Pharmacological and therapeutic activity of Cissus quadrangularis: an overview. Int. J. PharmTech Res. 2 (2), 1298–1310.

Mittal, R., Singh, S.P., 2007. Shifting from agriculture to agribusiness: the case of aromatic plants. Agric. Econ. Res. Rev. 20 (2007), 541–550.

Mohanta, B., Chakraborty, A., Sudarshan, M., Dutta, R.K., Baruah, M., 2003. Elemental profile in some common medicinal plants of India. Its correlation with traditional therapeutic usage. J. Radioanal. Nucl. Chem. 258 (1), 175–179.

Mukherjee, P.K., Venkatesh, M., Kumar, V., 2007. An overview on the development in regulation and control of medicinal and aromatic plants in the Indian system of medicine. B. Latinoam. Caribe Plant. Med. Aromát. 6 (4), 129–136.

Nanyunja, R.K., Baguma, T.I.M.O.T.H.Y., 2005. In: The role of indigenous knowledge in forestry assessment and monitoring in Uganda. Non-Timber Forest Products—Traditional Products and Technologies. World Forest Congress, Brisbane Australia, pp. 76–85.

Nikam, P.H., Kareparamban, J., Jadhav, A., Kadam, V., 2012. Future Trends in Standardization of Herbal Drugs. J. Appl. Pharm. Sci. 2 (6), 38–44.

Odisha Forest Development Corporation Ltd (OFDC, 2006). https://www.odishafdc.com/UploadMaque/Susant.pdf.

Olsen, C.S., Bhattarai, N., 2005. A typology of economic agents in the Himalayan plant trade. Mt. Res. Dev. 25 (1), 37–43.

Olsen, C.S., Helles, F., 2009. Market efficiency and benefit distribution in medicinal plant markets: empirical evidence from South Asia. Int. J. Biodivers. Sci. Manag. 5 (2), 53–62.

Padalia, K., Bargali, K., Bargali, S.S., 2015. How does traditional home-gardens support ethnomedicinal values in Kumaun Himalayan Bhabhar belt, India? Afr. J. Tradit. Complement. Altern. Med. 12 (6), 100–112.

Pan, S.Y., Litscher, G., Gao, S.H., Zhou, S.F., Yu, Z.L., Chen, H.Q., Ko, K.M., 2014a. Historical perspective of traditional indigenous medical practices: the current renaissance and conservation of herbal resources. Evid. Based Complement. Alternat. Med. 2014, 525340.

Pan, S.Y., Litscher, G., Gao, S.H., Zhou, S.F., Yu, Z.L., Chen, H.Q., Ko, K.M., 2014b. Historical perspective of traditional indigenous medical practices: the current renaissance and conservation of herbal resources. Evid. Based Complement. Alternat. Med. 2014, 525340.

Panda, H., 2004. Handbook on Herbal Medicines. Asia Pacific Business Press Inc., New Delhi.

Pandey, A.K., 2015. 11 Sustainable bark harvesting of important medicinal tree species, India. In: Ecological Sustainability for Non-Timber Forest Products: Dynamics and Case Studies of Harvesting. Routledge, Abingdon, Oxon, p. 163.

Pandey, M.M., Rastogi, S., Rawat, A.K.S., 2013. Indian traditional ayurvedic system of medicine and nutritional supplementation. Evid. Based Complement. Alternat. Med. 2013, 376327.

Pangriya, R., 2015. Study of aromatic and medicated plants in Uttrakhand, India: with focus on role in employment generation and supply chain management. Int. J. Soc. Sci. Manag. 2 (2), 148–156.

Patra, K.C., Pareta, S.K., Harwansh, R.K., Kumar, K.J., 2010. Traditional approaches towards standardization of herbal medicines—a review. J Pharm Sci Technol 2 (11), 372–379.

Patwardhan, B., 2000. Ayurveda: the designer medicine. Indian Drugs 37, 213–227.

Patwardhan, B., Vaidya, A.D., 2010. Natural products drug discovery: accelerating the clinical candidate development using reverse pharmacology approaches. Indian J. Exp. Biol. 48 (3), 220–227.

Patwardhan, B., Warude, D., Pushpangadan, P., Bhatt, N., 2005. Ayurveda and traditional Chinese medicine: a comparative overview. Evid. Based Complement. Alternat. Med. 2 (4), 465–473.

Patwardhan, R.B., Dhakephalkar, P.K., Chopade, B.A., 2015. Antibacterial and plasmid curing activities of root extracts of Plumbago zeylanica. Int. J. Herbo Med. 2 (1), 13–25.

Phondani, P.C., Bhatt, I.D., Negi, V.S., Kothyari, B.P., Bhatt, A., Maikhuri, R.K., 2016. Promoting medicinal plants cultivation as a tool for biodiversity conservation and livelihood enhancement in Indian Himalaya. J Asia Pac. Biodivers. 9 (1), 39–46.

Phondani, P.C., Negi, V.S., Bhatt, I.D., Maikhuri, R.K., Kothyari, B.P., 2011. Promotion of medicinal and aromatic plants cultivation for improving livelihood security: a case study from West Himalaya, India. Int. J. Med. Arom. Plants 1 (3), 245–252.

Poudel, K.L., 2007. Trade Potentility and Ecological Analysis of NTFPs in Himalayan Kingdom of Nepal. Berkeley Electronic Press, Berkeley, USA.

Pradeep, T., 2016. Traditional ethnomedicinal knowledge of Indian tribes. Curr. Sci. 110 (4), 486.

Prakash, O., Jyoti, A.K., Kumar, P., Manna, N.K., 2013. Adulteration and substitution in Indian medicinal plants: an overview. J. Med. Plant. Stud. 1 (4), 127–132.

Purkayastha, J., 2016. Emerging trends in sustainable bioprospecting of bioresources. In: Bioprospecting of Indigenous Bioresources of North-East India. Springer, Singapore, pp. 3–19.

Rajpurohit, D., Jhang, T., 2015. In situ and ex situ conservation of plant genetic resources and traditional knowledge. In: Plant Genetic Resources and Traditional Knowledge for Food Security. Springer, Singapore, pp. 137–162.

Ram, M., Abdin, M., 2008. SCAR markers for authentication of herbal drugs. MAPSB 2, 79–85.

Ramawat, K.G., Goyal, S., 2008. The Indian herbal drugs scenario in global perspectives. In: Bioactive Molecules and Medicinal Plants, pp. 325–347. Bingley, United Kingdom.

Rao, K.S., Saxena, K.G., Tiwari, B.K., 2015. Biodiversity, Climate Change and Socio-Economic Development in the Indian Himalaya. Bishen Singh Mahendra Pal Singh, Dehra Dun, India.

Rao, M.R., Palada, M.C., Becker, B.N., 2004. Medicinal and aromatic plants in agroforestry systems. In: New Vistas in Agroforestry. Springer, Netherlands, pp. 107–122.

Rath, B., 2005. Globalisation, Global Trend in Herbal Market, and the Impact Thereof on Medicinal Plants in Orissa. Vasundhara, Bhubaneswar.

Rather, M.A., Bhat, B.A., Qurishi, M.A., 2013. Multicomponent phytotherapeutic approach gaining momentum: is the "one drug to fit all" model breaking down? Phytomedicine 21 (1), 1–14.

Rawat, A.K.S., Tewari, S.K., 2015. Quality assurance of medicinal and aromatic plants-good agricultural and collection practices (GAP & GCP). In: Medicinal and Aromatic Plants of the World. Springer, Netherlands, pp. 273–303.

Rawat, S., Jugran, A.K., Bahukhandi, A., Bahuguna, A., Bhatt, I.D., Rawal, R.S., Dhar, U., 2016. Anti-oxidant and anti-microbial properties of some ethno-therapeutically important medicinal plants of Indian Himalayan Region. 3 Biotech 6 (2), 1–12.

Regmi, K.R., Joshi, M., 2008. Emerging food security challenges and sustainable management strategies in Nepal. In: Agricultural Research for Poverty Reduction for Poverty Alleviation and Livelihood Enhancement. Nepal Agricultural Research Council (NARC) Society of Agricultural Scientists, Nepal, 27, pp. 376.

Rentala, S., Anand, B., Shaban, M., 2014. Technological capabilities and firm resources as determinants of export competitiveness: evidence from Indian pharmaceutical industry using quantile regression approach. J. Med. Mark. 14 (2-3), 133–144.

Sadhu, A., Upadhyay, P., Singh, P.K., Agrawal, A., Ilango, K., Karmakar, D., Dubey, G.P., 2015. Quantitative analysis of heavy metals in medicinal plants collected from environmentally diverse locations in India for use in a novel phytopharmaceutical product. Environ. Monit. Assess. 187 (8), 542.

Sahil, G., 2016. Characterization of Antimicrobial and Antioxidant Metabolites From *Pinus Gerardiana* and *Picea Smithiana* Growing in North Western Himalayas. INFLIBNET's Institutional Repository.

Sahoo, N., Manchikanti, P., 2013. Herbal drug regulation and commercialization: an Indian industry perspective. J. Altern. Complement. Med. 19 (12), 957–963.

Sahoo, N., Manchikanti, P., Dey, S., 2010. Herbal drugs: standards and regulation. Fitoterapia 81 (6), 462–471.

Sahoo, N., Manchikanti, P., Dey, S.H., 2011. Herbal drug patenting in India: IP potential. J. Ethnopharmacol. 137 (1), 289–297.

Sahoo, S., Ghosh, G., Das, D., Nayak, S., 2013. Phytochemical investigation and in vitro antioxidant activity of an indigenous medicinal plant Alpinia nigra BL Burtt. Asian Pac. J. Trop. Biomed. 3 (11), 871–876.

Saini, N., 2011. Establishment of an herbal industry in India. Int. J. Pharm. Biol. Arch. 2 (4).

Saks, M., 2015. Health policy and complementary and alternative medicine. In: The Palgrave International Handbook of Healthcare Policy and Governance. Palgrave Macmillan, Basingstoke, UK, pp. 494–509.

Samal, P.K., Palni, L.M.S., Agrawal, D.K., 2003. Ecology, ecological poverty and sustainable development in Central Himalayan region of India. Int. J. Sust. Dev. World 10 (2), 157–168.

Sati, V.P., 2005. Natural resource conditions and economic development in the Uttaranchal Himalaya, India. J. Mt. Sci. 2 (4), 336–350.
Satyanarayana, T., 2010. Indian medical tourism—international health care destination. Indian J. Public Adm. 56 (4), 901–919.
Scalera, V.G., Perri, A., Mudambi, R., 2015. Managing innovation in emerging economies: organizational arrangements and resources of foreign MNEs in the Chinese pharmaceutical industry. In: Emerging Economies and Multinational Enterprises. Emerald Group Publishing Limited, Bingley, UK, pp. 201–233.
Schippmann, U.W.E., Leaman, D., Cunningham, A.B., 2006. A comparison of cultivation and wild collection of medicinal and aromatic plants under sustainability aspects. Frontis 17, 75–95.
Schmidt, M., Butterweck, V., 2015. The mechanisms of action of St. John's wort: an update. Wien. Med. Wochenschr. 165 (11-12), 229–235.
Sen, S., Chakraborty, R., 2017. Revival, modernization and integration of Indian traditional herbal medicine in clinical practice: importance, challenges and future. J. Tradit. Complement. Med. 7 (2), 234–244.
Sen, S., Chakraborty, R., De, B., 2011. Challenges and opportunities in the advancement of herbal medicine: India's position and role in a global context. J. Herb. Med. 1 (3), 67–75.
Sengar, A., Sengar, A., Sharma, V., Sharma, V., Agrawal, R., Agrawal, R., 2017. Market development through integrating value chains—a case of Patanjali food and herbal park. EEMCS 7 (4), 1–22.
Shahidullah, A.K.M., 2016. Community-Based Developmental Entrepreneurship: Linking Microfinance with Ecosystem Services. Doctoral Dissertation, University of Manitoba.
Shahidullah, A.K.M., Haque, C.E., 2010. Linking medicinal plant production with livelihood enhancement in bangladesh: implications of vertically integrated value chain. J. Transdiscipl. Environ. Stud. 9 (2), 1–18.
Shameem, S.A., Kangroo, I.N., 2011. Comparative assessment of edaphic features and phytodiversity in lower Dachigam National Park, Kashmir Himalaya, India. Afr. J. Environ. Sci. Technol. 5 (11), 972–984.
Shankar, D., 2010. Conceptual framework for new models of integrative medicine. J Ayurveda Integr Med 1 (1), 3.
Sharma, M., Garg, S., 2013. Functional foods: marketing 'health' to modern India. IJIRD 2 (5), 720–739.
Sharma, R., 2013. Demand Analysis of Medicinal Plants in Himachal Pradesh Based on Selected Pharmaceutical Units. Doctoral Dissertation.
Sharma, S., Rathi, N., Kamal, B., Pundir, D., Kaur, B., Arya, S., 2010. Conservation of biodiversity of highly important medicinal plants of India through tissue culture technology—a review. Agric. Biol. J. N. Am. 1 (5), 827–833.
Sharma, U.R., Malla, K.J., Uprety, R.K., 2017. Conservation and management efforts of medicinal and aromatic plants in Nepal. Banko Janakari 14 (2), 3–11.
Sheikh, M.A., 2014. Phytochemical investigation and uses of some of the medicinal plants of Kashmir Himalaya—a review. IJBPAS 3, 462–489.
Sher, H., Bussmann, R.W., Hart, R., 2017. Promoting sustainable use of medicinal and aromatic plants for livelihood improvement and biodiversity conservation under global climate change, through capacity building in the Himalaya Mountains, Swat District, Pakistan. Ann. Mo. Bot. Gard. 102 (2), 309–315.
Sheth, P.P., 2005. Global opportunities and challenges for medicinal uses of ayurveda, herbal products, nutraceuticals and alternatives. Health Adm. 1, 74–75.
Shrivastava, A., Jain, R., Agrawal, R.K., Ahirwar, D., 2008. Clinical trial of herbal drugs and products in India: past and current status and critical issues. RJPT 1 (2), 69–74.
Shukla, A.C. (Ed.), 2014. Advances in Medicinal and Aromatic Plants. Agrobios, Jodhpur, India.
Shukla, N., Pandey, S., 2015. A study on marketing of forest produce of Chhattisgarh State. IRJET 2 (8), 1665–1671.

Singh, A., 2014. Role of women entrepreneurs in India: a SWOT analysis. IJMIBS 2277-31774 (2), 231–238.
Singh, A.K., 2017a. Medicinal and aromatic plants. In: Wild Relatives of Cultivated Plants in India. Springer, Singapore, pp. 165–176.
Singh, B.P., 2017b. Biodiversity, tribal knowledge and life in India. Environ. Soc. Psychol. 2 (1), 1–10.
Singh, H., 2006. Prospects and challenges for harnessing opportunities in medicinal plants sector in India. Law Environ. Dev. J. 2, 196.
Singh, A., Singh, P.K., 2009. An ethnobotanical study of medicinal plants in Chandauli District of Uttar Pradesh, India. J. Ethnopharmacol. 121 (2), 324–329.
Singh, H., Husain, T., Agnihotri, P., Pande, P.C., Khatoon, S., 2014b. An ethnobotanical study of medicinal plants used in sacred groves of Kumaon Himalaya, Uttarakhand, India. J. Ethnopharmacol. 154 (1), 98–108.
Singh, K.M., Shankar, T., Jha, A.K., Kumar, A., 2014a. Scope and possibilities for cultivation of medicinal and aromatic plants in Bihar: some evidences. Environ. Ecol. 32 (4B), 1642–1646.
Singh, K.M., Swanson, B.E., Singh, J.P., 2005. In: Development of supply chains for medicinal plants: a case study involving the production of vinca rosa by small farmers in the Patna District of Bihar India. Workshop on Building New Partnerships in the Global Food Chain, Chicago.
Singh, P.S., Vidyasagar, G.M., 2015. Cultivation, marketing of medicinal and aromatic plants from Telangana: a review. J. Med. Plant Res. 3 (5), 76–79.
Singh, V., Raidoo, D.M., Harries, C.S., 2004. The prevalence, patterns of usage and people's attitude towards complementary and alternative medicine (CAM) among the Indian community in Chatsworth, South Africa. BMC Complement. Altern. Med. 4 (1), 3.
Singh, M., Govindarajan, R., Nath, V., Rawat, A.K.S., Mehrotra, S., 2006. Antimicrobial, wound healing and antioxidant activity of Plagiochasma appendiculatum Lehm. et Lind. J. Ethnopharmacol. 107 (1), 67–72.
Singh, N., Kaushik, N.K., Mohanakrishnan, D., Tiwari, S.K., Sahal, D., 2015. Antiplasmodial activity of medicinal plants from Chhotanagpur plateau, Jharkhand. India. J. Ethnopharmacol. 165, 152–162.
Skaria, B.P., 2007. Aromatic Plants. Vol. 1. New India Publishing, Delhi.
Spielman, D.J., Hartwich, F., Grebmer, K., 2010. Public-private partnerships and developing-country agriculture: evidence from the international agricultural research system. Public Adm. Dev. 30 (4), 261–276.
Sridharan, S., Viswanathan, M., 2008. Marketing in subsistence marketplaces: consumption and entrepreneurship in a South Indian context. J. Consum. Mark. 25 (7), 455–462.
Srinivasan, R., Srinivasan, R., 2016. Himalaya: head-to-heel herbal healthcare. EEMCS 6 (3), 1–29.
Srivastava, R., 2016. How Indian pharmaceutical companies are building global brands: the case of the Himalaya Herbal brand. TIBR 58 (5), 399–410.
Subramoniam, A., 2014. Present scenario, challenges and future perspectives in plant based medicine development. Ann. Phytomed 3 (1), 31–36.
Sujatha, S., Bhat, R., Kannan, C., Balasimha, D., 2011. Impact of intercropping of medicinal and aromatic plants with organic farming approach on resource use efficiency in arecanut (*Areca catechu* L.) plantation in India. Ind. Crop. Prod. 33 (1), 78–83.
Taneja, S.C., 2009. Therapeutic products from botanical sources: the Indian scenario. Planta Med. *75* (04), S-25.
Thillaivanan, S., Samraj, K., 2014. Challenges, constraints and opportunities in herbal medicines—a review. Int. J. Herb. Med. 2 (1), 21–24.
Torri, M.C., 2010. Community-based enterprises: a promising basis towards an alternative entrepreneurial model for sustainability enhancing livelihoods and promoting socio-economic development in rural India. J. Small Bus. Entrep. 23 (2), 237–248.

Torri, M.C., 2011. Facilitating the entry of communities in the herbal medicine sector: GMCL in India. Dev. Pract. 21 (2), 282–294.

Torri, M.-C., 2008. Linking tribal communities with the medicinal plant sector for better rural livelihood and sustainable use of resources: the case study of GMCL (Gram Mooligai Limited Company). Innogen Working Paper No. 60 JULY 2008. ESRC Genomics Network.

Umashanker, M., Shruti, S., 2011. Traditional Indian herbal medicine used as antipyretic, antiulcer, anti-diabetic and anticancer: a review. IJRPC 1 (4), 1152–1159.

Uniyal, S.K., Awasthi, A., Rawat, G.S., 2002. Current status and distribution of commercially exploited medicinal and aromatic plants in upper Gori valley, Kumaon Himalaya, Uttaranchal. Curr. Sci. 82, 1246–1252.

Upadhyay, R.K., Verma, R.S., Singh, V.R., Bahl, J.R., Sharma, S.K., Tewari, S.K., 2016. New agrotechnology for quality planting material production of rose-scented geranium (*Pelargonium graveolens* L. Herit.). JARMAP 3 (3), 128–130.

Van de Kop, P.E.T.R.A., Alam, G., De Steenhuijsen Piters, B., 2006. Developing a sustainable medicinal-plant chain in India. In: Agro-Food Chains and Networks for Development. Springer, Netherlands.

Ved, D.K., Goraya, G.S., 2007. Demand and Supply of Medicinal Plants in India. NMPB; FRLHT, New Delhi; Bangalore, India.

Verma, N., Shukla, S., 2015. Impact of various factors responsible for fluctuation in plant secondary metabolites. JARMAP 2 (4), 105–113.

Verma, P., Mathur, A.K., Jain, S.P., Mathur, A., 2012. In vitro conservation of twenty-three overexploited medicinal plants belonging to the Indian sub continent. Sci. World J. 2012, 929650.

Verma, R.S., Padalia, R.C., Goswami, P., Upadhyay, R.K., Singh, V.R., Chauhan, A., Tiwari, A.K., 2016. Assessing productivity and essential oil quality of Himalayan thyme (*Thymus linearis* Benth.) in the subtropical region of north India. Ind. Crop. Prod. 94, 557–561.

Verma, R.S., Rahman, L.U., Chanotiya, C.S., Verma, R.K., Chauhan, A., Yadav, A., Yadav, A.K., 2010. Essential oil composition of *Lavandula angustifolia* Mill. cultivated in the mid hills of Uttarakhand, India. J. Serb. Chem. Soc. 75 (3), 343–348.

Verma, S., Singh, S.P., 2008. Current and future status of herbal medicines. Veterinary World 1 (11), 347–350.

Verma, U., Sharma, R., Gupta, P., Gupta, S., Kapoor, B., 2007. Allopathic vs. Ayurvedic practices in tertiary care institutes of urban North India. Indian J. Pharm. 39 (1), 52.

Vijayan, N., Swapna, T.S., Kiran, G.S., Abdulhameed, S., 2017. Medicinal plants of india: implications in modern medicine. In: Bioresources and Bioprocess in Biotechnology. Springer, Singapore, pp. 421–443.

Vyuptakesh, S., 2013. International Business: Concept, Environment and Strategy. Pearson Education India, Noida, India.

Wal, P., Singh, A., Mehra, R., Rizvi, S., Vajpayee, R., 2016. Pharmacovigilance: need and future prospective in herbal and ayurvedic medicines. Pharm. Innov. 3 (7, Part A), 18–22.

Wani, S.P., Bergvinson, D., Raju, K.V., Gaur, P.M., Varshney, R.K., 2016. Mission India for transforming agriculture (MITrA). Research Report IDC-4.

WHO, 2011. Global initiative for the elimination of avoidable blindness. http://www.who.int/blindness/Vision2020_report.pdf.

Yadav, N.P., Dixit, V.K., 2008. Recent approaches in herbal drug standardization. Int. J. Integr. Biol. 2 (3), 195–203.

Yadav, V.G., 2013. Biosynthetic Engineering for the Assembly of Better Drugs. Doctoral Dissertation Massachusetts Institute of Technology.

Yadav, V., Goyal, P., 2015. User innovation and entrepreneurship: case studies from rural India. J. Innov. Entrep. 4 (1), 5.

Zhang, J., Wider, B., Shang, H., Li, X., Ernst, E., 2012. Quality of herbal medicines: challenges and solutions. Complement. Ther. Med. 20 (1), 100–106.

Further Reading

Chauhan, B., Kumar, G., Kalam, N., Ansari, S.H., 2013. Current concepts and prospects of herbal nutraceutical: a review. J. Adv. Pharm. Technol. Res. 4 (1), 4.

Máthé, Á., 2015. Botanical aspects of medicinal and aromatic plants. In: Medicinal and Aromatic Plants of the World. Springer, Netherlands, pp. 13–33.

Srirama, R., Kumar, J.S., Seethapathy, G.S., Newmaster, S.G., Ragupathy, S., Ganeshaiah, K.N., Ravikanth, G., 2017. Species adulteration in the herbal trade: causes, consequences and mitigation. Drug Saf. 40 (8), 651–667.

Appendix

TABLE A.1 Medicinal Tree Species From Indian Himalaya and Their Status

Tree species	Family	Altitude range	Status
Abies pindrow	Pinaceae	2900–3100 m 3350–3550 m	Vulnerable
Aesculus indica	Hippocastanaceae	1550–1750 m	Alien
Carpinus viminea	Corylaceae	2000–2200 m	Vulnerable
Callicarpa arborea	Verbenaceae	1550–1750 m	–
Hippophae salicifolia	Elaeagnaceae	2450–2650 m	
Lyonia ovalifolia	Ericaceae	1550–1750 m	Rare
Myrica esculenta	Myricaceae	1550–1750 m	Vulnerable
Neolitsea pallens	Lauraceae	1550–1750 m	Vulnerable
Pinus roxburghii	Pinaceae	–	Vulnerable
Rhododendron arboreum	Ericaceae	–	Rare
Sapindus mukorossi	Sapindaceae	–	–
Sorbus aucuparia	Rosaceae	3350–3550 m	Threatened
Sorbus cuspidata	Rosaceae	2450–2650 m	Threatened
Taxus baccata	Taxaceae	2900–3100 m	Endangered
Viburnum cotinifolium	Caprifoliaceae	2450–2650 m	Rare

TABLE A.2A Medicinal Shrub Species From Indian Himalaya and Their Status

Tree species	Family	Altitude range	Status
Arachne cordifolia	Euphorbiaceae	1550–1750 m	Vulnerable
Buddleja asiatica	Buddlejaceae	1550–1750 m	Alien
Clematis buchananiana	Ranunculaceae	2450–2650 m	Vulnerable
Coriaria nepalensis	Coriariaceae	2900–3100 m	Rare
Cotoneaster microphyllus	Rosaceae	2900–3100 m	Vulnerable

(Continued)

TABLE A.2A Medicinal Shrub Species From Indian Himalaya and Their Status—cont'd

Tree species	Family	Altitude range	Status
Debregeasia salicifolia	Urticaceae	1550–1750 m	Alien
Lonicera angustifolia	Caprifoliaceae	3350–3550 m	Vulnerable
Leptodermis lanceolata	Rubiaceae	1550–1750 m	–
Leycesteria formosa	Caprifoliaceae	2900–3100 m	
Inula cappa	Asteraceae	2450–2650 m	Rare
Indigofera heterantha	Fabaceae	2450–2650 m	Vulnerable
Hypericum choisianum	Hypericaceae	1550–1750 m	Rare
			Vulnerable
Holmskioldia sanguinea	Verbenaceae	1550–1750 m	Rare
Eupatorium odenophorum	Asteraceae	1550–1750 m	–
Elsholtzia fruticosa	Lamiaceae	2450–2650 m	Threatened
Deutzia compacta	Hydrangeaceae	2450–2650 m	Threatened
Desmodium elegans	Fabaceae	2000–2200 m	–
Persicaria polystachya	Polygonaceae	2900–3100 m	Rare
Prinsepia utilis	Rosaceae	1550–1750 m	
Rhamnus virgatus	Rhamnaceae	–	
Rhus javanica	Anacardiaceae	–	
Rosa sericea	Rosaceae	2900–3100 m	Rare
Roylea cinerea	Lamiaceae	1550–1750 m	Vulnerable
Rubia manjith	Rubiaceae	2450–2650 m	Alien
Rubus niveus	Rosaceae	2450–2650 m	Vulnerable
Sarcococca saligna	Buxaceae	1550–1750 m	–
Sorbaria tomentosa	Rosaceae	2450–2650 m	
Viburnum erubescens	Caprifoliaceae	2450–2650 m	Rare
Viburnum grandiflorum	Caprifoliaceae	2900–3100 m	Vulnerable
Viburnum nervosum	Caprifoliaceae	2900–3100 m	–
Woodfordia fruticosa	Lythraceae	1550–1750 m	–
Zanthoxylum armatum	Rutaceae	–	Vulnerable

TABLE A.2B Medicinal Herb Species From Indian Himalaya and Their Status

Herb species	Family	Altitude range	Status
Aconitum heterophyllum	Ranunculaceae	3350–3550 m	Critically endangered
Ainsliaea aptera	Asteraceae	2900–3100 m	Alien
Ainsliaea latifolia	Asteraceae	2900–3100 m	Vulnerable
Anaphalis contorta	Asteraceae	1550–1750 m	Rare
Anaphalis margaritacea	Asteraceae	2450–2650 m	Vulnerable
Anaphalis triplinervis	Asteraceae	2000–2200 m	Alien
Anemone obtusiloba	Ranunculaceae	3350–3550 m	Vulnerable
Anemone rivularis	Ranunculaceae	2450–2650 m	–
Arisaema jacquemontii	Araceae	2450–2650 m	
Artemisia japonica	Asteraceae	2000–2200 m	Rare
Artemisia roxburghiana	Asteraceae	2450–2650 m	Vulnerable
Asparagus filicinus	Asparagaceae	–	Rare
Aster peduncularis	Asteraceae	–	Vulnerable
Barleria cristata	Acanthaceae	2000–2200 m	Rare
Begonia picta	Begoniaceae	2900–3100 m	–
Bergenia ciliata	Saxifragaceae	2450–2650 m	Threatened
Bidens bipinnata	Asteraceae	2000–2200 m	Threatened
Bidens biternata	Asteraceae	2000–2200 m	–
Bidens pilosa	Asteraceae	2900–3100 m	Rare
Bistorta amplexicaulis	Polygonaceae	2900–3100 m	
Bistorta vaccinifolia	Polygonaceae	3350–3550 m	Rare
Blumea lanceolaria	Asteraceae	1550–1750 m	
Bupleurum falcatum	Apiaceae	2000–2200 m	Rare
Calanthe tricarinata	Orchidaceae	3350–3550 m	Vulnerable
Cannabis sativa	Cannabinaceae	2000–2200 m	Alien
Clematis montana	Ranunculaceae	1550–1750 m	Vulnerable
Clinopodium umbrosum	Lamiaceae	2000–2200 m	–
Corallodiscus lanuginosus	Gesneriaceae	2450–2650 m	
Cyathula capitata	Amaranthaceae	2000–2200 m	Rare

(Continued)

TABLE A.2B Medicinal Herb Species From Indian Himalaya and Their Status—cont'd

Herb species	Family	Altitude range	Status
Cyathula tomentosa	Amaranthaceae		Vulnerable
Cynoglossum glochidiatum	Boraginaceae	2000–2200 m	–
Cynoglossum lanceolatum	Boraginaceae	1550–1750 m	–
Delphinium vestitum	Ranunculaceae	3350–3550 m	Vulnerable
Dicliptera bupleuroides	Acanthaceae	2000–2200 m	Vulnerable
Dipsacus inermis	Dipsacaceae	2450–2650 m	Alien
Elephantopus scaber	Asteraceae	1550–1750 m	Vulnerable
Elsholtzia strobilifera	Lamiaceae	2900–3100 m	Rare
Euphorbia chamaesyce	Euphorbiaceae	1550–1750 m	Vulnerable
Euphorbia hypericifolia	Euphorbiaceae	2900–3100 m	Alien
Euphorbia pilosa	Euphorbiaceae	2900–3100 m	Vulnerable
Fagopyrum dibotrys	Polygonaceae	2450–2650 m	–
Fragaria nubicola	Rosaceae	1550–1750 m	
Galinsoga parviflora	Asteraceae	2450–2650 m	Rare
Galium aparine	Rubiaceae	2900–3100 m	Vulnerable
Galium asperifolium	Rubiaceae	3350–3550 m	Rare
Geranium wallichianum	Gerianiaceae	2450–2650 m	Vulnerable
Gerbera gossypina	Asteraceae	1550–1750 m	Rare
Girardinia diversifolia	Urticaceae	1550–1750 m	–
Gonatanthus pumilus	Araceae	1550–1750 m	Threatened
Gonostegia hirta	Urticaceae	1550–1750 m	Threatened
Impatiens scabrida	Balsaminaceae	2900–3100 m	–
Jurinea dolomiaea	Asteraceae	3350–3550 m	Endangered
Lamium album	Lamiaceae	2000–2200 m	
Leucas lanata	Lamiaceae	2450–2650 m	
Lindenbergia indica	Scrophulariaceae	2450–2650 m	
Maianthemum purpureum	Liliaceae	3350–3550 m	Rare
Morina longifolia	Morinaceae	2900–3100 m	Rare
Nepeta ciliaris	Lamiaceae	2000–2200 m	Alien
Nomocharis oxypetala	Liliaceae	3350–3550 m	Vulnerable

TABLE A.2B Medicinal Herb Species From Indian Himalaya and Their Status—cont'd

Herb species	Family	Altitude range	Status
Origanum vulgare	Lamiaceae	3350–3550 m	–
Paeonia emodi	Paeoniaceae	1550–1750 m	
Parnassia nubicola	Saxifragaceae	2900–3100 m	Rare
Pedicularis hoffmeisteri	Scrophulariaceae	3350–3550 m	Vulnerable
Phalaris minor	Poaceae	1550–1750 m	–
Picrorhiza kurroa	Scrophulariaceae	3350–3550 m	Critically endangered
Pimpinella acuminata	Apiaceae	2000–2200 m	Vulnerable
Pimpinella diversifolia	Apiaceae	3350–3550 m	Vulnerable
Plantago depressa	Plantaginaceae	3350–3550 m	Alien
Plantago himalaica	Plantaginaceae	3350–3550 m	Vulnerable
Podophyllum hexandrum	Podophyllaceae	3350–3550 m	Rare
Polygonatum verticillatum	Liliaceae	2450–2650 m	Vulnerable
Primula denticulata	Primulaceae	3350–3550 m	Alien
Ranunculus hirtellus	Ranunculaceae	3350–3550 m	Vulnerable
Reinwardtia indica	Linaceae	2000–2200 m	–
Roscoea alpina	Zingiberaceae	2900–3100 m	Rare
Rubus nepalensis	Rosaceae	2900–3100 m	Rare
Rumex hastatus	Polygonaceae	1550–1750 m	Vulnerable
Rumex nepalensis	Polygonaceae	2900–3100 m	Rare
Salvia hians	Lamiaceae	3350–3550 m	Rare
Salvia nubicola	Lamiaceae	3350–3550 m	Rare
Saussurea albescens	Asteraceae	2000–2200 m	–
Saussurea auriculata	Asteraceae	3350–3550 m	Rare
Saxifraga diversifolia	Saxifragaceae	2900–3100 m	Threatened
Selinum candollii	Apiaceae	3350–3550 m	–
Senecio graciliflorus	Asteraceae	3350–3550 m	Rare
Silene edgeworthii	Caryophyllaceae	2450–2650 m	
Solanum suratteuse	Solanaceae	1550–1750 m	

(Continued)

TABLE A.2B Medicinal Herb Species From Indian Himalaya and Their Status—cont'd

Herb species	Family	Altitude range	Status
Solidago virgaurea	Asteraceae	2900–3100 m	
Swertia chirayita	Gentianaceae	2900–3100 m	Rare, vulnerable
Swertia ciliata	Gentianaceae	3350–3550 m	Vulnerable
Synotis alatus	Asteraceae	2450–2650 m	Rare
Taraxacum officinale	Asteraceae	3350–3550 m	Vulnerable
Triumfetta rhomboidea	Tiliaceae	1550–1750 m	–
Urena lobata	Malvaceae	–	
Urtica ardens	Urticaceae	–	Rare
Urtica dioica	Urticaceae	–	Vulnerable
Valeriana hardwickii	Valerianaceae	–	–
Verbascum thapsus	Scrophulariaceae	–	–
Vernonia anthelmintica	Asteraceae	2450–2650 m	Vulnerable
Vernonia cinerea	Asteraceae	2000–2200 m	Vulnerable
Veronica anagallis-aquatica	Scrophulariaceae	2450–2650 m	Alien
Viola canescens	Violaceae	2900–3100 m	Vulnerable

Artemisia absinthium

Aster amellus

Celosia cristata (1)

Celosia cristata (2)

Celosia cristata

Cynara cardunculus

Datura stramonium

Dioscorea deltoidea

Lagerstroemia indica

Lagerstroemia

Lavandula angustifolia (1)

Lavandula angustifolia

Malva sylvestris

Melissa officinalis

Mentha spicata

Nerium oleander

Origanum marjoram

Origanum vulgare

Origanum vulgare (1)

Pelargonium quercifolium

Rosmarinus officinalis

Salvia sylvestris

Tagetes minuta

Taraxacum officinale

Thymus serphyllum

Index

Note: Page numbers followed by *f* indicate figures and *t* indicate tables.

CPI Antony Rowe
Chippenham, UK
2018-10-11 20:06